About Island Press

Island Press is the only nonprofit organization in the United States whose principal purpose is the publication of books on environmental issues and natural resource management. We provide solutions-oriented information to professionals, public officials, business and community leaders, and concerned citizens who are shaping responses to environmental problems.

In 2003, Island Press celebrates its nineteenth anniversary as the leading provider of timely and practical books that take a multidisciplinary approach to critical environmental concerns. Our growing list of titles reflects our commitment to bringing the best of an expanding body of literature to the environmental community throughout North America and the world.

Support for Island Press is provided by The Nathan Cummings Foundation, Geraldine R. Dodge Foundation, Doris Duke Charitable Foundation, Educational Foundation of America, The Charles Engelhard Foundation, The Ford Foundation, The George Gund Foundation, The Vira I. Heinz Endowment, The William and Flora Hewlett Foundation, Henry Luce Foundation, The John D. and Catherine T. MacArthur Foundation, The Andrew W. Mellon Foundation, The Moriah Fund, The Curtis and Edith Munson Foundation, National Fish and Wildlife Foundation, The New-Land Foundation, Oak Foundation, The Overbrook Foundation, The David and Lucile Packard Foundation, The Pew Charitable Trusts, The Rockefeller Foundation, The Winslow Foundation, and other generous donors.

The opinions expressed in this book are those of the author(s) and do not necessarily reflect the views of these foundations.

About The Nature Conservancy

The Nature Conservancy is a leading, international nonprofit organization that preserves plants, animals, and natural communities representing the diversity of life on Earth by protecting the lands and water they need to survive. To date, the Conservancy and its more than one million members have been responsible for the protection of more than 14 million acres in the United States and have helped preserve more than 102 million acres in Latin America, the Caribbean, Asia, and the Pacific. The Nature Conservancy has developed a strategic, science-based planning process that helps the organization identify the highest-priority places—landscapes and seascapes—that, if conserved, promise to protect biodiversity over the long term. The Nature Conservancy works to conserve large landscapes, consistent with the need to ensure that protected lands and waters retain their ecological integrity, and works across traditional geographic, political, and intellectual lines, because ecological systems rarely coincide with human-drawn boundaries. Visit The Nature Conservancy on the Web at nature.org.

DRAFTING A CONSERVATION BLUEPRINT

DRAFTING A CONSERVATION BLUEPRINT

A Practitioner's Guide
to Planning for Biodiversity

CRAIG R. GROVES

CONTRIBUTING AUTHORS
Michael W. Beck, Jonathan V. Higgins, and Earl C. Saxon

FOREWORD BY
Malcolm L. Hunter Jr.

THE NATURE CONSERVANCY

ISLAND PRESS

Washington   Covelo   London

Library of Congress Cataloging-in-Publication data

Groves, Craig.
   Drafting a conservation blueprint : a practitioner's guide to planning for biodiversity / Craig R. Groves ; contributors, Michael W. Beck, Jonathan V. Higgins, Earl C. Saxon.
      p. cm.
Includes bibliographical references and index.
   ISBN 1-55963-938-5 (cloth : alk. paper)—ISBN 1-55963-939-3 (pbk.: alk. paper)
1. Biological diversity conservation. II. Title.
   QH75.G73 2003
   333.95′16—dc21   2003001910

British Library Cataloguing-in-Publication data available.

Printed on recycled, acid-free paper ✪

Design by Teresa Bonner.

Manufactured in the United States of America

10 9 8 7 6 5 4 3 2 1

CONTENTS

Foreword   *xi*

Preface   *xv*

Acknowledgments   *xix*

About the Authors   *xxi*

PART ONE **Groundwork**   *1*

1. The Challenge of Conserving Biological Diversity   *3*

2. Foundations: Ecoregions, Guiding Principles,
   and a Planning Process   *24*

3. Building Blocks for Regional Conservation Planning   *41*

PART TWO **The Seven Habits of Highly Effective Planning**   *79*

4. What to Conserve? Selecting Conservation Targets   *81*

5. Evaluating Existing Conservation Areas and Filling
   Information Gaps   *114*

6. How Much Is Enough? Setting Goals for Conservation Targets   *146*

7. Will Conservation Targets Persist? Assessing Population Viability
   and Ecological Integrity   *178*

8. Drafting Nature's Blueprint: Selecting and Designing
   a Network of Conservation Areas   *216*

9. Safeguarding Nature's Investments: Setting Priorities
   for Action among Conservation Areas   *260*

PART THREE **Conservation Planning for the Biosphere**   *289*

10. Maintaining the Ebbs and Flows of the Landscape:
    Conservation Planning for Freshwater Ecosystems   *291*
    Jonathan V. Higgins

vii

11. The Sea Around: Conservation Planning in Marine Regions   *319*
     Michael W. Beck

12. Adapting Ecoregional Plans to Anticipate the Impact
     of Climate Change   *345*
     Earl C. Saxon

PART FOUR **From Planning to Practice**   *367*

13. Putting the Pieces Together: Implementing Conservation Plans
     for Biodiversity at Multiple Scales   *369*

14. Conservation Planning at the Crossroads   *398*

Literature Cited   *405*

Index   *447*

For Jay E. Anderson—scientist, mentor, conservationist, and friend

FOREWORD

Although the term *biodiversity* emerged from the pool of obscure jargon quite a few years ago, it is still enshrouded with significant ambiguity. At one extreme, some people use it as a loose synonym for nature. At the other extreme, some people reduce biodiversity to simplistic parameters, such as the number of species. Conservation organizations, both private and public, must navigate these waters with great care. They need to engage a public that loves nature as embodied in beautiful places and charming creatures—elk in a Yellowstone meadow or snow leopards on a Himalayan snowfield—while expounding the values of species and ecosystems that have significantly less appeal. It is no small task to promote projects that have "swamp," "rat," or "spider" in their titles. At the same time, to undertake their work in a systematic and rigorous manner, conservation organizations need to be able to quantify their goals and accomplishments. Some things are easy to measure, such as the areas of new reserves, but biodiversity writ large is not one of them. For example, take one of the commonest metrics of biodiversity, species richness. First of all, species richness is essentially impossible to measure comprehensively because our knowledge of taxonomy is so limited, especially for microorganisms and invertebrates. Second, it can be badly misconstrued, as by one U.S. government official who argued that we should not be overly concerned by the long list of endangered species in the United States because they were greatly outnumbered by new, exotic species, thus yielding a net increase in biodiversity.

Among the many organizations sailing these waters, The Nature Conservancy stands out as a leader. Decades ago it moved beyond the syndrome of pretty places and cute creatures, but in doing so it has not forsaken the public: Roughly a million nature lovers provide their backbone of support. More significantly, it has articulated a science-based approach to conserving biodiversity that is both comprehensive and efficient. It is comprehensive because it embraces all forms of life at all levels of organization, explicitly for species and ecosystems, implicitly for the genetic component of biodiversity. It is efficient because it targets conserving those places where the needs and opportunities are greatest, and it uses an array of conservation techniques tailored to the particular situation. The keystone to these undertakings is conservation planning and measuring the successes or failures of conservation action, what the Conservancy calls "Conservation by Design."

In an organization as large and diverse as The Nature Conservancy, there is a danger of becoming too insular, of developing elaborate systems that are not widely shared outside the organization. Efforts such as this volume, *Drafting a Conservation Blueprint,* avoid such tendencies. Here we have Craig Groves, one of the architects and leaders of Conservation by Design, articulating everything he has learned about conservation planning, especially the selection of sites for conservation action, for an audience that reaches far beyond the ranks of the Conservancy to include a diverse array of private and governmental agencies.

The diversity of Groves's audience extends in three other directions besides the private-public axis. First, Groves has leapt over the pitfalls of parochialism to examine conservation planning far beyond the United States. He is well aware that the literature of conservation planning is replete with both cutting-edge innovations and case histories from Australia, South Africa, and all corners of the globe. Second, his primary audience is all the conservation practitioners who travel under many names— natural resource managers, wildlifers, foresters, fisheries biologists, restoration ecologists, coastal zone managers, environmental planners, landscape architects, and more—and they will appreciate his practical, step-by-step approach. However, other readers will enjoy the book as well, notably academics and students who will applaud the book for both its rigorous scholarship and its coherent synthesis of disciplines, such as population biology, community ecology, and landscape ecology. Finally, with the help of two key contributors (Jonathan V. Higgins and Michael W. Beck), this book moves beyond the terrestrial bias of most conservation practitioners to embrace freshwater and marine conservation. A third contributor, Earl C. Saxon, offers practical advice and new insights into the challenges of conserving biodiversity in the face of global climate change.

Most of the conservationists I know are optimists; one has to be optimistic to continue working in the face of widespread loss and degradation of ecosystems. We routinely cite the litany of doom and gloom statistics to convince others that our cause merits their attention, but balanced against that are our cherished success stories. We think about the species and places that are thriving now despite being on the brink of annihilation a century ago. We can continue to generate more success stories, but it will not be easy. It will require careful marshaling of the limited resources available for conservation, which in turn will require the kind of thoughtful conservation planning that is cultivated by this book.

Malcolm L. Hunter Jr.
President of Society for Conservation Biology
Libra Professor of Conservation Biology
University of Maine

PREFACE

It has been over 13 years since a group of Australian biologists published a series of papers on conservation evaluation in the journal *Biological Conservation*, and 15 years since the inception of the Gap Analysis program in the United States. Both of these events were major benchmarks in the evolving field of systematic planning for the conservation of biological diversity. In the 1990s, several of the world's major biodiversity conservation organizations undertook systematic planning and priority-setting efforts, and the United Nations Convention on Biological Diversity mandated national planning for biodiversity conservation for those countries that were signatory to this important treaty. What would have once been described as a limited and opportunistic use of scientists and scientific information for setting priorities for biodiversity conservation has now evolved into a cottage industry of conservation planning that is steadily growing in stature within both nongovernmental and governmental spheres.

Although The Nature Conservancy has used scientific information to help determine its conservation priorities for nearly 30 years, it embarked on a much more comprehensive effort to systematically plan for biodiversity conservation in the mid-1990s. Despite spending tens of millions of U.S. dollars to protect important natural areas and endangered species, the Conservancy realized that its efforts were not nearly enough in the face of overwhelming worldwide losses of natural habitats and biodiversity. As a result, the Conservancy initiated a more leveraged strategy: develop biodiversity conservation plans whose target audience would be the entire conservation community. In 1997, the Conservancy published its first guidebook on these planning efforts—*Designing a Geography of Hope: Guidelines for Ecoregion-Based Conservation*—and its programs for developing ecoregional plans for biodiversity conservation were already well under way. By

2003, the Conservancy will have completed comprehensive biodiversity conservation plans for nearly all the ecoregions of the United States and several ecoregions internationally from Yunnan, China, to Central and South America. Many other organizations, from Conservation International and the World Wildlife Fund to state and national governments around the globe, will have undertaken or completed similar planning efforts.

World Conservation Union (IUCN) biologist Jeffrey McNeely was the first to suggest to The Nature Conservancy that it should make the guidance outlined in *Geography of Hope* available to the larger conservation community. With the encouragement of Conservancy Board of Governor Dr. Peter Kareiva and the experience of participating in the writing of two editions of *Geography of Hope* behind me, I began the research and writing of this book in the fall of 2000. My hopes were to write a book that would be useful to conservation practitioners in the United States and abroad, building on what I had learned as a conservation biologist and planner for the Conservancy for over a decade, but adding the richness and diversity of experiences and methods in conservation planning from around the globe. Although I have endeavored to achieve that goal, the single greatest challenge has been the vast number of published papers and books relevant to the subdisciplines of conservation planning, ranging from the recovery of endangered species and classification of ecosystems to population viability analyses, assessments of ecological integrity, and computer algorithms for the selection and design of nature reserves. While I have attempted to be comprehensive in my efforts to incorporate the important literature from all of these fields that contribute to effective conservation planning, I have no doubt fallen short of being exhaustive.

I have been fortunate to spend the better part of my adult years living and working in the Rocky Mountains of the western United States, particularly the wilderness areas of central Idaho. That has been both a blessing and a curse from the perspective of conservation planning. It has provided me the opportunity to observe patterns and processes of biological diversity that are relatively little disturbed by human activities. As I have worked on conservation plans around the world, it has also made me appreciate how increasingly difficult it is to find large blocks of unfragmented natural lands or intact watersheds that can serve as benchmarks for how communities and ecosystems should look and function. I now recognize how much more difficult it is to develop these sorts of plans in regions of the world where it is necessary to rely on historical ecology or, in some cases, imagination, to picture what the landscapes, watersheds, and seascapes might once have looked like and how they may have functioned before being altered so extensively by humans.

Ask any group of biologists and planners what conserving biological diversity means, and you are likely to get a wide range of opinions. For some, it means restoring the last remnants of the most threatened natural communities and ecosystems in a region. For others, it is recovering endangered species from the brink of extinction. For still others, it is ensuring that intact functional landscapes remain that way. In large part, one's worldview on such matters depends on both the geography of place and the biologist's prism through which the world is viewed—populations, communities, ecosystems, or landscapes. An effective conservation planning framework needs to be robust enough to accommodate the varied biogeographic contexts that biologists and planners worldwide face today. It should also draw from the best thinking that the fields of population, community, ecosystem, and landscape ecology have to offer to conservation. Accordingly, the methods and strategies outlined in this book are focused not only on recovering and restoring endangered species and degraded habitats, but on maintaining those landscapes where native species and communities are intact and natural disturbances and ecological processes are still functional. Although a world in which most native species persist is far better than a biologically impoverished one, a world in which at least many of these species can continue to play out their roles on the ecological and evolutionary stage is far better.

Craig R. Groves
Boise, Idaho
August 2002

ACKNOWLEDGMENTS

I owe a debt of gratitude to many people who assisted me with this book, but none deserves more thanks than Peter Kareiva, who provided encouragement and advice from the beginning until completion and kindly reviewed every chapter. Writing this book would not have been possible without the support of a fellowship from the John Simon Guggenheim Memorial Foundation and the continued financial and professional support of my principal employer during the writing of this book, The Nature Conservancy. In particular, I want to especially thank Amy Lester, Deborah Jensen, Jonathan Reed, and Alan Holt for generously allowing me significant blocks of time to devote to research and writing. Thanks also to the Wildlife Conservation Society, especially Bill Weber, who supported me taking the time to finish this book while taking on a new staff position with WCS. In 2001, I wrote early drafts of several chapters while a sabbatical fellow at the National Center for Ecological Analysis and Synthesis (NCEAS), a center funded by National Science Foundation Grant number DEB-0072909, at the University of California, Santa Barbara. I particularly appreciate the efforts of Sandy Andelman at NCEAS in helping arrange this sabbatical opportunity.

Renée Mullen helped obtain several references and prepare the literature cited. Laura Landon (Valutis) went well beyond the call of duty in preparing many figures and soliciting previously published figures. The book may have never reached print without her invaluable help. Special thanks also to Tim Boucher, Carlos Carroll, Cherie Moritz, Zach Ferdana, Ben Halpern, and Ree Brannon for providing several figures.

Conversations with and materials provided by biologists David Brunckhorst, Richard Thackway, Chris Margules, Julianne Smart, and Simon Ferrier during a research trip through Australia greatly broadened the scope and improved the content of the book. Lisa Johnstone, Deanna

Chilson, Eric Atkinson, Melonie Atkinson, and Kathy O'Hara spent many hours proofing the literature cited, which is greatly appreciated. Ed and Alison Wilson helped set up several interviews with WWF and IUCN staff in Switzerland and kindly provided me lodging while doing some of the research for the book.

The following individuals each reviewed one or more chapters and their reviews greatly improved the quality of the manuscript: Robin Abell, J. David Allan, Mark Anderson, Paul Angermeier, Jeff Baumgartner, Terry Cook, Patrick Crist, Francis Crome, Zach Ferdana, Tom Fitzhugh, Paula Gagnon, Ben Halpern, Lee Hannah, Jeff Hardesty, Malcolm Hunter, Deborah Jensen, Nels Johnson, Mary Lammert, Laura Landon, Bernhard Lehner, Colby Loucks, Stewart Maginnis, Renée Mullen, Betsy Neely, Reed Noss, Michael O'Connell, Sheila O'Conner, Dave Olson, Sam Pearsall, Karen Poiani, Bob Pressey, Helen Regan, Mary Ruckelshaus, Mike Scott, Craig Shafer, Mark Shaffer, Frederick Steiner, David Stoms, Tim Tear, David Wilcove, and Rob Wilder. Lynn Macquire kindly allowed me to borrow heavily from her writings on expert judgment. Barry Baker and Chris Zganjar prepared the maps for the climate change chapter. Special thanks go to Kent Redford for reviewing most of the book and serving as a constant source of inspiration.

Many of the ideas for this book first surfaced while writing a handbook for The Nature Conservancy on ecoregional planning. To my co-authors of that handbook, *Designing a Geography of Hope,* I am very grateful for their intellectual contributions, which I was able to use as a starting point for this book: Laura Landon, Betsy Neely, Bruce Runnels, Jerry Touval, Diane Vosick, and Kimberly Wheaton. Too numerous to mention are the members of 50 plus ecoregional planning teams of The Nature Conservancy and several ecoregional or regional planning efforts of the World Wildlife Fund and Conservation International, which also contributed many significant ideas to regional conservation planning that found their way into this book. Charlie Ott and Alan St. John generously allowed the use of their photographs.

Barbara Dean, Barbara Youngblood, and Laura Carrithers, all editors at Island Press, provided many useful suggestions throughout the writing of the book and their editing greatly improved each and every chapter. Last, but far from least, I want to thank my wife, Victoria, who tolerated countless evenings and weekends of my writing and focus on this project.

ABOUT THE AUTHORS

Craig R. Groves was the Director of Conservation Planning for The Nature Conservancy during 1997–2002. Prior to that time, he worked for the Conservancy as a conservation biologist in several capacities and as a nongame biologist for the Idaho Department of Fish and Game. He is currently a research biologist and conservation planner for the Wildlife Conservation Society in the Greater Yellowstone area.

Jonathan V. Higgins is the Senior Aquatic Ecologist for The Nature Conservancy, where he leads conservation planning for freshwater biodiversity. He previously managed the sampling and analyses for monitoring the condition of the Laurentian Great Lakes as a contractor for the U. S. Environmental Protection Agency's Great Lakes National Program Office. He received his Ph.D. in population and evolutionary ecology from the University of Illinois–Chicago.

Michael W. Beck is a Senior Scientist with the Marine Initiative of The Nature Conservancy and a research associate at the University of California, Santa Cruz. His research examines factors that control the diversity and abundance of animals in seagrass, mangrove, rocky intertidal, and salt marsh habitats. He received his Ph.D. from Florida State University.

Earl C. Saxon is the Chief Scientist in The Nature Conservancy's Climate Change Initiative. He was previously the Conservancy's Asia-Pacific Regional Ecologist and has worked extensively as a conservation ecologist throughout Australia. His Ph.D. (Cambridge University) research focused on human adaptation to the impacts of climate change at the end of the last ice age.

PART ONE

GROUNDWORK

Chapters 1–3 explore the
meaning of biological diversity and the
challenge of conserving it, outline a conservation
planning framework and seven-step planning process,
and discuss the building blocks to successful planning that
biologists and planners should consider prior to launching
into a full-scale planning project. Collectively, these first
three chapters provide conservation biologists and plan-
ners with critical background information that will
enable them to more successfully use the guid-
ance that is provided in Parts II–IV.

CHAPTER ONE

The Challenge of Conserving Biological Diversity

Still, too many land conservation efforts are haphazard and reactive in nature. They deal with whatever comes over the transom. The result is haphazard conservation and haphazard development.

—MARK BENEDICT
AND EDWARD MCMAHON (2002)

Seventeen and a half billion dollars—that is the amount state and local governments in the United States alone directed toward open space preservation from 1999 through 2001 (Benedict and McMahon 2002).Yet much of this money has not been spent wisely from a conservation perspective. According to Benedict and McMahon, the authors of a recent report on "green infrastructure," smart land conservation of the future needs to be more proactive and less reactive, more systematic and less haphazard, multifunctional and not single-purpose, and larger in spatial scale.

Achieving smarter conservation in the future will be challenging, in part because many important decisions about land and water conservation are made locally, especially with regard to privately owned lands. All over the world, local governments, from counties and provinces to cities, townships, and villages, make decisions every day about land use. These decision makers, through no fault of their own, have a limited view of the world, usually defined by political boundaries of their particular jurisdiction. Unfortunately, the natural world does not operate along these geopolitical lines. Most species, ecological communities, and ecosystems depend upon

3

a much larger domain for their long-term survival. The result has been what could only be described as the "tyranny of the local." Lacking information with which to make better decisions, local governments, often unwittingly, reach conclusions that result in the degradation or destruction of some of the best remaining examples of the world's ecosystems (Dale et al. 2000). Clearly, many important decisions about natural resources are also made at levels above local government. Natural resource agencies at the state or national level, although usually operating at larger spatial scales, are guilty as well of planning along inappropriate boundaries and making poorly informed decisions. Taken as a whole, incremental decisions, from the local to the national level, often result in driving species to the edge of extinction—and even over the edge.

What we need are a road map and appropriate planning boundaries for making more informed decisions, decisions that benefit the natural world and, in turn, through the services nature provides, the human world. That is what this book is about—developing regional-scale road maps or plans that, if implemented, will help ensure that the world's species, communities, and ecosystems, and the underlying ecological processes that sustain them, will not only persist, but continue to evolve and adapt for generations to come. To devise such a road map, this book lays out a step-by-step planning process for conserving the biological diversity of entire regions. To best appreciate the value of this planning framework, we first need to have a good understanding of what problems conservation planning is attempting to address, how biological diversity is best conceptualized and defined, and what contributions the disciplines of ecology and conservation biology can make to developing better planning processes and credible conservation plans.

The Biological Diversity Problem

Over 3500 vertebrate species, nearly 2000 invertebrate species, and over 5600 species of plants from around the world made the 2000 IUCN Red List of Threatened Species (Hilton-Taylor 2000). All these species face a high risk of extinction in the wild. In the United States alone, the number of species listed as Threatened or Endangered under the Endangered Species Act (ESA) has increased sixfold from 174 in 1976 to 1244 species as of November 2001 (http:endangered.fws.gov, U.S. Fish and Wildlife Service). These ever-lengthening lists of threatened and endangered species worldwide are symptomatic of an even more serious problem—the extinction crisis—whereby species are currently going extinct at a rate conservatively estimated to be 100 to 1000 times greater than rates recorded through recent geological time (Lawton and May 1995). Furthermore, the extinction crisis is likely worse than the current data indicate. Recent studies suggest that today's fragmentation and destruction of natural habitats may result in extinctions that will not be apparent to us for several generations, creating what has been termed an "extinction debt" (Tilman et al. 1994).

The principal causes of this march toward extinction are well documented (World Resources Institute, World Conservation Union, and United Nations Environment Programme 1992, Noss and Cooperrider 1994, McNeely et al. 1995, Vitousek et al. 1997). They include the conversion and fragmentation of natural habitats, the introduction of non-native species, pollution, the direct exploitation of species, the disruption of natural ecological processes, industrial-scale agriculture and forestry, climate change, and overall human domination of Earth's ecosystems. Habitat loss, primarily from urbanization and agriculture, is the single largest cause of species endangerment (Wilcove et al. 1998, Czech et al. 2000, Hilton-Taylor 2000). Although the ESA has been an effective piece of legislation for abating the rate of extinction in the United States, it is largely a reactive tool, used only for species that are already well down the road to critical levels of endangerment. Too often, ESA implementation results in situations popularly described to the media by former U.S. Secretary of the Interior Bruce Babbitt as environmental "train wrecks" (Reid and Murphy 1995). An example close to my home in the Pacific Northwest demonstrated this point emphatically. Sockeye salmon (*Oncorhynchus nerka*) historically returned from the Pacific Ocean to spawn by the thousands in central Idaho, yet they were not listed as Threatened or Endangered until their returning numbers were reduced to less than ten individual fish.

By itself, single-species approaches such as the ESA are necessary but insufficient tools for effectively addressing the extinction crisis and stemming the tide of overall losses of biological diversity worldwide. Entire ecosystems are being lost at alarming rates (World Conservation Monitoring Centre 1992). Hardly a day passes without reports about the dire status of the world's coral reefs, wetlands, or tropical forests. The "homogenization" of many ecosystems through the spread and invasion of non-native species is a serious issue in itself, as it complicates efforts to conserve biological diversity. These points corroborate the notion that conservation actions, to be successful, need to operate at multiple levels of biological organization, from populations and species to landscapes and ecosystems.

Poor planning in the identification of important areas for conserving biological diversity has exacerbated the extinction crisis. Most areas in the world that have been designated for conservation purposes were set aside in an ad hoc manner, and not specifically for conserving biological diversity (Pressey 1994). Recent analyses in the United States (Scott et al. 2001a) and Australia (Pressey et al. 1996a, Mendel and Kirkpatrick 2002) have shown that the locations of areas set aside for conservation are strongly biased toward the most unproductive soils, the steepest slopes, and the highest elevations. Similar trends exist for many other parts of the world (Scott et al. 2001b). One of the most oft-cited examples of this trend is the "rocks and ice" national parks of the western United States, many of which were established for their scenic grandeur in the Rocky Mountains, Sierra Nevadas, or Cascade Mountain ranges, but as a whole are poorly representative of the region's different ecosystems. Conversely, other recent analyses on the distribution of endangered species and threatened habitats in the United States and elsewhere have shown that the majority of endangered species, and, indeed, the majority of biological diversity (as measured by the number of species), tends to occur in lower elevations, warmer climates, and coastal areas that are more attractive to human occupation and use (Dobson et al. 1997, 2001).

Natural resource managers, conservation practitioners, and scientists from around the globe have recognized the serious nature of the problems that must be addressed to effectively conserve biological diversity, and they have reacted to these problems on several fronts. Several examples of these reactions are noted here. First, in June 1992, at the United Nations Conference on the Environment and Development in Rio de Janeiro (also known as the Earth Summit), a record 150 countries signed a global Convention on Biological Diversity (CBD), a landmark treaty that takes a comprehensive approach to the conservation and sustainable use of Earth's biological resources (Glowka et al. 1994). This treaty has already had a significant conservation impact globally, and it has potential for substantial influence in the future (see the section on the CBD later in this chapter). Second, the World Conservation Monitoring Centre (1992) compiled the first global review and sourcebook on biological diversity. Third, the United Nations Environment Programme (1995) commissioned and published the *Global Biodiversity Assessment* to provide a state-of-the-art understanding of society's knowledge of biological diversity and the nature of human impact upon this diversity.

Fourth, a renowned group of scientists from agencies, academia, and nongovernmental organizations met in a workshop in 1995 sponsored by the U.S. Marine Mammal Commission and issued a set of seven conservation principles along with guidelines for their implementation (Mangel et al. 1996). Aimed at natural resource managers, some of the important principles included maintaining biological diversity at genetic, species, population, and ecosystem levels; assessing the ecological and sociological effects of resource use before both proposed use and proposed restriction; using the full range of knowledge and skills from the natural and social sciences to address conservation problems; and understanding and taking into account the motives, interests, and values of all users and stakeholders.

Finally, recognizing that habitat loss and degradation are the leading culprits in the loss of diversity and that many decisions resulting in such losses occur at a local level, the Ecological Society of America issued a set of guidelines to better inform local land-use decision making (Dale et al. 2000). Among the guidelines were examining the impact of local decisions in a regional context, planning for long-term change and unexpected events, preserving rare landscape elements and associated species, retaining large contiguous or connected areas that contain critical habitats, avoiding the introduction of exotic species, and avoiding or mitigating for negative effects of development on ecological processes.

In the chapters ahead, I will explore each of these principles and guidelines, and others in more detail, as I develop a comprehensive planning framework for conserving biological diversity. But before doing that, we need to examine more closely what is meant by the term *biological diversity* or its shortened form, *biodiversity*. An important part of the challenge to conserving biological diversity is getting society to better understand and appreciate the concept.

Biodiversity: Definition, Perceptions, and Values

Popularized by the scientific community over the last 15 years, biodiversity and its conservation have been the focus of numerous books, notably those by Harvard biologist E. O. Wilson (e.g., Wilson 1992). Not surprisingly, a variety of definitions for biodiversity have been advanced (see review by Baydack and Campa 1999). Early definitions focused nearly exclusively on the diversity or variety of life-forms or living organisms. In practice, many biologists have interpreted this to mean that areas with high levels of biodiversity have relatively high numbers of species. In part, this interpretation of the term has led to an emphasis on conserving biological diversity in the world's tropical regions that harbor the vast majority of described species on Earth. More recent definitions view biodiversity as the variety of living organisms, the ways in which they organize themselves (genes, populations, species, communities, ecosystems), and the ways in which they interact with the physical environment and with one another (Redford and Richter 1999).

Whether biodiversity is actually defined primarily as the variety of living organisms, or whether it also includes the ways these organisms organize themselves and interact with the environment, makes little difference from a theoretical standpoint. However, from a practical standpoint of conserving biodiversity, it matters a great deal. To actually conserve biodiversity, natural resource managers and conservation biologists must pay attention to its three components (Noss 1990, Redford and Richter 1999): composition, structure, and function (Figure 1.1). From this point

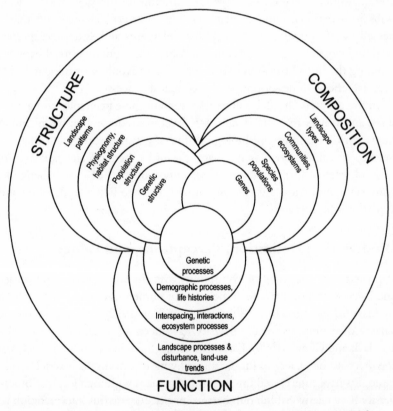

Figure 1.1 The three components of biodiversity: composition, structure, and function. Each component can be described at different levels of biological organization, from genes to landscapes. The components interact to maintain biological diversity. (From Noss, R. F. 1990. Indicators for monitoring biodiversity: a hierarchical approach. *Conservation Biology* 4:355–364. Copyright, Blackwell Publishing, Inc.)

of view, the definition offered by Redford and Richter (1999) is preferable to definitions that focus solely on numbers of species. *Composition* refers to the identification of elements within the different levels of biological organization, from genes and species to communities and ecosystems. The description of a particular area of tropical forest that harbors 400 species of birds is a reference to the composition of biological diversity at that site. *Structure* refers to how these different biological elements are physically organized. For example, walking through a stand of trees, we might observe some bird species on the forest floor, others in shrubs and trees of the forest understory, and still others high in the forest canopy. These different places within the forest describe the structure of biological diversity. *Function* refers to ecological processes that sustain composition and structure. Many forest types, for example, need periodic disturbances from natural fires or storm events to maintain structure and initiate reproduction.

In addition to having components of composition, structure, and function, biological diversity occurs at different spatial scales. The renowned ecologist Robert Whittaker (1975) first advanced this notion. He termed these scales alpha, beta, and gamma (Figure 1.2). *Alpha diversity* refers to the number and types of species that occur at a particular site or area. For example, at La Selva Biological Station in the Atlantic Forest lowlands of Costa Rica, a wide variety of bird species can frequently be observed. If

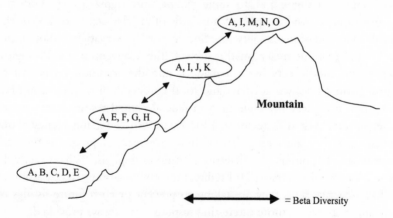

Figure 1.2 Spatial scales of species diversity. The letters inside the circles represent different bird or plant species. The diversity at a given site or locale (within circles) is referred to as alpha diversity. As one hikes up a mountain, the diversity of species changes. A measure of that change, or turnover in species, is referred to as beta diversity. In this simplified diagram, one species is broadly distributed along the entire elevational gradient, while most species are restricted to certain bands of elevation, often represented by distinct habitats. Areas with greater levels of beta diversity require more intensive efforts to conserve their biological diversity. The diversity of plants and birds encompassing the entire mountain from bottom to top is referred to as gamma diversity.

one treks upslope from La Selva toward the continental divide of Costa Rica, different types of bird species and plants can be observed along the way. High on the divide itself, which is characterized by the cloud forests of Braulio Carrillo National Park, even the most casual naturalist would again observe the changes in bird and plant composition. These changes or turnover in species composition along a gradient, such as an elevational one, are referred to as a measure of *beta diversity,* or between-area diversity. The diversity of plants and birds encompassing this entire landscape, from the coastal lowlands to continental divide, is collectively known as *gamma diversity.* In many sites around the world, alpha diversity is actually increasing from the spread of non-native species, while beta and gamma diversity are generally declining.

Anyone who has hiked in the Rockies, Himalayas, Andes, or Alps, or in other mountain ranges with relatively rapid changes in relief, has likely made similar observations. The concept of beta diversity is especially important from a conservation standpoint. In areas with high turnover in species composition, such as many tropical forests, more intensive conservation efforts may be required than in regions where species are more broadly distributed. I will explore this concept and its ramifications for conserving biological diversity in more detail in later chapters.

Despite the popularity of its use in the scientific community, biodiversity remains a vague and, to some extent, unrecognized term with the general public. And in spite of hundreds of articles, scores of books, the birth of a professional society (the Society for Conservation Biology), and a United Nations treaty on the subject (the Convention on Biological Diversity), biodiversity has not made it into the common vernacular of most natural resource agencies and local conservation organizations. For example, if I query my colleagues in federal natural resource agencies in the western United States or in a local state fish and game agency about what steps they are taking, books they have read, or programs they are involved in to conserve biodiversity, a common reply is, "What exactly do you mean by biodiversity?" If I redirect the conversation to a discussion of what proactive steps they are taking to prevent or curb future listings of endangered species, more interesting responses are always provoked.

Recent surveys of the American public concerning the recognition and understanding of the term *biological diversity* are encouraging (Belden et al. 2002). Compared to 1996 surveys, more Americans now recognize the term, more understand its meaning, and more appreciate the importance of species and habitat conservation. Polls also reveal that although the vast majority of Americans understand that species are going extinct and feel a responsibility toward environmental protection, most do not understand how their individual actions can influence species and their habitats. Whether there is similarly an increasing recognition and understanding of biodiversity in other parts of the world is unclear.

Although there have been elegant and articulate explanations of the value of biodiversity by E.O. Wilson (1988, 1992) and others (World Conservation Monitoring Centre 1992, Callicott 1997), conservation biologists, practitioners, and policy makers have struggled to explain why it is important to conserve biodiversity. Noss and Cooperrider (1994) summarized four general sets of values for conserving biodiversity: direct utilitarian values, indirect utilitarian values, recreational and esthetic values, and intrinsic, spiritual, and ethical values. Direct utilitarian values refer to those uses of a species that are of some direct benefit to humans. Commonly cited examples of such uses are the medicinal values of many plants and the direct food values of some species groups, such as commercial fisheries. As Noss and Cooperrider (1994) point out, arguments for conservation based on direct utility are limited and potentially dangerous. What are we to say for all those species that appear to have no direct benefits to society?

Indirect utilitarian benefits refer to what are more commonly now known as ecosystem services and benefits (Daily 1997). These are benefits provided by natural ecosystems that most of us take for granted, ranging from climate stabilization to flood control and the maintenance of air and water quality. This valuation of ecosystem benefits and services has spawned an entire new subdiscipline of ecology and economics (e.g., Folke et al. 1996, Costanza et al. 1997). Much like direct utilitarian benefits, the limitation in arguing for biodiversity conservation on the basis of ecosystem services and functions is that we know very little about which species within any given ecosystem are most critical to its functions. In addition, we do not know how many species an ecosystem can lose before its functions become impaired. Considerable evidence indicates that the loss of species from an ecosystem can negatively impact its functioning (Naeem et al. 1999), but there are also persuasive studies to the contrary. The nature of the relationship between biodiversity and ecosystem function remains the topic of a contentious and acrimonious debate in the scientific community (Kaiser 2000).

Maintaining places for their natural beauty and as places for outdoor recreation is another explanation often given for conserving biodiversity. Many parks, refuges, and nature reserves were established for such reasons. Although it is clear that biological considerations entered into the rationale for establishing some U.S. national parks and monuments (Shafer

1999a), they were never a driving force in that decision-making process. Setting aside lands primarily for recreational purposes also carries some liabilities for biodiversity conservation (Knight and Gutzwiller 1995, Czech et al. 2000), especially if that recreation involves the use of roads and motorized vehicles (Trombulak and Frisell 2000, Gucinski et al. 2001, Havlick 2002).

The final set of values for conserving biodiversity rests upon the argument that the natural elements and processes of biodiversity have intrinsic values that make them worth conserving, and that as humans we have a moral and ethical responsibility to conserve all living things. Taken at face value, this argument suggests that all species deserve an equal opportunity to persist and evolve, humans included. Despite the persuasiveness of this argument for some people (Ehrenfeld 1981), the human domination of Earth's ecosystems and estimated current rates of extinction suggest that it has had little significance for post–Industrial Revolution society as a whole.

Given the difficulties of defining and conserving biological diversity, how do we most effectively communicate about it, within the community of scientists and natural resource managers as well as with the general public? Among the different levels of biological organization (see Figure 1.1), scientists can generally agree, with a few exceptions, on what constitutes a gene or species. We all know a robin, a rose, a palm tree, a cactus, or a tiger when we see one. On the other hand, plant ecologists will argue endlessly over how to define a mixed deciduous-coniferous forest. The general public, although perhaps a little less sophisticated in recognizing the components of biodiversity, reacts similarly. People relate to species and the different kinds of habitats in which they occur. Most people understand that a swamp is very different from an alpine meadow or from a sandy desert, and they appreciate that different assemblages of species occur in these different habitats. Although this book is aimed at improving our efforts to conserve biodiversity, conservation biologists may most effectively communicate about their work by focusing on the need to maintain different habitats to help ensure that all native species can survive and continue to evolve. In the vernacular of conservation biology and planning, this focus implies representing and maintaining high-quality examples of different ecosystem types and the processes and disturbances that sustain them in conservation areas. As the various chapters of this book will detail, we need not champion solely a species-by-species approach to achieve effective conservation. Indeed, the modern-day Noah must employ a variety of tools and approaches in both conservation planning and conservation practice, if he or she hopes to be effective.

As to why it is important to conserve biodiversity, discussions that focus on the economic status of individuals and/or their public health generally carry the most weight. In many parts of the world, but especially the United States, the addition of a species to the Endangered Species list carries with it the perception and sometimes the reality of an entangled bureaucracy and potential economic hardships for some sectors of society. Working to avoid such listings has proven to be a powerful incentive for taking conservation action. Conservation biologists and natural resource managers also need to be more supportive of arguments that articulate the need to conserve ecosystems (i.e., different types of habitats) based on their services and benefits to society. Harte (1997) recently made this point in eloquent fashion: "If we do not understand for ourselves, and then educate the public about, the essential role of ecosystems in sustaining the human economy, we have not a ghost of a chance of stanching the worldwide biodiversity hemorrhage." Recent biodiversity polls in the U.S. support this argument; most Americans want to conserve species and their habitats because of the ecosystem services they provide to society (Belden et al. 2002).

Ecological economist Carl Folke and colleagues (1996) have referred to biodiversity as "natural insurance capital" for securing the ecological services the natural world provides to society. As the pressures from human population growth and their manifestation on such ecosystem products as clean air and potable water supplies mount, arguments for land and water conservation should be increasingly persuasive. Making these and any other arguments for biodiversity conservation lead to successful conservation action will require thoughtful and strategic conservation planning at a variety of spatial and temporal scales.

The Emergence of Conservation Planning as a Discipline

Systematic efforts to plan for the conservation of biodiversity were initiated independently in a number of different countries during the 1970s. For example, The Nature Conservancy (TNC), a conservation organization formed in the United States as an outgrowth of scientists from the Ecological Society of America concerned about the disappearance of natural areas, hired its first scientist and established the first state Natural Heritage Program in the early 1970s. These programs were designed to systematically collect, manage, and disseminate information about the status and distribution of rare plants, animals, and natural communities (Groves et al. 1995). In turn, that information is used to identify high-priority sites for conservation purposes. A network of Natural Heritage Programs and Conservation Data Centers now exists throughout most states and provinces of the United States and Canada, as well as a number of Caribbean and Latin American countries.

During approximately the same time period as TNC's initiation of conservation planning efforts in the United States, the United Nations Educational, Scientific, and Cultural Organization (UNESCO) launched the Man and the Biosphere Program (MAB) in 1974 (Spellerburg 1992). The purpose of the MAB is to establish an international network of biosphere reserves that conserve important biological resources, develop environmentally sound economic growth, and support environmental research, monitoring, and education (see www.unesco.org/mab, Man and the Biosphere Program). To qualify as a biosphere reserve, an area should meet the criteria of representation of a major biogeographic region; contain landscapes, ecosystems, or plant and animal species in need of conservation; be of sufficient size to accomplish the first two criteria; and provide an opportunity to experiment in sustainable development approaches. As of October 2000, 368 biosphere reserves covering more than 300 million hectares (740 million acres) have been established in 91 countries.

In 1977, *A Nature Conservation Review* was published in the United Kingdom (Spellerburg 1992). This publication described and assessed ten criteria for evaluating sites for their ability to serve as representative examples of biological natural areas. The most important criteria were size, species diversity, degree of naturalness, rarity, and fragility. In addition to evaluating these criteria, the nature review identified 784 potential sites for nomination as National Nature Reserves in Britain. Today, English Nature, a government agency established to "champion wildlife, geology, and wild places in England," manages over 200 National Nature Reserves in collaboration with other approved organizations, such as Wildlife Trusts (www.English-Nature.org, English Nature).

Concurrently with TNC and MAB efforts, Australian scientists began to focus on establishing a comprehensive and biologically representative set of nature reserves throughout the country as early as 1974 (Spellerburg 1992). Since that time, Australia has been recognized as a scientific center for the research and development of site-selection methods for nature reserves (Margules 1989). The Australian national government has played a leading role in establishing a biologically representative set of nature reserves state by state through the systematic assessment of current reserves on a bioregional basis and the establishment of new reserves through a National Reserves

Program (Thackway and Cresswell 1997). As will be evident throughout this book, Australian scientists continue to be among the leading thinkers in systematic planning for the conservation of biological diversity.

As conservation planning emerged as a discipline, many countries began to recognize the need for more systematic planning efforts. As mentioned earlier, a United Nations agreement known as the Convention on Biological Diversity (CBD) formally recognized this need and has provided a vehicle for the development of country-wide conservation plans.

The Convention on Biological Diversity

At the 1992 Earth Summit in Rio de Janeiro, world leaders reached a number of agreements for "sustainable development" of Earth's resources. Those agreements that focus on reducing greenhouse gases to combat the deleterious effects of global climate change have received a great deal of attention in the media. A less well-known but equally important agreement was the CBD, a legally binding instrument under the auspices of the United Nations, that obligates signatory countries to take a variety of actions to stem the tide of biological diversity losses. The CBD's objectives are "the conservation of biodiversity, the sustainable use of its components, and the fair and equitable sharing of benefits arising from genetic use of resources" (Glowka et al. 1994, Secretariat of the Convention on Biological Diversity 2000, 2001). Some of the requirements of the CBD are for governments to:

- Develop national biodiversity plans and strategies and integrate these plans into broader environmental and development plans.
- Identify and monitor important components of biodiversity.
- Establish protected areas to conserve biodiversity.
- Restore degraded ecosystems and promote the recovery of threatened and endangered species.
- Preserve and maintain traditional knowledge and the sustainable use of biodiversity with the involvement of indigenous peoples and local communities.
- Prevent and control the introduction of alien species.
- Control the risks posed by organisms modified through biotechnology.
- Promote public participation in environmental evaluations of projects that threaten biodiversity.
- Educate people about the importance of biodiversity and its conservation.

• Report periodically on the progress of each country toward its biodiversity goals.

Over 175 countries, with the notable exception of the United States, have ratified the Convention on Biological Diversity. Activities by developing countries under the auspices of the CBD are eligible for financial support from the Global Environmental Facility (GEF). Support for GEF projects come from the United Nations Environment Programme (UNEP), the United Nations Development Programme (UNDP), and the World Bank. As of 2000, the GEF had contributed over US$1 billion in funding to convention-related biodiversity projects in more than 120 countries. Two requirements of the CBD—developing National Biodiversity Strategies and Action Plans (NBSAPs) and establishing a system of protected areas—are especially relevant to the theme of this book. Chapter 2 will focus on information available through the Convention on Biological Diversity that can contribute to the development of conservation plans. Fortunately, the science of conservation planning continues to advance the development of tools and methods for helping signatory countries prepare NBSAPs and identify the most significant remaining areas in their countries that can contribute to the conservation of biodiversity.

The Influence of Ecology and Conservation Biology on Planning for Biodiversity

During the 1990s, a number of advances in ecology and conservation biology had a significant impact on our thinking about conserving biological diversity in the twenty-first century. First, ecologists adopted a new way of thinking about the natural world. Termed the nonequilibrium paradigm, it places greater emphasis on the interconnectedness and unpredictability of natural ecosystems, and on the role that natural disturbances and ecological processes play in maintaining nature's patterns (Baker 1992, Smith et al. 1993, Balmford et al. 1998). The expansive wildfires in the western United States during the summer of 2002 certainly underscored the importance of considering disturbances as a natural feature of ecosystems. Second, as mentioned at the beginning of this chapter, the growing list of endangered species highlighted the need for approaches to conservation that go beyond single-species management. Gap Analysis (Scott et al. 1993, Jennings 2000), a scientific program in the United States to identify "gaps" in the network of nature reserves, was initiated, in part over frustration with the ineffectiveness of single-species approaches. Third, we gained a greater appreciation for the fact that biodiversity occurs at multiple spatial and temporal scales as well as different levels of biological organization (Schwartz 1999, Poiani et al. 2000). Finally, the growth of the field of landscape ecology has taught us that the patchiness of the natural world is really important in determining how many species may occupy an area, in the transmission of pests and pathogens, in the dynamics of local populations, and consequently in the selection and design of nature reserves or conservation areas (Wiens 1997).

These advances in science, coupled with the deepening of the extinction crisis and the recognition of limited conservation dollars with which to abate this crisis, have resulted in significant changes to planning methods and conservation strategies by both nongovernmental organizations (NGOs) and governmental agencies. Over the last decade, numerous institutions have been engaged in planning efforts that have established regional, country-wide, or continental priorities for biodiversity conservation (Redford et al., in press). For example, international conservation organizations such as the World Wildlife Fund (WWF) and Conservation International (CI) are setting global priorities for biodiversity conservation through assessments that pinpoint the most biologically important ecoregions and biodiversity "hotspots," respectively (Mittermeier et al. 1998, Olson and Dinerstein 1998, Myers et al. 2000). BirdLife International has identified hotspots of bird endemism throughout the world (Stattersfield et al. 1998). The European Union is creating a network of protected areas across Europe known as Natura 2000 to conserve endangered species and vulnerable habitats (European Commission 2000).

In a similar fashion in the United States, the National Gap Analysis Program (Jennings 2000) is assessing, on a state-by-state basis, the degree to which vegetation cover types and vertebrate species are adequately represented in a system of biodiversity management areas. Numerous states are undertaking planning efforts for biodiversity conservation (e.g., California Resources Agency 2000, 2002), and several of these are multiyear efforts that are already being implemented in states such as Oregon (Oregon Biodiversity Project 1998) and Florida (Hoctor et al. 2000, Kautz and Cox 2001). The Environmental Law Institute (2001) recently sponsored a symposium to review and synthesize the numerous ongoing state-based biodiversity initiatives. In fiscal year 2002, the U.S. Congress provided US$85 million in grants to state wildlife agencies, with the requirement that each state will prepare by 2005 a comprehensive wildlife conservation plan that considers the broad range of wildlife habitats and species with the greatest conservation needs. The Conservation and Recovery Act

(CARA) Committee of the International Association of Fish and Wildlife Agencies is now in the process of generating guidelines for the preparation of those plans (T. Johnson, Arizona Game and Fish Department, personal communication).

Most of these planning and priority-setting exercises identify priority regions or ecoregions that range in size from tens to hundreds of thousands of square kilometers, or they are focused on entire nations (NBSAPs under the CBD) or states. These types of planning jurisdictions are conceivably inhabited by thousands of species distributed over scores of natural communities and landscapes. Brazil's Cerrado region, for example, identified as a hotspot by Conservation International, extends over approximately 360,000 km^2 (139,000 mi^2) and harbors at least 10,000 species of vascular plants (Mittermeier et al. 1999). Overlaying this biological complexity are varying patterns of human use, land ownership, and land management. It is critical that we be able to identify which lands and waters across these vast areas are most essential for conserving biodiversity.

We now need a practical, science-based framework for planning the conservation of biodiversity *within* these priority ecoregions or hotspots, nations, or states (Groves et al. 2002). Such a planning framework is critically required for many reasons. First and foremost, there are limited conservation funds available. We want to ensure that those dollars are invested in places where they will reap the most benefit for conserving biodiversity. A recent conference attended by a number of renowned ecologists and conservationists concluded that "knowing which areas within hotspots are especially important could reduce costs considerably" (Pimm et al. 2001). Investing in places that harbor species or ecosystems that are not really in need of conservation, or places in which the threats to biodiversity are so severe that their abatement is unlikely, is bad for business and bad for conservation. And like many development projects, once an investment has been made in a conservation project, good or bad, it is difficult to walk away from it.

I mentioned a second reason for planning a framework at the beginning of this chapter: Conservation decisions and actions are increasingly taking place at local levels. Municipalities and county governments make decisions about whether and where to establish new parks, open spaces, and nature reserves. Similarly, in the United States, local and regional land trusts, which have mushroomed from 887 trusts in 1990 to over 1200 in 2000 (Figure 1.3), can increasingly play a role in biodiversity conservation (Tarlock 1993, Press et al. 1996). At the end of 2000, these land trusts had protected a total of 2,520,334 hectares (6,225,225 acres) (www.lta.org,

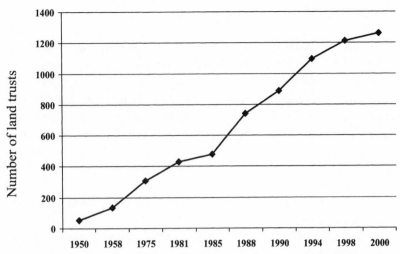

Figure 1.3   The growth of local and regional land trusts in the United States, 1950–2000. As of 2000, land trusts have conserved over 2.5 million hectares (more than 6 million acres) of land. (Data from Land Trust Alliance, www.lta.org.)

Land Trust Alliance). Yet, given the relatively small areas that local government or local institutions have jurisdiction over, how are they to know which species may be most vulnerable or which types of communities and ecosystems are in greatest decline and therefore in greatest need of conservation? Moreover, local governments that have statutory authority to make land-use decisions often lack expertise in conservation planning and are seldom able by themselves to design or implement conservation programs based on sound scientific information and thinking. In addition, they often lack the capacity to provide the regional coordination necessary to meet effective conservation goals or reliable access to biodiversity data to enable them to prepare credible plans and implement effective programs in the first place (Reid and Murphy 1995, Theobald et al. 2000). There is little doubt that local decisions and actions with regard to conserving biodiversity will be most effective when made and taken within the context of a wider, regional-scale conservation plan.

A third critical reason for developing a practical yet scientifically credible planning framework for biodiversity is to provide guidance for important global efforts such as the development of National Biodiversity Strategies and Action Plans (NBSAPs) under the Convention for Biological Diversity. Although the identification of priority areas for establishing new protected areas is a key component of these plans, little useful guidance has been provided to countries on how this might best be accomplished (Miller and Lanou 1995, Hagen 1999). Finally, and specifically for the United States, neither the major federal natural resource agencies (U.S. Forest Service, U.S. Bureau of Land Management, U.S. Fish and Wildlife Service, U.S. National Park Service) nor the states, with a few exceptions, have a collective or individual strategy for conserving biological diversity beyond that of implementing the Endangered Species Act. The Forest Service, with its species viability requirements under the National Forest Management Act (Andelman et al. 2001), was until recently the only federal agency with a legal mandate that includes biodiversity conservation. Now the U.S. Fish and Wildlife Service has broadened the legal scope of the purpose of the National Wildlife Refuge system to include a biodiversity component (Gergely et al. 2000). The National Park Service, through the writings of one of its scientists, Dr. Craig Shafer, has advanced a framework with many similar elements to those outlined in this book for how a national park system can be used to conserve biodiversity (Shafer 1999c), but this framework has yet to be transformed into policy. Although there are efforts in the right direction from all of these federal agencies and many state agencies, all would profit from incorporating a systematic approach to biodiversity conservation within their planning frameworks and land management actions.

How This Book Is Organized

The *Global Biodiversity Strategy* (World Resources Institute, World Conservation Union, and United Nations Environment Programme 1992) identified 85 different actions that could be taken to conserve biodiversity, ranging from providing universal access to family planning services to establishing new conservation areas to strengthening the conservation role of zoological parks and botanical gardens. In this book, I will be concerned primarily with identifying the most important lands and waters whose on-the-ground conservation will have the greatest impact on conserving biodiversity. Conservation biologists who specialize in the identification and design of such areas often refer to this process as area selection or site selection. We will attempt to identify these areas through a systematic planning framework. Planning for biodiversity conservation can be a complicated process, because the patterns of biodiversity we observe on the landscape and the underlying ecological processes that sustain them are, in themselves, complex. When we overlay those patterns with the distribution of human land use, the picture becomes even more complicated.

One of the goals of this book is to demystify and simplify this complex picture and planning process.

The book is organized in four parts. Part I (Chapters 1–3) is intended to provide essential background information for conducting a successful planning project and a big-picture view of the planning framework. It defines biodiversity, articulates some of the challenges to conserving it, and reviews some of the history of conservation planning as a discipline. Then it outlines a planning framework and the basic steps in the planning process for carrying out that framework. Finally, it provides guidance on how to get the planning project off to a good start by considering such topics as the audience, putting together a strong team, the overall vision and goals of the project, important sources of information, and so forth. Part II (Chapters 4–9) is the heart of the planning methodology—the seven habits of effective planners. Each chapter represents a distinct step (or series of steps) in the planning process, from selecting conservation targets (the "things" we are trying to conserve) to assessing the viability and integrity of these targets and designing a network of conservation areas. Each chapter concludes with a list of key points important to that particular planning step.

Part III (Chapters 10–12) focuses on conservation planning in freshwater and marine ecosystems and some of the nuances particular to the conservation of aquatic systems. Although the methods outlined in Part II are applicable to aquatic as well as terrestrial ecosystems, the relative paucity of biological information on aquatic ecosystems, the inability to easily obtain this information from remote-sensing methods, and our collective inexperience in planning for conservation in these environments make them worthy of special attention. Equally worthy is the timely topic of the concluding chapter in Part II—the impact of climate change on biodiversity and what we can do about that in a conservation planning context. Part IV (Chapters 13 and 14) is devoted to the implementation of conservation plans and future directions for conservation planning. The chapters give advice about how to best move forward in turning plans into conservation action, thereby avoiding the "dusty bookshelf" fate that awaits so many planning efforts. Included in this part are some outstanding examples of plans that have been translated into effective "on-the-ground" or "on-the-water" conservation efforts.

Identifying a set of priority places to conserve biodiversity within a regional planning framework is, of course, just one part of the picture of what it takes to actually accomplish biodiversity conservation. The Nature Conservancy (2001) has articulated a dynamic four-part Conservation

Process that succinctly describes the broader set of activities involved in achieving biodiversity conservation (Figure 1.4). These include setting site-based or area-based priorities, developing strategies to conserve biodiversity and abate threats at appropriate scales, taking conservation action to abate current and future threats, and then measuring the degree to which conservation actions are actually effective at achieving a project's goals. The different components of this conservation process take place at different spatial and temporal scales. For example, the first component of setting priorities takes place over broad regions through a systematic planning framework, while the other three occur largely at the scale of individual conservation projects but also at a range of scales appropriate to the entities we are trying to conserve and the threats to them. The first component—setting propreties—will be the focus of this book, although the Conservation Process introduced here will be explored in more detail in Chapter 13.

As the book's subtitle suggests, the intended audience for this book is conservation practitioners, biologists, and planners in both governmental and nongovernmental organizations worldwide who have the responsibil-

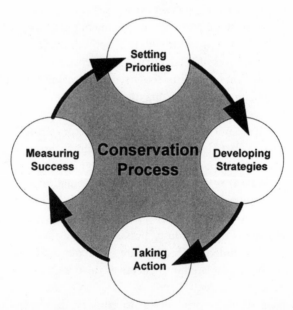

Figure 1.4   The Conservation Process: four components of an overall process for conserving biodiversity, as described by The Nature Conservancy (2001). The focus of this book is on the first component, setting priorities, a reference to the identification of important areas for conservation in a scientifically based regional planning process.

ity to develop, use, or simply understand planning processes to conserve biodiversity. Whether you work for a small local land trust or a large multilateral development organization, my hope is that you will find the guidance in this book to be useful. I have strived to explain scientific concepts and illustrate them with examples as often as possible, while at the same time not going too deeply into the technical detail of any one concept or step. For those who want more detail, the book is rich with references that will provide that detail, as well as additional examples.

CHAPTER TWO

Foundations: Ecoregions, Guiding Principles, and a Planning Process

Save some of everything, save enough to last.
—MARK SHAFFER AND BRUCE STEIN (2000)

Although humans tend to organize the world along geopolitical lines, environmental problems and biodiversity pay scant attention to such boundaries. Information on biological and environmental resources is collected and managed at a variety of scales along many different types of boundaries, some ecological and some geopolitical. For natural resource managers and planners who are concerned with many resources (e.g., soils, wildlife, vegetation cover, hydrology), there is a real need for a mapping and classification system that integrates these different natural resources. Because what happens on one piece of land or one body of water is related to what happens on adjacent areas, such a classification system needs to be spatially hierarchical.

Different geographers and biologists have developed multipurpose classifications, usually referred to as ecosystem classifications, for many parts of the globe. The study of the distribution pattern, structure, and process of differentiating ecosystems at different spatial scales has been termed *ecosystem geography* (Bailey 1996). The hierarchy begins with ecosystems that extend over only a few hectares (or acres) and are relatively homogeneous in cover. Local marshes and wetlands are good examples of such ecosystems that are often referred to as *sites* (Bailey 1996). Adjacent sites can be linked together in units called *landscapes* that cover from ten to several thousand square kilometers (or miles). In the northern Rocky Mountains in the United States, a common type of landscape comprises small wetlands or riparian areas surrounded by shrub steppe on valley floors, which in turn is enclosed by adjacent mountainsides with dry coniferous forests dominated by Ponderosa pine (*Pinus ponderosa*) or Douglas fir (*Pseudotsuga menziesii*). On an even larger scale, landscapes are connected together to form regions or *ecoregions*. The break between the grasslands of the Great Plains in the central United States and the adjacent Rocky Mountain landscapes is a good example of the differentiation between two ecoregions.

In southern South America, two adjacent ecoregions—the Patagonian grasslands and the Magellanic subpolar forests—illustrate a similar contrast. The Magellanic forests are dominated by beech (*Nothofagus*) trees, the topography is mountainous, and the climate is wet, temperate, and cold. To the north and east, the Patagonian grasslands consist of mesas and plains dominated by perennial grasslands, interspersed with shrubs and wetlands (www.worldwildlife.org/wildlworld/profiles/terrestrial_nt.html, World Wildlife Fund, United States). In central Australia, the Finke and MacDonnel Ranges bioregions (i.e., ecoregions) form an obvious but less distinct contrast (Thackway and Cresswell 1995). The former is dominated by arid plains with spinifex hummock grasslands, whereas the latter grades into foothills and mountains with sparse acacia shrublands and woodlands along riparian corridors.

Bailey (1998) defined ecoregions as major ecosystems, resulting from large-scale predictable patterns of solar radiation and moisture that, in turn, affect the distribution of local ecosystems and their component plant and animal species. From a biodiversity perspective, Dinerstein et al. (1995) provided a more pertinent definition: "[Ecoregions are] relatively large areas of land and water that contain geographically distinct assemblages of natural communities." These natural communities "share a large majority of their species, dynamics, and environmental conditions, and function together effectively as a conservation unit at global and continental scales." Because ecoregions contain relatively distinct assemblages of flora and fauna, they make useful planning units for conserving biological diversity. In the next section, I explore their use as planning units in more detail.

Ecoregions as Planning Units

There have been many efforts to develop ecosystem or ecoregional classification systems (see Grossman et al. 1999 for review). Some of these classifications are based largely on biotic factors (e.g., Dasmann 1973, 1974,

Udvardy 1975), whereas others place more emphasis on environmental and physical features (e.g., Bailey 1996, 1998). Still others have merged the concept of ecosystem with the cultural and social landscape to define *bioregions*—units of governance that match social and ecological functions to promote sustainability (Branekhorst 2000). No matter what classification system is used, conservation planners should keep three caveats in mind (Olson et al. 2001). First, no ecoregional classification will be optimal for all taxa or biological features that a particular planning effort is attempting to conserve. For example, Wright et al. (1998) found that existing vegetation patterns in the Pacific Northwest of the United States did not correspond well with the patterns of two different ecoregional classifications. Second, the boundaries between ecoregions are usually gradual transitions from one major ecosystem type to another and are only rarely represented by distinct edges. Third, most ecoregions will contain patches of habitats that are more distinctive of adjacent ecoregions.

In the United States, The Nature Conservancy (TNC) has used Bailey's (1995) ecoregional classification (Figure 2.1, in color insert) combined with the work of a U.S. Forest Service Ecosystem Classification and Mapping Team (ECOMAP 1993) to define its conservation planning units, while relying on World Wildlife Fund (WWF) characterizations of ecoregions for planning outside the United States (see below). The Bailey and ECOMAP classifications divide ecoregional provinces into sections, subsections, landtype associations, landtype, and landtype phases on the basis of similarities of geomorphic process, surficial geology, soils, drainage networks, and regional climate patterns. These section and subsection units can be useful delineators for identifying and planning for the conservation of species, communities, and ecosystems across the different environmental gradients in which they are distributed. Chapter 6 on setting conservation goals will explore the use of such subdivisions in more detail.

Eric Dinerstein, David Olson, and their colleagues at WWF spent much of the 1990s creating a global framework for conserving biodiversity based on ecoregions (Figure 2.2, in color insert; Olson et al. 2002) and setting conservation priorities for many regions of the world (Dinerstein et al. 1995, Ricketts et al. 1999, Abell et al. 2000, Wikramanayake et al. 2001). In North America, their maps are based upon the work of Omernik (1987, 1995) for the United States, the Ecological Stratification Working Group for Canada (1995), and Gallant et al. (1995) for Alaska. Ricketts et al. (1999) pointed out that some natural resource managers and conservationists have raised concerns that there are several different ecoregional classifications in use in the United States (McMahon et al. 2001). They also

noted, however, that there is a great deal of overlap in all of these ecoregional maps, especially the center points of ecoregions, and that careful
scrutinizing of the maps reveals more similarities than differences. As
unfortunate as it is that different organizations and agencies are using different classification systems, any anxiety over these incongruities should be
tempered by two points. First, any well-thought-out ecoregional classification will be a big improvement over using geopolitical boundaries as planning units. Second, for organizations such as TNC and WWF that are
developing ecoregional conservation plans across contiguous regions, the
priority-setting and planning work that occurs *within* ecoregional boundaries is of far greater import to conservation than energy spent debating
the exact location of these boundaries (e.g., Jeppson and Whittaker 2002).
Too many times, I have witnessed biologists and planners devoting substantial time and energy to debating the placement of ecoregional boundaries while seeming almost indifferent to the lack of information on the
viability and integrity of biodiversity features inside those boundaries. At
best, this is a graphic example of biologists' own failure to "see the forest
for the trees."

I observed in the first chapter that a number of different governmental
agencies and conservation organizations are undertaking priority-setting
exercises for biodiversity. Many of these, although not conducted along
ecoregional lines, are occurring at similar spatial scales. Figure 2.3 provides
a comparison of the different spatial scales at which samplings of these
planning efforts are being conducted. For example, the biodiversity hotspots identified by Conservation International (Myers et al. 2000) typically
range in size from a few hundred to over a million square kilometers.
Birdlife International (Stattersfield et al. 1998) has identified global centers
of endemism for bird species that generally range from tens of thousands
to several hundred thousand square kilometers. Most states are completing
Gap Analysis projects (Jennings 2000), which provide the foundation for a
conservation plan by identifying ecosystems and species that are not adequately protected in existing conservation areas or reserves. Finally, as I
noted in Chapter 1, many countries that are signatory to the Convention
on Biological Diversity are preparing national strategies and action plans
for conserving biodiversity (e.g., Namibian National Biodiversity Task
Force 1998).

The underlying assumption of planning within an ecoregional framework is that ecoregions are based on environmental and ecological variables known to influence the distribution of plants and animals; therefore,
they will do a better job of serving as a template for conservation plans than

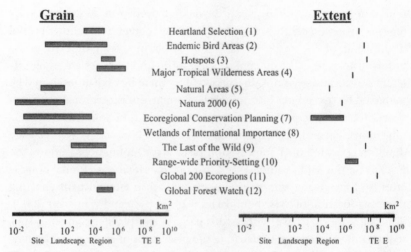

Figure 2.3   Priority-setting exercises for conserving biodiversity. Many of these planning efforts are occurring at similar spatial scales. Grain refers to the unit of analysis (such as a conservation area identified in a Nature Conservancy ecoregional plan), and extent refers to the total area under consideration (such as an entire ecoregion). References: (1) African Wildlife Foundation 2000; (2) Stattersfield et al. 1998; (3) Myers et al. 2000; (4) Mittermeier et al. 1999; (5) English Nature 2001; (6) European Commission 2000; (7) Groves et al. 2000a; (8) Ramsar 2001; (9) Sanderson et al. (submitted); (10) Sanderson et al. 2002b; (11) Olson and Dinerstein 1998; (12) World Resources Institute 2001. (From Redford et al. Mapping the conservation landscape. *Conservation Biology*, in press. Copyright, Blackwell Publishing Inc.)

geopolitical boundaries. Accordingly, the planning principles and framework detailed in this book will emphasize conservation planning on an ecoregional basis. However, these principles and guidelines are sufficiently robust for application to other types of planning units of similar spatial scale. It is worth reiterating that the most important outcome of these large-scale planning projects is the identification of priority lands and waters, referred to hereafter as *conservation areas,* and the ecological processes and disturbances that sustain these areas, the management of which will conserve a substantial majority of a region's biodiversity. The principles and planning framework developed in this book are all designed to help planners, managers, biologists, and conservation practitioners achieve that end.

Conservation Planning Principles

John Prendergast and colleagues (1999) published a provocative paper that pondered the question of why science-based, systematic approaches to identifying priority areas for conserving biodiversity have not been more widely adopted by governmental agencies and conservation organizations. They suggested that a lack of awareness, low levels of funding, a lack of understanding of the purpose of sophisticated software tools, and a generally negative attitude toward prescriptive approaches to conservation were all to blame. Although these are all plausible explanations, we must also recognize that most successful conservation planning efforts are both very public and very political processes. Strong scientific frameworks and tools will improve the conservation outcome and bring credibility to the process. However, sophisticated tools and the best scientific information to guide a planning process are insufficient by themselves.

Recognizing that conservation planning needs to be effective within a scientific, social, and political framework, Nels Johnson and the World Resources Institute (1995) advanced a useful set of principles for setting conservation priorities for biodiversity:

Principle One: Link biodiversity priorities with clear conservation goals and objectives.

Principle Two: Use a replicable, transparent process to develop credible priorities.

Principle Three: Clarify local, national, and global biodiversity priorities.

Principle Four: Evaluate the advantages and disadvantages of relevant priority-setting schemes.

Principle Five: Make full use of relevant and available information.

Principle Six: Involve those responsible for implementing conservation action.

Principle Seven: Involve communities and other stakeholders.

Principle Eight: Consider how priorities fit into a policy and institutional context.

Principle Nine: Link conservation priorities to other planning and policy processes.

Principle Ten: Establish a process to revise or reassess priorities at regular intervals.

The final section of this chapter outlines a seven-step planning process for developing a regional conservation plan. Each of these steps is more fully developed in Part II of this book, which, in turn, incorporates either explicitly or implicitly each of the above principles. Conservation goals (Principle One), for example, are the topic of Chapter 6. Relevant information (Principle Five) is a major theme of Chapters 3 and 5. Various approaches for setting priorities are discussed at length in Chapters 8 and 9. How conservation priorities fit into a policy and institutional context is addressed in Chapter 13.

The Four-R Framework

The most important product of this planning process is identifying a network of conservation areas that will conserve the full array of biological diversity found within a region, ecoregion, or similarly scaled planning unit. Chapter 8 will focus in more detail on the exact meaning of the term *conservation area*. For now, it is sufficient to say that conservation areas are areas in which the primary management concern is the conservation of specific biotic or environmental features. Many scientists and conservationists have proposed criteria or protocols for identifying conservation areas. Shaffer and Stein (2000) suggested three that I find to be the most useful for their clarity and simplicity, and I have added a fourth. A portfolio of conservation areas for biodiversity should be representative, resilient, redundant, and restorative. I have termed these criteria the Four-R Framework, as they should provide the structural underpinnings of any proposed system of biodiversity conservation areas.

> *Representative. Conservation areas within the planning region should represent the biological features and the range of environmental conditions under which they occur.* The idea that a system of conservation areas should represent the biological variation in a region has been advocated since at least the beginning of the twentieth century (Scott 1999). As early as the 1920s, Victor Shelford and the Ecological Society of America suggested the need for a network of conservation areas (Croker 1991). The Man and the Biosphere program of UNESCO, initiated in 1974, was probably the first major conservation initiative focused on representing the range of global biotic diversity in an international system of reserves (Austin and Margules 1986). Noss and Cooperrider (1994) defined representation as the full spectrum of biological and environmental variation in a region. They further suggested that conservation programs should attempt to represent all genotypes, species, ecosystems, and landscapes in a network of reserves. Although this is a laudable goal, there is never sufficient information to know with certainty whether these elements are actually being represented. In reality, some subset of species, communities, and ecosystems will be the focus of attention in most planning projects, although it will often be necessary to rely on additional surrogate measures of biodiversity. For example, there will never be sufficient information on most invertebrate or plant species to plan for their conservation. As a result, we must rely on surrogate measures, predictive modeling, and post hoc validation steps in the selection of conservation areas to ensure that these species do not fall through the fig-

urative cracks of our planning framework. Chapter 4 addresses this topic of what features of biodiversity, referred to hereafter as *conservation targets,* should be the focus of regional planning efforts in considerable detail.

Resilient. Conservation targets occurring within lands and waters identified as priority conservation areas should be resilient to both natural and human-caused disturbances. This notion of resilience incorporates both the concepts of population viability (Shaffer 1981) and the ecological integrity of communities and ecosystems (Angermeir and Karr 1994). This criterion means that individual populations or occurrences of conservation targets—be they species, communities, ecosystems, or other types of targets—should be of sufficient quality to persist for long periods of time. Quality implies that populations of target species are large enough and reproducing sufficiently to remain viable. For communities, ecosystems, and other surrogate measures, quality implies that natural ecological processes and disturbance regimes such as fires and floods are operating within their historical range of variability (Landres et al. 1999), and that the size of an area is sufficient to allow conservation targets to recover from natural disturbances.

Ecological processes have traditionally received short shrift from conservation planners (Balmford et al. 1998, Margules and Pressey 2000), primarily because it is easier to determine the presence or absence of a species or community than it is to assess its viability and biotic integrity. A credible conservation plan must focus on conserving the patterns and the processes that sustain biodiversity. Changes in global climate are influencing both pattern and the process (Hughes 2000). The impact of these changes represents a significant challenge to conservation planners who must attempt to ensure that efforts to design regional portfolios of conservation areas are resilient to both natural and human-related disturbances, including those stemming from the difficult-to-predict effects of climate change. Chapter 7 explores this issue of resilience (viability and integrity) at length, and Chapter 12 focuses in more detail on how conservation planners can take the potential impact of climate change into consideration when developing regional biodiversity plans.

Redundant. To avoid extinction or endangerment caused by both naturally occurring stochastic events (e.g., disease, predation, floods, fires) and human-related threats, conservation targets should be represented multiple times within a system of conservation areas. This principle is a simple reiteration of the wise old adage, "Don't put all of your eggs in one basket." Sufficient numbers of populations need to be conserved so that target species have a high likelihood of persisting for long periods of time. Similarly, targets at other levels of biological organization (e.g., communities and ecosystems) need to be conserved multiple times across the environmental gradients in which they are distributed. This is especially critical for communities and ecosystems because these targets are serving as surrogates for conserving other species. Ensuring that conservation targets are adequately represented in conservation areas leads planners to set specific goals. Conservation goals should have two components: (1) a representation component, which refers to either the number of occurrences or the percentage of each target that should be represented within conservation areas, along with some indication of how those targets should be distributed or stratified across a planning region; and (2) a quality component, which addresses the level of viability or ecological integrity thought necessary for persistence. Chapter 6 provides helpful guidance in setting conservation goals for different types of conservation targets.

Restorative. Conservation planners should evaluate where, when, and how occurrences of conservation targets that are not viable or lack ecological integrity may be feasibly restored to appropriate levels of viability and integrity within the planning region. In many of the more developed regions of the world, remaining natural lands are highly fragmented, populations of many species within these fragments are below viability thresholds, and natural disturbance regimes and ecological processes no longer function properly. Consequently, there may be opportunities to restore conservation targets to viable levels and reestablish natural disturbance regimes. In a few cases, targets may occur at viable levels but not at ecologically functional levels, and it may be possible to restore some populations to a level of functionality. Restoration activities can be thought of as occurring along a continuum (Hobbs and Norton 1996), from the re-creation of lost habitats at one end to improving degraded ecosystems within largely intact landscapes at the other (see Figure 7.10).

Traditional ecological restoration efforts have been modest in scale (Soulé and Terborgh 1999a), usually focused on the reintroduction of an ecological process (e.g., prescribed fire), the removal of exotic species, or the revegetation of highly disturbed sites, all in relatively small areas. Some restoration projects are now taking place at larger spatial scales. For example, the restoration of the Kissimmee River/Everglades ecosystem in southern Florida, slated to cost around US$7.8 billion over 20 years, is the largest single ecological restoration project that has ever been undertaken (U.S. Army Corps of Engineers 2000). Biodiversity planning at the scale of ecoregions should examine restoration opportunities that span the continuum of possibilities, from restoring natural disturbance regimes to expanding existing conservation areas through a reconnection of remnant natural habitats. Chapter 7 delves more deeply into selecting areas for restoration, including a framework for evaluating the pros and cons of different types of restoration projects at different scales.

★ ★ ★

The Four-R Framework is just that—a framework to support a more detailed process. Planners, managers, biologists, and conservation practitioners need specific guidance and direction on a regional or broad-scale planning process aimed at conserving biodiversity. Although there is likely to be no single planning process that is universally appropriate, The Nature Conservancy has developed, implemented, and improved, over time, an ecoregional planning process that works well in terrestrial, freshwater, and coastal marine systems and that is applicable to both data-rich and data-poor regions. Between 1996 and 2002, over 50 regional and ecoregional conservation plans from Yunnan, China, to Central America to the United States were completed using this planning process. The process itself has emerged from the collective experience of hundreds of conservation planners, biologists, and practitioners.

The seven-step planning process outlined below forms the underlying foundation for Part II of this book (Figure 2.4). Although the basic steps parallel those being used by TNC in its ecoregional planning efforts (Groves et al. 2000a, 2002), the details of how each step is carried out go well beyond the work of TNC. They draw in depth upon literature from the fields of ecology, conservation biology, professional planning, and landscape architecture as well as the work of many experts in conservation organizations, planning institutions, and government natural resource agencies worldwide. In fact, there are several parallel lines of thought in the seven steps outlined below and the steps used by planners who employ the ecological planning method as described by landscape architect and planner Frederick Steiner in his book, *The Living Landscape: An Ecological Approach to Landscape Planning* (2000).

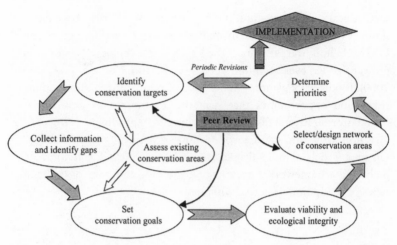

Figure 2.4   A seven-step process for planning for the conservation of biological diversity. The actual planning process is not as linear as the numbered steps in the text suggest, because several steps can occur simultaneously. Feedback loops within steps indicate that the entire process is dynamic; it is constantly being updated and revised. Peer review and evaluation should occur throughout the planning process.

The Seven Habits (Steps) of Effective Planning for Biodiversity Conservation

In his best-selling book, *The Seven Habits of Highly Effective People,* Stephen Covey (1990) mixes experience and wisdom into lessons learned from individuals who are successful in both their professional and their personal lives. The seven steps outlined below are similarly a blend of scientific wisdom and practical experience in the development and implementation of conservation plans.

Step One: Identify Conservation Targets.

Conservation targets are those features or elements of biodiversity that planners seek to conserve within a system of conservation areas. These targets may be biologically based features, such as species and communities, or they may be environmentally derived targets (based on such factors as soils, climate, geology, and elevation) that serve as surrogates for biological features. Some types of conservation targets have legal standing under various laws. For example, under the Endangered Species Act, some subspecies and distinct populations are legally recognized and protected. Con-

siderable debate has taken place over which levels of biological organization are most appropriate to serve as targets for conserving biodiversity
(e.g., Franklin 1993, Orians 1993a, Wilcove 1993). Because biological
diversity occurs at a variety of spatial scales and levels of organization
(Noss 1990, Poiani et al. 2000), conservation planners should, whenever
feasible, identify targets at multiple levels of biological organization and
spatial scales. In addition, planners should ensure that they identify and
focus attention on the unique or distinct biological features of a planning
region (e.g., high degree of plant endemism, outstanding ungulate migration processes). Identifying these sorts of features requires the planning
team to be able to place the planning region within a broader biogeographic context.

Step Two: Collect Information and Identify Information Gaps.

A wide variety of data, ranging from socioeconomic trends and land ownership information to target-specific biological data, is necessary for developing a regional conservation plan for biodiversity. The more important
categories of information include land use and land ownership; physical
and environmental attributes (e.g., soils, geology, elevation, hydrography);
socioeconomic variables (e.g., human population trends); and biological
(e.g., species occurrences) and ecological data (from remotely sensed vegetation maps). The best regional conservation plans will use information
from all available sources, including conservation organizations, public
natural resource agencies (local, state, provincial, federal), academia, planning commissions and agencies, indigenous communities, research institutions, and individual experts. In many cases, information critical to the
development of a conservation plan may be lacking. In these situations, it
is important for planning teams to identify the most significant information gaps and to begin filling them. It is also critical to identify the level of
confidence or uncertainty associated with various data sets used in conservation planning.

Step Three: Assess Existing Conservation Areas
for Their Biodiversity Values.

A logical step in any planning process for conserving biodiversity is to
determine what biological features are already under appropriate management within existing conservation lands and waters. This is an especially important step, given the limited conservation funds available for

establishing new conservation areas and the paucity of biological survey data for most of the world's existing parks, reserves, and protected areas. Identifying "gaps" in a system of conservation areas formed the conceptual basis for the U.S. Gap Analysis Program (Scott et al. 1993, Jennings 2000). These state-based programs analyze which biological features are found within existing conservation areas and to what extent these areas are actually managed for conserving these features. Gap analysis projects, similar to those conducted in the United States, have now been completed in numerous regions of the world (see Chapter 5). In many developing countries, these analyses must be carefully scrutinized to ascertain whether existing protected areas are being adequately managed to conserve the biological features within them. Conservation planning methods developed principally in Australia have incorporated a similar "gap-checking" procedure (Margules and Pressey 2000).

Step Four: Set Conservation Goals.

Establishing conservation goals is important for several reasons. First, goals allow an evaluation of how adequate existing conservation areas are for conserving the biodiversity of a region. Second, they provide guidance to planners who may also be balancing a number of competing demands for various lands and waters in a region. Third, they essentially set the upper limit to the number and size of conservation areas that are needed in a planning unit. Finally, goals provide a means for measuring the success of how well a portfolio of conservation areas will perform in conserving a region's biological diversity. Unfortunately, there is only a limited amount of both theoretical and empirical work upon which to base conservation goals. As a result, goal setting involves a substantial amount of informed guesswork and a heavy reliance on expert opinion. In most cases, planning teams should establish a range of reasonable numeric goals and use them to determine what conservation alternatives may be possible in the planning unit.

Step Five: Evaluate the Viability and Integrity of Conservation Targets.

Although conservation planners increasingly recognize that factors responsible for ensuring the persistence of biological diversity receive little attention in most planning processes, only minimal practical guidance and methods for assessing the viability and ecological integrity of conservation targets have been available. The Nature Conservancy, in conjunction with

NatureServe (formerly known as the Association of Biodiversity Information) and the network of Natural Heritage Programs, has developed an expert opinion–based approach for evaluating viability and ecological integrity in a conservation planning context (Anderson et al. 1999, Groves et al. 2000a, Stein and Davis 2000). This approach uses a qualitative ranking scheme to evaluate the viability and integrity of conservation targets on the basis of three criteria: size, condition, and landscape context. Other promising methods include the overlay of multiple variables (e.g., road density, human population density, percentage of natural cover remaining) in a Geographic Information System (GIS) environment as an index to the "suitability" of a site or area for conservation purposes, along with the mapping of spatial components known to represent important ecological processes (e.g., sand dunes, wildlife migration corridors, floodplains). Suitability analyses have their origins in schools of landscape planning and architecture (McHarg 1969, 1997, Steiner 2000). Where occurrences of conservation targets are determined to lack sufficient viability or integrity, planning teams should consider the feasibility of restoration.

Step Six: Select and Design a Network of Conservation Areas.

Often referred to as assembling a network of conservation areas, this step in the process synthesizes all of the information from previous steps to result in the most important product of regional planning—a portfolio of conservation areas (Figure 2.5, in color section). Because of the complexity of selecting conservation areas to represent large numbers of conservation targets, often with different goals, the use of computerized algorithms or programs with GIS is recommended as a tool to assist planners. The primary advantage of using these algorithms is that they enable planners to select conservation areas with an explicit set of "rules" and then assess alternative networks by making adjustments to these rules. The selection of conservation areas should place an emphasis on the landscape scale (Poiani et al. 2000). On average, landscape-scale conservation areas will "capture" more targets, and individual occurrences of those targets are likely to be larger and therefore more viable. There is also a greater likelihood that natural disturbance regimes and ecological processes will be intact and functional.

The entire area selection process should be as transparent, explicit, and repeatable as possible. It is critical that key decision makers and stakeholders be involved in this process so they buy in to the results. Ultimately, planners should strive to design not just a portfolio—a collection of independent conservation areas—but a *network* of conservation areas that pays due diligence to relationships among conservation areas, including the juxtaposition of new conservation areas next to existing ones, linkages among conservation areas where appropriate, and the limiting of habitat fragmentation. More than any other step in the planning process, designing the network of conservation areas is the best opportunity for the planning team to express its vision and ambitions.

Step Seven: Assess Threats and Setting Priorities within the Planning Unit.

In practice, regional conservation plans may identify one hundred or even several hundred potential conservation areas within a planning region. Some of these areas are in more urgent need of conservation action than other areas. Therefore, a final step in the conservation planning process prior to implementation is to set priorities for action within the network of areas. Five criteria are useful for setting these priorities: degree of existing protection, conservation or biodiversity value, threat, feasibility, and leverage. The most important criterion among these is the degree of threat to conservation areas and to the targets contained in them. Evaluating threats is important for two reasons. First, the severity and scope of threats help determine which conservation areas are in most urgent need of attention. Second, for threats that recur across many conservation areas, it may be possible to design multiarea strategies to abate these threats. Graphing the degree of threat or vulnerability of an area versus the extent to which conservation targets in the area are irreplaceable is a helpful technique in setting conservation priorities (Pressey 1999).

★ ★ ★

For discussion purposes, this planning process is described in a stepwise fashion. In the real world of planning, many steps take place concurrently, and there are built-in feedback loops among the steps (see Figure 2.4). For example, gathering information needs to begin even prior to selecting the conservation targets, as their selection will be influenced by the data gathered. Assessing the value of existing conservation areas is often done concurrently with the gathering of information about conservation targets and other data relevant to the planning process. Peer review and evaluation of the planning process need to occur throughout the life of the project, but especially at the target selection and area selection stages. Finally, implementation need not and should not wait until the plan is completed.

The most effective teams are thinking about and moving the planning process toward implementation from the onset of the project.

A final point about the planning process is the need for it to be iterative and dynamic. There will always be data gaps, and with scientists' unquenchable thirst for data, there will always be a need for more information. The first iteration or version of a plan may not include every detail or component that is discussed in this book. The best plans are living documents, and they are improved over time as new or better information becomes available and as improved methods are developed.

Australian ecologists Chris Margules and Bob Pressey have long advocated taking a systematic approach to planning for the conservation of biodiversity and have been responsible for developing many of the methods for this approach. In a recent review paper (Margules and Pressey 2000), they identified six characteristics of systematic conservation planning, which overlap extensively with the seven-step process outlined here. Margules and Pressey argue that planners must first be clear about what features or elements they are trying to represent within reserves (step 1). Second, they should set explicit goals for these features (step 4). Third, existing reserves should be examined to determine the extent to which goals can be met in these areas (step 3). Fourth, new reserves should be selected based on the principle of complementarity (see Chapter 8). Fifth, planners should be explicit about which areas are of highest priority for on-the-ground conservation action (step 7). Finally, specific management objectives and monitoring programs should be adopted and implemented, respectively, for reserve areas to ensure that the targeted features are maintained (see Chapter 13).

Several issues related to this seven-step framework require special consideration when developing regional conservation plans for biodiversity. Among these are some of the nuances in selecting targets, setting goals, and identifying conservation areas for freshwater and coastal marine systems. Chapters 10 and 11 in Part III of the book will address these issues. Similarly, the effects of global climate change are likely to influence all aspects of conservation planning. In Chapter 12, we consider the implications of climate change on the conservation of biodiversity and what can be done about it in the context of conservation planning. Finally, the worst possible outcome of an otherwise quality conservation planning effort is a plan that is consigned to the dusty bookshelves of history. Part IV of this book will focus on the most effective strategies for implementing these plans, including a number of examples of how different institutions are going about this process.

There are several key ingredients to the development of a successful conservation plan. Chief among these are agreement from the outset on an ambitious overall vision and goals for the project, a talented project director and team, a good plan for managing the project to deliver quality products on time and within budget, and the best possible set of science-based information with which to draft a conservation blueprint. The next chapter focuses on these and other critical building blocks.

CHAPTER THREE

Building Blocks for Regional Conservation Planning

One of the fallacies of project management is that once a project plan has been assembled, the world will stand still while the plan is executed.

—JAMES P. LEWIS (2000)

Many conservation biologists have formal education and work experience in some discipline of ecology, the use of statistics, and perhaps the application of natural resource economics or policy. Yet no amount of scientific education and training is sufficient by itself to succeed at conservation planning. Developing a conservation plan that will have conservation impact goes well beyond the tedious tasks of selecting conservation targets and using sophisticated computer mapping technology to identify potential conservation areas. Another set of tasks is equally important to the ultimate success of a conservation planning project: identifying what conservation problems the plan is attempting to address, assembling an effective team, managing the planning process, including critical partners and stakeholders in that process, and accessing and managing the best information. In addition, both internal and external peer review throughout the life of the project will improve its quality and credibility. Although few books in ecology or conservation biology address these tasks, they are so critical to the success of a planning project that I am devoting a separate chapter to

41

them. I refer to these essential tasks as the building blocks of a successful planning project.

Building Block One: Defining Conservation Problems and Envisioning Solutions

There is no better time to think about the desired outcomes of a conservation planning project than at its beginning. It is a critical period in which to give thoughtful consideration to the scope of the project, the audience for the plan, how much time and money should be invested in it, the participation of major stakeholders and potential partners, and, most importantly, the primary goals or objectives of the project. It is equally important to be clear about which conservation problems will be addressed by the planning project and which will not. Nearly all problems related to conserving biodiversity have a scientific, social, and political side to them. Yet most of the conservation plans that are being developed by major conservation organizations like The Nature Conservancy (TNC) have a tendency to focus primarily on the scientific components of a plan, often without much consideration of the social and political aspects that are key to implementation. Policy scientists have developed a process for better defining problems (Clark 1998, 2002). They have outlined five tasks that seem appropriate to any team undertaking a planning effort to conserve a region's biodiversity: clarifying the goals of the project, describing the trends and history of the problem with respect to the goals, analyzing the conditions under which these trends have taken place, projecting possible future trends in the problem, and inventing alternatives to solve the problem and reach the goals.

A biodiversity strategy and action plan developed for the Cape Floral Kingdom of South Africa and coordinated by the World Wide Fund for Nature–South Africa (WWF–South Africa) provides a good example of this process of problem definition (CAPE Team 2000). The Cape Floral Kingdom is a global hotspot of vascular plant endemism and has a diverse set of marine and freshwater ecosystems as well. Problems, however, are as diverse as the biodiversity, which is under intensive pressure from an array of threats, including habitat loss to agriculture, rapid development, commercial exploitation of marine resources and wildflowers, and invasions of exotic species. Conservation organizations in the region have generally acted independently, with few shared strategies or priorities, and there is no integrated legal framework to conserve biodiversity. In addition, these problems are compounded by the restructuring and reorganization of post-apartheid society in South Africa.

The overall vision of the Cape Action Plan for the Environment (CAPE) is:

> We, the people of South Africa, as proud custodians of the Cape Floral Kingdom, will protect and share its full ecological, social, and economic benefits now and in the future.

The goal of CAPE is as follows:

> By the year 2020, the natural environment and biodiversity of the Cape Floral Kingdom will be effectively conserved, restored wherever appropriate, and will deliver significant benefits to the people of the region in a way that is embraced by local communities, endorsed by government and recognized internationally.

Three overarching themes of CAPE are conserving biodiversity in priority areas, promoting sustainable use, and strengthening institutions. Each of these themes, in turn, has a set of strategic goals and objectives associated with them. For example, an action under the theme of conserving biodiversity is to identify and establish an effective network of nature reserves via a conservation planning process.

This type of strategic evaluation of the overall dimensions of a conservation problem, and envisioning the solutions, is critical components of any successful conservation plan. Team members should periodically revisit the problems being addressed by the planning process—as well as the plan's vision, goals, and objectives—to help ensure that they stay on track. The success of the team will depend a great deal on its leadership and composition. A periodic reevaluation of roles and responsibilities may be needed, because they may tend to evolve during the life of the project.

Building Block Two: Assembling a Team

My first attempts at assembling a team and launching one of TNC's first ecoregional planning projects were notably inauspicious. I brought together a diverse group of scientists and nonscientists from several state programs. Because our methods for developing these plans were in their infancy, the scientists devoted their time at this first meeting to delving into every detail of the biological information that we would need for the project. The reaction of the nonscientists in the room was predictable—disinterest and boredom. To make matters worse, our vision and overall goal for the project at that point were fuzzy at best, and there was not yet a recognized leader for the project. Unwittingly and unknowingly, I broke nearly every rule of good project management at that first meeting.

Not everyone needs to travel down this painful road to learn how to assemble an effective team or launch a project effectively. A great deal of knowledge and expertise about how to build effective teams is available. Katzenbach and Smith (1993) defined a team as a small number of people with complementary skills who are committed to a common purpose, to goals, and to an approach for which they hold themselves mutually accountable (Figure 3.1). Teams are a means to an end, not an end in themselves, and the members are bound together by one mutual objective—performance. High-performance teams require both individual and mutual accountability, and they strive to have an impact that is greater than what its members could achieve individually.

A number of lessons on team building that apply equally well to governmental or nongovernmental organizations can be learned from the business sector (Katzenbach and Smith 1993). First, keep it small. Most effective *teams* have fewer than ten members. Beyond that size, communication and the logistics of holding meetings become serious barriers to success. If more members are necessary, it may be useful to maintain a core planning team and create additional working groups or teams. Working groups contribute to specific parts of the planning process but generally

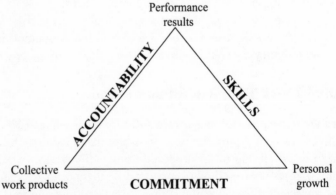

Figure 3.1 The key elements of a successful team: individual and mutual accountability, commitment to purpose and goals, and multidisciplinary and interpersonal skills. Teams strive for high performance, allow for personal growth, and have collective products that are of higher quality than any single individual could produce. (Reprinted by permission from Harvard Business School Press. Modified from *The Wisdom of Teams* by J. R. Katzenbach and D. K. Smith, Boston, MA [1994], p. 8. Copyright © 1993 by McKinsey & Company, Inc., all rights reserved.)

not to the extent that the core team does, nor do they have the level of accountability for the project's completion that the core team carries. Many of the ecoregional planning teams of TNC comprise numerous working groups related to areas of expertise in such subjects as rare plants, vulnerable animals, terrestrial communities, freshwater and marine systems, socioeconomic trends, and information management. Working groups are dynamic entities, and their number and type will vary throughout the life of the project.

Choosing a team based on a mix of skills is the second important lesson in assembling a team. Most teams need a variety of skills, including technical, problem-solving, decision-making, and interpersonal skills. For example, many of TNC's ecoregional planning teams have at least one team member with strong writing skills, another member who might work well with partner organizations or stakeholders, a member who focuses on the collection of socioeconomic data, and still another who takes on the critical task of mapping and information management. All team members should, of course, be able to work well with others in a group environment, and they must be able to commit their time to the project. This last point should not be taken lightly. The lack of investment in time and effort to produce a credible plan is, at best, inefficient and, at worst, a fatal flaw in the planning process. Inevitably, more time and money will be spent improving an inadequate plan that could have been satisfactorily completed the first time. In the worst scenarios, interested parties will recognize a poor-quality plan and have little incentive to invest further in its implementation.

Developing and agreeing on a common sense of purpose and shared goals, as well as a common approach for effectively working together, is another important team-building lesson. The most successful conservation planning teams have *members* who clearly understand their respective roles and responsibilities in the planning process. Ecoregional planning teams in The Nature Conservancy have found that relatively short written team charters can help clarify these roles and responsibilities (Table 3.1).

The final and critical ingredient to building a successful team is finding members who have a commitment to and trust in one another and are willing to hold themselves accountable as a team, not just as individuals. Much to the detriment of the team and an organization, some planning teams struggle to the point that they cannot bring a credible plan to completion, or do so only with great difficulty. More often than not, the failure of a team to produce is due in large part to a lack of mutual accountability, although weak leadership can also be part of the problem.

Table 3.1 *Team Charter for the Central Shortgrass Prairie Ecoregional Planning Team, The Nature Conservancy (revised February 1998)*

Purpose and Role: The purpose of the Central Shortgrass Prairie Ecoregional Planning Team is to create a conservation program within the Central Shortgrass Prairie by actively employing the principles espoused in Conservation by Design. The team will develop an ecoregional plan for the Central Shortgrass Prairie utilizing guidelines outlined in *Designing a Geography of Hope: Guidelines for Ecoregion-Based Conservation in The Nature Conservancy* (The Nature Conservancy 1997). This plan will identify a portfolio of conservation sites and strategies that will ensure the long-term survival of all viable native species and community types for the Central Shortgrass Prairie. The results of this planning effort will serve as a first credible iteration of a portfolio to determine where the Conservancy and/or others should work on the ground to protect biodiversity within the ecoregion.

Expected Products: The team will create an ecoregional plan for conservation of biodiversity targets in the Central Shortgrass Prairie Ecoregion. The major products to be completed include:

1. Documentation of methods
2. Summary of conservation targets that are the focus of conservation activity
3. Landscape identification and assessment consisting of GIS coverage of large blocks of natural vegetation delineated from TM Imagery
4. Identification of key ecological processes influencing major types within the ecoregion
5. Conservation lands assessment consisting of maps of conservation areas and lists of elements by conservation area and management
6. Identification of research needs, data gaps, and additional inventory and planning needs
7. System of sites composing the portfolio
8. GIS maps of conservation sites
9. Threats and feasibility assessments of multiple sites
10. Strategies to reduce region-wide threats and to fill data gaps

Timeline: The first iteration of the ecoregional plan, initiated in October 1996, will be completed by March 1998.

Structure and Composition: A core team with the ability to work quickly is essential. Also essential are input, advice, and guidance from a large number of people with an interest and a stake in the shortgrass ecoregional process. To achieve both needs, we propose a structure featuring a core team of seven people, supported by five functional groups. The core team will request advice and input from the functional groups throughout the process, and the functional groups will provide guidance and review products as the core team produces them. Each functional group will have a liaison from the core team, and the liaison will manage the relationship with the functional group. Most of the work will be completed by working groups, with lead staff serving on the core team.

Functional Advisory Groups:
1. Portfolio Assessment
2. Portfolio Design
3. Steering Committee and Regional Directors
4. Partner Organizations
5. Communications

Portfolio Assessment: A core team, consisting of representatives from states with the most land area within the ecoregion as well as regional office staff, will do the bulk of the data compilation and scientific analysis, which includes determining data needs, developing methods, and compiling and analyzing data. Texas and New Mexico field offices will be on the mailing list and kept informed of activities, but not expected to participate fully in the planning team due to the small size of area within the ecoregion.

Portfolio Design Team: Selected staff from the Portfolio Assessment Team will serve on the Portfolio Design Team. Other team members will include practitioners in protection, stewardship, and planning from CO, NE, KS, and OK. Proposed team members include Directors of Conservation Programs within these states, Protection staff, Heritage representatives, and state office Stewardship Scientists and Planners.

Steering Committee: This committee consists of directors of Conservancy Field Offices and Coordinators of Heritage Programs, and Director of the Great Plains Program. The objectives of the steering committee are to endorse the cooperative effort to identify a portfolio of conservation sites within the Central Shortgrass Prairie (CSP), commit varying degrees of staff, financial, and other resources to the planning effort, provide strategic leadership, and endorse the ecoregional plan for the CSP. Steering Committee members will review plans and work together on development of methods, team composition, and financial and human resource commitments. Team members will attend one meeting during the spring of 1997 to review progress made by the planning team. They will also identify partners and funding sources for the ecoregional project. The team will also be responsible for communicating with the Great Plains Program, Great Plains Partnership, and Great Plains Capital Campaign. Mark Burget and David Harrison are primary contacts.

Regional directors of the Western Region and Midwest Region also provide support and funding for the plan. The regional directors will keep in regular communication with state directors and Heritage coordinators regarding ecoregion-based conservation. Regional directors will provide staff to assist with special projects relating to Conservation by Design as it relates to ecoregional planning. Mark Burget is the primary contact.

Partner Organizations: Successful implementation of the plan will depend heavily on other institutions. The team will develop a strategy for gaining the input and advice of potential partners during the planning effort to encourage their interest and ownership in the plan. We anticipate creating a rough draft plan as a core team with input and guidance from the functional and working groups, then collecting substantial input from partners before revising and finalizing the plan. Biological Resources Division/USGS, Dept. of Natural Resources, and Division of Wildlife are primary partner organizations, but a number of organizations will be involved in the project through an ad hoc advisory group. An expert panel will be held to obtain input from partners and outside experts in the spring/summer of 1997. Heidi Sherk will be the primary contact with partner organizations.

Contributing Partners:
1. National Center for Atmospheric Research
2. University of Colorado, Boulder
3. Colorado Division of Wildlife
4. University of Michigan
5. Biological Resources Division
6. Colorado Department of Natural Resources

Communication: The heightened awareness of TNC's shortgrass investments and increased interest among staff, volunteers, and donors offers an ideal opportunity to use this planning effort to generate a great amount of internal and external support. We would like to capitalize on this opportunity by creating a steady flow of information from the planning team through internal publications, brown bags and other live media, and other opportunities decided by the core team.

Without a doubt, the most important member of a team is its leader. Teams can be effective with poor leadership, but the job is far more difficult. Leaders have a number of important functions to perform (modified from Katzenbach and Smith 1993):

- Keep the team focused on its purpose and goals, completing the project on time and within budget.
- Ensure that the team has the right mix of skills to get the job done.
- Develop meeting agendas and facilitate meetings.
- Delegate tasks to appropriate team members and clarify individual roles and responsibilities.
- Build the commitment and confidence of individuals and of the team as a whole, and provide opportunities for individual team members to shine.
- Communicate effectively internally among team members and externally to parties interested in the work of the team.
- Manage relationships with the rest of the organization and contacts from outside the organization.
- Remove obstacles and barriers to the team's success.
- Do real work.

It is unlikely that most team leaders will perform all of these functions exceedingly well, but the most effective leaders will ensure that the team fulfills them to some degree and that he or she excels at several of them. My own experience on teams in both governmental and nongovernmental organizations is that good leaders lead by example. They take on challenging work tasks, give credit to team members, insulate the team from bureaucratic matters that can unnecessarily divert attention from the primary purpose, and successfully confront barriers that are impeding a team's progress. Such leaders gain the respect of their team and are most likely to complete successful projects in which team members are satisfied with the product and their role in achieving the project's goals.

Assembling an effective team with strong leadership ensures that the planning project will get off to a good start. A productive first team meeting also helps set the stage for a successful project. Important activities for this first meeting include listing problems likely to be encountered during the life of the project, beginning to formulate solutions to these problems, describing the information the team will need to accomplish its tasks, developing a simplified and detailed breakdown of the tasks (see below), developing a project budget, proposing initial assignments, and developing a list of products (modified from Thomsett 1990).

There are a number of other issues in conservation planning whose resolution at the beginning of a project will help ensure that the project runs smoothly and achieves its goals and objectives. The next section will consider these issues under the broader topic of project management.

Building Block Three: Project Management

Whether developing a conservation plan for a national forest, a national park, a state, a country, or an ecoregion, a planning team should consider and answer several key questions at the outset that will make project management more efficient and effective:

• *What level of investment in time and money should be made in the planning project?*

Some important factors to consider in this investment decision are to what extent conservation opportunities remain in the region, how much information exists for analyzing these opportunities, the extent of existing conservation lands and waters, and what conservation institutions and organizations are active in the region. For example, in the Northern Tall-grass Prairie Ecoregion (Great Plains of the United States), only 4% of the ecoregion has not been converted to agriculture or urban settings. It may not be strategic to make the same level of investment in conservation planning in this ecoregion, where conservation opportunities are now very limited, as in other regions where there are substantially greater opportunities. On average, TNC has invested approximately US$234,000 per ecoregional plan ($n = 24$ plans), which includes staff salaries, travel, meetings, subcontracts, and printing. The average time to completion is about 2 years, with the caveat that team members are always involved in numerous other concurrent projects.

• *What organizational or institutional expectations exist that may affect the planning process?*

In some situations, there are clear and, at times, even legally mandated expectations of planning activities. For example, in TNC, there is now a short list of standards to which all ecoregional planning teams must adhere (P. Kareiva, TNC, personal communication). For governmental organizations, policies and regulations often guide the development of resource management plans. The important point here is that any planning team must be cognizant of organizational responsibilities and expectations with regard to the product of its efforts.

• *What products will be derived from the planning process?*

At the inception of a project, the team should determine what products will be generated from the planning process. The nature of those products will depend partially on the audience for the plan. Presumably the plan will contain a set of maps identifying important conservation areas, but how detailed should those maps be? How much text should accompany the maps in terms of explaining the maps and the process? Should the final product be a printed report, or should the team consider producing a CD-ROM version of the plan? How much data should be included in the plan? Should there be an abridged version of the plan that is meant to communicate to specific audiences and a more technical version for other interested parties? There is a whole host of possible products for a team to consider and make decisions about in the early stages, as those decisions will have ramifications for the work of the team.

• *Who is the audience for the plan?*

Just as it is imperative for a successful business to have a good understanding of the market sector it is trying to reach, in the conservation business, planners must know the audience a planning effort is attempting to reach. The first few years of ecoregional planning in TNC were marked by a poor understanding of the audience for its planning efforts. Some senior managers thought that the audience was primarily an internal one, while others interpreted it to be broader. The result was confusion over products, messages, and investment in planning. The important conclusion that the Conservancy eventually reached was that the target audience for these conservation plans was the entire community of interested public and private organizations on whom conservation success in any given region would depend. In this sense, ecoregional plans and their products could be viewed as leveraging instruments to achieve far more conservation than TNC could ever accomplish on its own.

• *Should a plan be prepared cooperatively by one or more institutions?*

In some cases, it may be advantageous for institutions to work together to develop a regional conservation plan. In the Cape Floral Kingdom of South Africa, WWF–South Africa worked closely with the government of South Africa and numerous local institutions, ranging from academic institutions to ecotourism forums to agricultural development institutes (CAPE Team 2000). The Nature Conservancy and WWF–US cooperatively prepared an ecoregional conservation plan for the Bering Sea (Banks et al. 2000). In the province of Yunnan, China, TNC and the Yunnan Provincial Government have undertaken a cooperative venture to conserve biodiversity and cultural resources of northwestern Yunnan, while fostering compatible and sustainable development in the region (The Nature Conservancy 2002). In the state of Oregon in the northwestern United States, Defenders of Wildlife took the lead, in association with TNC and a private environmental consulting firm, CH2M Hill, to develop a statewide biodiversity plan (Defenders of Wildlife 1998). In the state of Utah, the U.S. Forest Service and TNC are cooperatively gathering biological information that will benefit planning activities of both organizations in the Utah High Plateau ecoregion. These projects demonstrate that there are likely to be many situations in which different organizations with different skill sets and political constituencies will find it profitable to work together in conservation planning (see the section below on building collaborative relationships).

• *Are there ongoing biological assessment or planning activities in a region from which a biodiversity conservation planning project would benefit?*

I noted in Chapter 1 that many countries are preparing National Biodiversity Strategies and Action Plans (NBSAPs) under the auspices of the Convention on Biological Diversity(CBD). Within a country, many national and state/provincial natural resource agencies prepare resource management plans of various types. In a number of situations, it might be advantageous to develop a biodiversity conservation plan in parallel with, or on the heels of, other major planning projects, especially if those projects involve the collection or synthesis of new biological data. For example, many of the NBSAPs being prepared as a requirement under the CBD fall short of identifying a network of conservation areas that could conserve the biodiversity of that country (e.g., Biodiversity Support Program 1994 [Bulgaria]). At the same time, a great deal of information is collected and analyzed for the first draft of a NBSAP, and a more detailed conservation plan that identifies a network of conservation areas of a country is an appropriate follow-up exercise.

In the United States, the federal government is now conducting bioregional assessments over several parts of the country (Johnson et al. 1999). These assessments typically collect, develop, and synthesize a great deal of biological information about a region. As a result, there are considerable advantages to developing biodiversity conservation plans after such assessments have taken place, or in conjunction with them. In addition, all the major federal agencies in the United States (Forest Service, Bureau of Land Management, Fish and Wildlife Service, Army Corps of Engineers, Bureau of Reclamation) conduct an extensive amount of land and water management planning. It may be beneficial to synchronize regional biodiversity planning with federal natural resource agencies, or establish cooperative agreements and memoranda of understanding to collaborate on data collection and share information.

★ ★ ★

Planning teams that address these sorts of strategic questions at the outset are not only more likely to complete a planning project that will identify the requisite conservation areas of a region, but more importantly, their plan is more likely to be put into action. Of course, how the project proceeds from the beginning to the end can vary tremendously, from a horrifying professional experience at one extreme to a gratifying experience at the other—depending upon the team, its leadership, and how well the project is managed. Fortunately, there is no shortage of useful advice on how to organize and execute the successful management of a project. Michael Thomsett's *The Little Black Book of Project Management* (1990) provides an especially clear, simplified, and practical set of guidelines.

General Advice on Project Management

All projects can be viewed as having four components: performance, time, budget, and scope (Lewis 2000). Projects need to be completed at a certain performance level, on time, within budget, and for a given scope of work (Figure 3.2). The first three factors are often referred to as "good, fast, and cheap." Conventional wisdom is that a program manager can dictate two of these components, but not all three. A project can be cheap and fast, but it will not be of high quality. Conversely, a project can be good and fast, but there will be a corresponding increase in cost.

Results—the completion of specific tasks—is the driving force behind a project (Thomsett 1990). The budgets of most projects fall outside routine departmental or programmatic budgets, and often affect multiple programs and departments. Consequently, it is especially important to adhere to budgeted expenses, because cost overruns are more difficult to backfill than routine programmatic budgets. Finally, projects have specific beginning points and end points. Because they often involve team members from several programs or organizations, expectations come from many sources that a project will get completed in the specified amount of time. As a result, one of the greatest challenges for the project manager is to keep the project on schedule and complete it within the allotted time.

There are two important elements to managing any project: definition and control. *Definition* refers to how the project is defined, including its overall purpose, the tasks to achieve that purpose, the schedule for completing the tasks, and the budget for funding all the work. *Control* has to do with the day-to-day ongoing activities required to complete the project, and it includes five parts: the team itself, coordinating team members and the various project tasks, monitoring the schedule and budget to keep the project on track, taking a variety of actions often required to bring the project back on course, and completing the project.

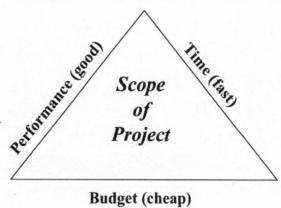

Figure 3.2 Four components of any project: performance, time, budget, and scope of work. The first three are also known as "good, fast, and cheap."

Several tools exist to help a team develop a schedule for project completion. One of the most useful tools is the Gantt chart, named after Henry Gantt, an industrial engineer who was responsible for developing a daily schedule of munitions production in World War I. A Gantt chart parses a project out into specific phases and develops a schedule for completing those various phases. Figure 3.3 shows a Gantt chart for the various phases of an ecoregional planning project in The Nature Conservancy.

Most projects are complicated enough to warrant the use of a Gantt chart for planning and scheduling all the various tasks necessary to complete the project. Developing a diagram or table that breaks work into different tasks will help planning teams determine what products will be generated from which tasks, estimate how long a particular phase and its associated tasks will take to complete, make assignments for the various tasks, and estimate how much the various phases will cost (Figure 3.4). Project management tools like Gantt charts and work breakdown structures, or work plans, are also available in commercial project management software packages. Once these work phases and products have been determined, the planner can develop an overall budget for the project (Table 3.2).

One of the most important jobs of the team leader and project manager is keeping the project on track in terms of its goals, timetable, and budget. Late delivery of products, cost and time overruns, and low morale or high turnover of team members are all warning signs that the project is veering off course. There are many causes of such difficulties, but some of the most common ones are poor communication among the team members; a lack of clarity on roles, responsibilities, and assignments; mid-course changes in

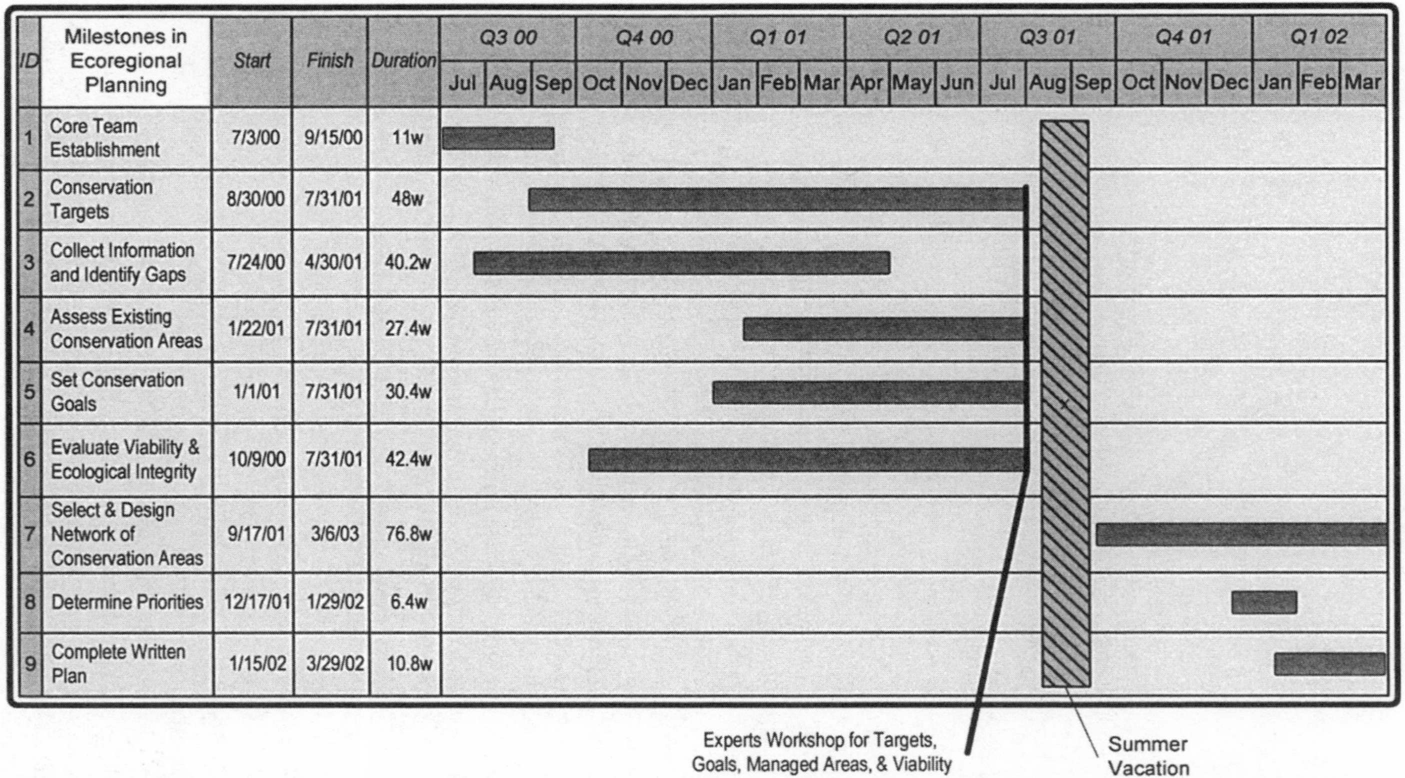

ID	Milestones in Ecoregional Planning	Start	Finish	Duration
1	Core Team Establishment	7/3/00	9/15/00	11w
2	Conservation Targets	8/30/00	7/31/01	46w
3	Collect Information and Identify Gaps	7/24/00	4/30/01	40.2w
4	Assess Existing Conservation Areas	1/22/01	7/31/01	27.4w
5	Set Conservation Goals	1/1/01	7/31/01	30.4w
6	Evaluate Viability & Ecological Integrity	10/9/00	7/31/01	42.4w
7	Select & Design Network of Conservation Areas	9/17/01	3/6/03	76.8w
8	Determine Priorities	12/17/01	1/29/02	6.4w
9	Complete Written Plan	1/15/02	3/29/02	10.8w

Timeline columns: Q3 00 (Jul, Aug, Sep), Q4 00 (Oct, Nov, Dec), Q1 01 (Jan, Feb, Mar), Q2 01 (Apr, May, Jun), Q3 01 (Jul, Aug, Sep), Q4 01 (Oct, Nov, Dec), Q1 02 (Jan, Feb, Mar)

Summer Vacation

Experts Workshop for Targets, Goals, Managed Areas, & Viability

Figure 3.3 A Gantt chart for project management, with tasks and timelines for a generic ecoregional conservation plan in The Nature Conservancy.

Action	Component Tasks	Date	Info/GIS Component	Lead	Status
Identify terrestrial ecological systems	•Engage Heritage program & key expert involvement on terrestrial team	June 02	Contact info	Bev	completed
	•Develop ecological systems classification	Aug 02	Metadata	Tracy	completed
	•Field test ecological system classification	Oct 02	ArcView	Tim	in progress
Identify aquatic communities & ecological systems	•Engage Freshwater Initiative in freshwater classification	June 02	none	Maria	completed
	•Develop freshwater ecological systems classification	Aug 02	Metadata	Tom	completed
	•Field test freshwater ecological system classification	Oct 02	ArcView	Kim	in progress
Identify species targets	•Identify imperiled, threatened, endangered species	Aug 02	Metadata	Carlos	completed
	•Identify declining, endemic, disjunct, wide-ranging, vulnerable, focal, and species aggregations	Aug 02	Metadata	Carlos	completed
	•Test to see if captured by terrestrial or freshwater ecological systems	Oct 02	ArcView	Renee	in progress
Experts workshop to test target list	•Invitations, logistics, preparatory materials, data management, GIS support	Nov 02	Metadata ArcView	Diane	in progress

Figure 3.4   A sample portion of a work breakdown structure, or work plan, for completing an ecoregional conservation plan. Work plans typically contain more detailed information than can be presented in a Gantt project management chart.

project priorities or poorly defined project goals; and scope creep—the tendency for the scope of the project to expand unchecked. The sooner these problems are identified, the greater number of options the team is likely to have to resolve them, and at relatively lower costs.

Bringing a planning project to closure is never easy. The temptation to write the perfect plan can be paralyzing, even for some of the best planning teams. And the unfinished plan, regardless of the reason, not only delays conservation action, but can leave a bad taste in the mouth of program managers, particularly those who might not have much of an

Table 3.2 *Budget for the Southern Rocky Mountains Ecoregional Plan, in U.S Dollars.*

Funding Source:* Funding Amount:	USFS $40,000	TNC–WWO $25,000	TNC–NM $5,000	CDOW $20,000	BLM $12,000	TNC–CO $77,443
Expense Type						
Salary	$27,930	$ 3,228		$16,690	$ 4,849	$53,390
Contractual		$20,000			$ 5,176	$19,641
Meetings	$ 2,912	$ 1,031	$1,492	$ 2,121		$ 3,000
Travel	$ 1,855		$3,540		$ 30	$ 1,398
Materials	$ 2,050	$ 496		$ 1,180	$ 11	
Communications	$ 393	$ 245		$ 8	$ 1,934	$ 13
Computer hard software	$ 4,228					
Other						
Total	$39,368	$25,000	$5,032	$19,999	$12,000	$77,442

*USFS = U.S. Forest Service, TNC–WWO = TNC–World Wide Office, TNC–NM = TNC of New Mexico, CDOW = Colorado Division of Wildlife, BLM = Bureau of Land Management, TNC–CO = TNC of Colorado.

From The Nature Conservancy (TNC).

appetite for planning in the first place. The Nature Conservancy has tried to walk a delicate line between bringing conservation plans to closure, yet treating them as dynamic "living" documents to be continuously improved and implemented. One successful "closing" tool TNC has used is peer review workshops, where ecoregional planning teams are required to make presentations about their planning project to colleagues from other programs. The date for these workshops is established months in advance, and teams are given ample opportunity to prepare both a presentation and complete a draft version of a written plan. Knowing that senior managers will be present to review their work, teams have a strong incentive to complete draft conservation plans for these workshops. Shortly after the workshop, teams are expected to revise their plans into a more final document. Like a closing pitcher in a baseball game, these workshops have largely been successful in bringing plans to completion. However, in retrospect, they have left a lot to be desired from a peer review standpoint, primarily because they are held too late in the planning process to make any major corrective actions or revisions (see the section later in the chapter on peer review).

Building Block Four: Collaborative Relationships with Stakeholders and Partners

In some situations, a planning team charged with developing a regional conservation plan may wish to undertake parts or all of that planning process in partnership with other organizations. Alternatively, the team may want to assess which organizations it will need to work with in the region to implement various aspects of the plan once it is completed. Either way, a first step is to define who the stakeholders are within the planning jurisdiction. Stakeholders are people, institutions, or any type of social group involved in or affected by the outcomes of a regional planning process for biodiversity conservation (World Wildlife Fund 2000). Here are the main different categories of stakeholders:

- *Primary stakeholders.* Because of their power, authority, and responsibilities, these stakeholders are critical participants in any conservation planning initiative. For example, because of the large amounts of land managed by the Bureau of Land Management and the U.S. Forest Service, two federal agencies in the western United States, they would be considered primary stakeholders for many of the ecoregional planning projects that occur in that region. Although the commonwealth government of Australia and federal government of Canada own only limited amounts of land, Australian state and Canadian provincial governments own and manage significant tracts of land and would be considered primary stakeholders in most ecoregional planning efforts. In many parts of the world, indigenous groups own or control large blocks of land and water that are important for conservation. In these situations, these groups should be considered primary stakeholders.

- *Secondary stakeholders.* This category of stakeholders has an indirect interest in the outcome of a regional conservation plan. In Yunnan, China, The Nature Conservancy is working with the provincial government to develop regional conservation plans for locating and establishing new nature reserves. At the same time, some county-level governments and private companies are working to increase tourism in some of the same areas. Those agencies and individuals promoting tourism would be considered secondary stakeholders in this region, whereas the provincial government would be considered a primary stakeholder.

- *Opposition stakeholders.* Some individuals and organizations are routinely opposed to many conservation efforts and have some capacity to influence the outcome. Private property advocacy groups in the United States and other organizations generally referred to as part of the AntiEnvironmental Movement often fall into this category. In Australia, there has long been intense debate between environmentalists and timber corporations over the disposition of forested parcels and concessions in New South Wales and other states (Pressey 1998). From a conservationist's perspective, the private timber companies could be considered opposition stakeholders in this situation.

A more detailed treatment of identifying various stakeholder groups and assessing their interest in collaboration is provided in World Wildlife Fund (2000).

Once the different types of stakeholder groups in a region have been identified, planning teams need to consider whether and how to involve them in the planning process. There are several questions to consider in any such collaborative process (World Wildlife Fund 2000). First, are there already existing collaborations in the region that can be built upon? Second, is the planning team willing to make the additional investment in time and money that it will take to involve other organizations in the planning process? Whether collaborating with governmental agencies or NGOs, bringing other institutions into the planning process almost always lengthens the time it takes to complete the project. Third, do the different organizations that are considering collaboration have sufficiently similar goals and visions for the project that the collaboration will be successful, or are there ideological differences that may preclude such success? Do any of the groups have their own planning processes that may preclude or limit participation in a separate effort? Finally, will the benefits realized from collaboration exceed the costs in time and money of involving multiple organizations and individuals?

The Biodiversity Support Program (BSP, a consortium of the World Wildlife Fund, The Nature Conservancy, and the World Resources Institute) examined the effectiveness of alliances for conservation based on its experiences with 20 conservation projects in the Asia and Pacific region (Margoluis et al. 2000). It differentiated three types of alliances: contractual agreements, partnerships, and a consortium. Contractual agreements occur when one primary organization hires another organization to complete specific tasks. In these cases, the contractors usually are not involved in decision making or liability for the project. Partnerships involve two organizations agreeing to work together on a specific project to achieve mutually beneficial goals, ideally sharing in decision making and liability for the project. Consortiums are similar to partnerships but involve three or more organizations.

The BSP study, although focused on questions related to successful site or project-based conservation, has significant application to the development of regional conservation plans, especially for teams considering working in partnerships or alliances with others. A set of principles for effective conservation alliances, based on results of the BSP study, provides helpful guidance for regional conservation planning efforts:

- *Create simple alliances.* Simple alliances are easier to manage and appear to achieve greater conservation impact.
- *Keep decision making as close to the ground as possible.* Organizations most involved in implementation should make decisions, and decision-making authority should be limited to as few organizations as possible.
- *Select strong leaders.* The most effective alliances have a single, capable leader.
- *Construct and maintain clear goals.* Alliances without a clear set of goals are unlikely to be successful.
- *Differentiate clear roles and responsibilities.* Each member of the alliance needs to clearly understand its responsibility, especially in alliances that involve multiple organizations.

Several ecoregional planning projects have been conducted successfully in alliance with other organizations. In Colombia, TNC has worked closely with the Instituto Alexander von Humbolt (part of the Colombian Ministry of the Environment) to develop ecoregional conservation plans in the Andes Region via a World Bank–funded project (Touval 2000). Similarly, in the Caribbean region of Colombia, TNC is working in an alliance with a local NGO (Fundación ProSierra Nevada de Santa Marta) to develop a biodiversity conservation plan as part of a larger project funded by the Global Environmental Facility (GEF) and conducted under the auspices of the Convention on Biological Diversity. In Brazil, three organizations (CI, TNC, WWF) are working together to produce a conservation plan for the Pantanal Ecoregion. Along the Canadian border, Nature Conservancy of Canada and TNC are cooperating in the development of several different ecoregional conservation plans.

In the Sonoran Desert Ecoregion in the southwestern United States, TNC entered into an alliance with both a U.S. NGO (Sonoran Institute) and a Mexican NGO (IMADES, Instituto del Medio Ambiente y el Desarrollo Sustentable del estado de Sonora) to produce a binational Sonoran Ecoregional Plan (Marshall et al. 2000). In this case, the Sonoran Institute had already developed expertise in the region in conducting outreach efforts to interested parties on conservation issues. As a result, it was well suited to play a critical role in assessing stakeholders and in informing a large number of institutions and organizations about the purpose and goals of the project. In part due to the Sonoran Institute's substantial public relations efforts in the planning process, the Sonoran Ecoregional Plan is being successfully implemented on a number of fronts, and new priority areas are coming under conservation management. In Chapter 13 on implementation, I will explore these successes in the Sonoran ecoregion in more detail.

The majority of these alliances reduced duplication of effort and resulted in higher quality plans because of the different talents and strengths the various organizations brought to the planning team. In addition, the involvement of local NGOs and local governmental organizations increased the credibility of these projects and the likelihood that they would be successfully implemented. This observation is consistent with the finding in the BSP study that local organizations are best suited to implement conservation projects, while international organizations like TNC and WWF are best suited to play a technical support and capacity-building role. These collaborations are also reinforced by both local and international donor organizations and foundations, which promote the funding of partnerships and alliances among conservation organizations and governmental agencies with similar interests and missions.

Building Block Five: Assembling Data and Information

Once a planning team is in place, a budget and time frame are established, and the overall purpose and goals of the project articulated, the actual work of conservation planning begins. The next step in the planning process starts with assembling the data and information that will be necessary to develop the plan. Table 3.3 lists some of the types of information most commonly used in biodiversity plans. There are many information sources for any ecoregion or region, and a great deal of information is now available on the Internet. However, for many regions and ecoregions around the world, there is a shortage of information on the status and distribution of native flora and fauna, as well as other ancillary information on environmental (physical) variables (e.g., soil maps) and socioeconomic trends. In freshwater and marine systems, this dearth of biological and environmental information is even more pronounced and creates additional challenges (see Chapters 10 and 11).

Table 3.3   *Useful Categories of Information for Conservation Planning.*

Category	Type of Information
Land use—ownership	Transportation
	Administrative boundaries
	Land cover
	Locations of dams and diversions
	Water quality monitoring stations
	Water flow monitoring stations
	Point sources for pollution
Physical	Soils
	Geology
	Climate
	Terrain and elevation
	Wave exposure
	Wave depth
	Watersheds and hydrography
Biological	Vegetation cover
	Wetlands
	Species distribution
	Ecoregions and bioregions
	Shellfish distributions
	Fisheries data
	Coral reef distribution
Socioeconomic	Human population density
	Human population trends
	Economic trends

From Groves et al. 2002. Planning for biodiversity conservation: putting conservation science into practice. *Bioscience* 52:499–512. Copyright © American Institute of Biological Sciences.

The best planning efforts will seek out information from a wide variety of sources, ranging from local or national natural resource agencies to universities, conservation organizations, and expert individuals. Although it is beyond the scope of this book to provide an exhaustive list of such sources, some of the more widely used and readily available ones are documented in the following section.

Natural Heritage Programs and Conservation Data Centers

During the early 1970s, The Nature Conservancy began working with state governments in the United States to establish statewide biological inventory programs. The primary purpose of these programs, known as state Natural Heritage Programs (NHPs), is to collect, manage, and disseminate information to a wide variety of users on the biological resources of a state (Groves et al. 1995, Adams et al. 2000). Since its inception, the network has grown to 75 individual programs, including Conservation Data Centers (CDCs, as they are known outside the United States) in most provinces of Canada and numerous countries in Latin America and the Caribbean. Most programs are now formally established within state, provincial, or country governments.

Elements of biodiversity (vertebrate and invertebrate species, vascular and nonvascular plant species, and vegetation communities) are the principal data units of NHP methodology (Noss 1987, Noss and Cooperrider 1994, Groves et al. 1995, Jenkins 1996). NHPs and CDCs use a standardized information management system to track both biological and nonbiological information, using manual or paper files, map files or Geographic Information System (GIS) files, and computerized database files. Typical biological information for species includes taxonomy, distribution, population trends, habitat requirements, and the condition or quality of individual occurrences or populations. Nonbiological information usually includes land ownership and land management status where elements of biodiversity are known to occur, as well as specific information on the location and management of protected areas and other conservation lands and waters.

NHP methodology assigns conservation ranks to all species and communities that are tracked in its databases (Master 1991, Bryer et al. 2000, Stein and Davis 2000). These ranks are assigned at three levels—globally, nationally, and subnationally (e.g., state)—and are based on several factors, including the number of occurrences or populations, population size and trends, threats, and the viability or integrity of individual occurrences. As a result, these ranks are useful in a planning context to determine which species are at greatest risk. Heritage programs also maintain detailed information on the status and distribution of both state and federally listed threatened and endangered species. In addition, most NHPs and CDCs also collect and manage information on the viability or integrity of individual populations of at-risk species or occurrences of at-risk communities (see Chapter 7).

In 1999, TNC and the network of Natural Heritage Programs worked to jointly establish a new, independent organization to advance the application of biodiversity information for conservation purposes. This organization, known as NatureServe (formerly known as the Association of Biodiversity Information), serves as an umbrella institution for the network of NHPs and CDCs. Its mission is to develop, manage, and distribute authoritative information critical to the conservation of the world's biological diversity. Much of the biodiversity information assembled by the network of NHPs, as well as general organizational information on NatureServe, is available at www.natureserve.org and www.natureserve.org/explorer (a searchable database on 50,000 plants, animals, and ecological communities of North America). However, individual state and provincial programs retain detailed information on the locations of populations and occurrences of individual species and communities. A separate Web site, www.infonatura.org, provides conservation information on more than 5500 common, rare, and endangered species in 44 countries of Latin America and the Caribbean.

Heritage-type data do have some weaknesses (Stein and Davis 2000). First, the inventories from which these data are derived are not systematic. In some cases, there are noticeable geographical biases, with heavy concentrations of data near road networks. Because the surveys are not systematic, the so-called white data on a map (areas with no data) are not necessarily an indication that a particular species or community is not present there. Second, there are some inconsistencies from program to program in how some occurrences of a given species or community are defined and mapped. Third, there are taxonomic gaps; some groups have been surveyed to a much greater extent than others. Freshwater fish, for example, are notably underrepresented in NHP databases. Finally, the frequency with which different species are surveyed is highly variable. In some cases, the information on a particular occurrence of a species or community may be current, whereas in other cases the information may be dated.

Gap Analysis Programs

As director of the recovery program to save the endangered California condor (*Gymnogyps californianus*) in the mid-1980s, Mike Scott (with the U.S. Fish and Wildlife Service at that time) grew increasingly frustrated over the emergency room tactics and large sums of money needed to bring species like the condor back from the edge of extinction. Knowing there had to be a more proactive way to keep species from becoming endangered, he conceived of the idea of "gap analysis" in collaboration with several colleagues (Scott et al. 1987, 1993). Gap Analysis Programs map endangered species' locations, predicted distributions of all native vertebrate species, and natural vegetation cover types. Subsequently, these maps are overlaid with information on land ownership and management patterns to determine which species and vegetation cover types are sufficiently represented within existing conservation lands and waters (Figure 3.5). Those biological features that are not adequately represented are identified as "gaps" in the network of conservation areas.

Gap Analysis Programs are administered on a state basis by the U.S. Geological Survey, Biological Resources Division, and are largely conducted through each state's Cooperative Fish and Wildlife Research Unit. A typical Gap Analysis Program involves cooperative representatives from

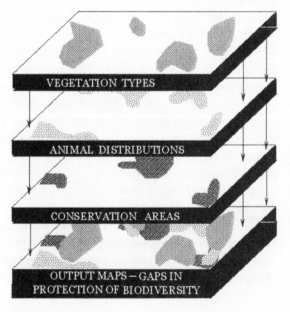

Figure 3.5 Data layers in a Geographic Information System (GIS) used in a typical Gap Analysis project. The purpose of Gap Analysis projects is to identify species, communities, and ecosystems that are not represented or are underrepresented in the existing network of conservation areas in a planning region (i.e., the "gaps" in the network). (Figure courtesy of R. Brannon, National Gap Analysis Program, Biological Resources Division, U.S. Geological Survey.)

several state, federal, and academic agencies and institutions; it crosses many disciplines, including ecology, botany, zoology, conservation biology, remote sensing, geography, and computer science (Jennings 2000). Prior to the inception of Gap Analysis Programs, there was little spatially explicit information available on natural resources. With advances in the use of GIS and remote sensing and the concurrent development of a vegetation classification for the United States (Grossman et al. 1998), Gap Analysis Programs can now deliver a number of highly useful digital data layers for conservation planning purposes. Among the most useful products of Gap Analysis are:

- Land cover maps based on the classification of Thematic Mapper satellite imagery.
- Predicted maps of vertebrate species distribution based on known locations, combined with species–habitat relationship information.
- Land stewardship maps that combine information on land ownership and management to classify all lands in a state based upon their value for conserving biodiversity.

Gap Analysis Programs have been completed for over 80% of the states. Efforts are now under way to improve the accuracy of vegetation and vertebrate species maps, and to develop an aquatic component of Gap Analysis Programs through the mapping of distributions of aquatic species, communities, and ecosystems. Chapter 5 will focus on the use of Gap Analysis in conservation planning, and Chapter 10 will address conservation planning in freshwater ecosystems in more detail. The Gap Analysis Web page (www.gap.uidaho.edu) provides a wealth of information about the program, including access to a detailed handbook on Gap Analysis methods and references to many other publications.

The World Wildlife Fund's Ecoregional Conservation Assessments

For much of the 1990s, Eric Dinerstein, David Olson, and their colleagues in the Conservation Science Program of WWF–US were occupied with mapping terrestrial and freshwater ecoregions on a continental basis and setting priorities for conservation among those ecoregions (Olson et al. 2001). Priorities are set primarily on the basis of two criteria: biological distinctiveness and conservation status (Dinerstein et al. 1995). Biological distinctiveness is based on broad measures of species richness (the number of species in a region), endemism (the degree to which species are largely found only within one ecoregion), unusual ecological or evolutionary phenomena (e.g., seasonal migrations of large ungulate herds across Africa's Serengeti Plains), and the global rarity of major habitat types (aggregations of ecoregions with similar climate regime, vegetation structure, and patterns of biodiversity). Conservation status is based on four landscape-level factors: habitat loss and degradation, the presence of large blocks of remaining natural habitat, the degree of habitat fragmentation, and the degree of existing protection. These two criteria (biological distinctiveness and conservation status) are combined into an integration score that is assigned to each ecoregion, ranking it in terms of its importance for conservation action.

Ecoregional assessments have been completed and published separately for North America, Latin America–Caribbean, and Asia-Pacific and for terrestrial and freshwater ecoregions (Dinerstein et al. 1995, Olson et al. 1998, Ricketts et al. 1999, Abell et al. 2000, Wikramanayake et al. 2001). Similar assessments are well under way for Africa. These assessments are valuable assets for the initial stages of regional-scale conservation planning. They establish priorities as to which regions or ecoregions in a country are most important to assess first (Olson and Dinerstein 1998) from a conservation perspective. The assessments also provide planners with helpful background information on each ecoregion, including important biological features, degree of both habitat loss and protection, types and severity of threats, priority conservation activities, potential conservation partners, and a few key references about the ecoregion. It is also important to understand what type of information these assessments do not have. They do not provide specific data on species, communities, or ecosystems that can be used as conservation targets in the regional planning process outlined in this book. In addition, a substantial portion of the information in these assessments comes from expert interviews and expert workshops, the strengths and weaknesses of which are discussed in Chapter 5.

The World Conservation Monitoring Centre

The World Conservation Monitoring Centre (WCMC) was first established in 1979 by the World Conservation Union (IUCN) for the purpose of monitoring endangered species. In 1998 it became an independent nonprofit organization jointly funded by the IUCN, WWF, and the United Nations Environment Programme (UNEP). Today, the WCMC is a UNEP program with three principal functions: facilitate access to information on biological diversity, conduct biodiversity assessments, and work with conventions (e.g., the Convention on Biological Diversity) to provide an information management service. The various programs within the WCMC publish and maintain a great deal of information related to biodiversity conservation, much of which is directly applicable to conservation planning.

Three main types of useful products are available from WCMC: publications, searchable databases, and linkages to Web sites that contain various types of information relevant to the conservation of biodiversity. Among the most useful publications are *The IUCN Red List of Threatened Species* (Hilton-Taylor 2000), a series of handbooks on biodiversity information management, a series of booklets examining various biodiversity issues ranging from the global trade in coral to priorities for biodiversity conservation in the tropics, and several documents describing the status of the world's protected areas.

Searchable databases on IUCN lists of threatened species include information on Red List status (extinct, endangered, rare, etc.), distribution, taxonomic status, and key references. The WCMC also maintains the most up-to-date global information on the status of protected areas. The World Conservation Union has assigned ranks of degree of legal protection to all of the world's protected areas (see Chapter 5). The database on protected areas is searchable on a country-by-country basis. It maintains information on the type of protected area (e.g., national park, wildlife refuge), the IUCN category of protection (I, II, III, etc.), the size of the area, location, and date established. All information available through the WCMC can be accessed through its Web site: www.unep-wcmc.org.

Conservation International (CI), Center for Applied Biodiversity Science (CABS)

Two programs within CABS house a considerable amount of information for biodiversity conservation planning. The first of these is the State of the Hotspots Program (www.biodiversityscience.org/xp/CABS/research/hotspots/hotspots.xml), a program designed to synthesize and disseminate current information on CI's biodiversity hotspots to decision makers, scientists, and conservation organizations (see Chapter 1). Current plans call for publishing a series of books on the State of the Hotspots, with the first book focused on the endangered Atlantic Forest ecosystem in South America.

The second program of interest within CI and CABS is the Conservation Priorities Setting Program. This program identifies critical areas for conservation action within the hotspots through a synthesis of existing information on that area and a participatory workshop (Conservation Priority Setting Workshops) that builds consensus with local partners and stakeholders about the most important areas and actions. Priority-setting workshops have been conducted and reports completed for the Amazon Basin, the Atlantic Forest of Brazil, Madagascar, West Africa, Papua New Guinea, Irian Jaya, and the Philippines (see www.biodiversityscience. org/xp/CABS/research/global_planning/globalplan.xml).

National Biodiversity Planning: The Convention for Biological Diversity

For signatories to the United Nations Convention on Biological Diversity (CBD), one of the requirements is the preparation of National Biodiversity Strategies and Action Plans (NBSAPs), as explained in Chapter 1. Many countries undertook initial efforts to prepare these plans. Early planning efforts produced mixed results, probably because there was little direction or guidance. More recently, such guidance has been forthcoming (Garcia Fernandez 1998, Hagen 1999) under the auspices of the Biodiversity Planning Support Programme (www.undp.org/bpsp), an initiative placed under the United Nations Development Programme's (UNDP) Global Environmental Facility (GEF).

A list server known as BioPlan, maintained by the UNDP and GEF, provides information on issues, methods, best practices, and tools for conservation planners worldwide who are engaged in developing NBSAPs under the Convention on Biological Diversity. These NBSAPs contain, among other items, a preliminary assessment of the elements of biodiversity within a country, an examination of threats to biodiversity and their causes, a prioritized list of native ecosystems in greatest need of conservation attention, and strategies for conserving biodiversity. The Web site for the CBD (www.biodiv.org) contains contact information (referred to as National Focal Points) about individuals who are responsible for preparing national biodiversity plans, downloadable versions of progress reports on the plans (see "National Reporting"), and completed plans. The World Resources Institute documented some of the early lessons learned in biodiversity planning under the CBD, including detailed information on approximately 15 case studies of country plans (Miller and Lanou 1995).

In addition to information on country plans, the CBD's Web site provides a variety of other useful information for conservation planners through the CBD's Clearinghouse Mechanism. For example, planners can obtain information about organizations that are involved in biodiversity planning in each country. There are also links to specific biodiversity topics, such as institutions engaged in freshwater or marine conservation. CBD recognizes that many of the world's countries do not have access to information and technology that would help them prepare biodiversity plans. The Clearinghouse Mechanism is a well-placed effort to rectify these information and technological disparities.

Bioregional Assessments in the United States

Over the last decade in the United States, several "bioregional assessments" have been conducted by the federal government (Johnson et al. 1999). Probably the best known of these—FEMAT, or Forest Ecosystem Management Assessment Team—was initiated because of concern for the spotted owl (*Strix occidentalis*) and the impact of logging of old-growth forest on the owl and other species that inhabited late-successional forests in the Pacific Northwest of the United States. It is difficult to define a bioregional assessment or place a particular label on it. Most are born in natural resource crises, and most address critical societal issues to which policy makers want answers (e.g., ecosystem degradation in the Everglades). The most important scientific outcome of assessments is not the data they gather, but the synthesis of that data into meaningful frameworks.

From a conservation planning perspective, bioregional assessments represent opportunities for collecting large amounts of information on the biological features of a region and assessing trends in those features. Such sets of data and information provide a strong foundation for developing regional plans focused on the conservation of biodiversity, and they are usually at a coarse scale of spatial resolution similar to that of ecoregions. Although these assessments are not plans in and of themselves, they often serve as a preliminary step in federal land management planning.

In the Sierra Nevada, Columbia Plateau, and Middle Rockies–Blue Mountains ecoregions, TNC was able to capitalize on bioregional assessments that had either been completed or were in progress by the federal government. These assessments provided critical data layers to ecoregional conservation plans, such as digital vegetation maps of the region, indices of biotic integrity for terrestrial and freshwater ecosystems, and detailed maps of land ownership and management, including the status and location of various types of designated conservation lands and waters. Assembling this type of information can cost hundreds of thousands (in some cases, millions) of US$ and is therefore usually beyond the scope of most regional conservation planning projects. Bioregional assessments have now been completed for numerous areas of the United States, and their results are generally available through a variety of federal government reports and Web sites (e.g., for the Interior Columbia River Basin Assessment, see www.icbemp.gov).

Millennium Ecosystem Assessment

Initiated in April 2001, the Millennium Ecosystem Assessment is a 4-year program aimed at providing decision makers and the public with scientific information on the condition of ecosystems around the world; the ability of these ecosystems to meet human needs in the face of major changes to these systems (climate change, economic growth); and technologies and tools for improving the management of these ecosystems. The assessment will take place at multiple scales including a global scale and two to four regional assessments in Southeast Asia, southern Africa, and other locations.

Information that is specifically relevant to conservation planners includes assessment of the "condition, pressures, trends, and changes in ecosystems." Other useful information likely to be found in these assessments includes the distributional extent of various ecosystems, ecological community types found within them, some species' distributional data, historical trends in the systems, threats to these ecosystems (including disruption of ecological processes), and changes in physical variables (e.g., nutrients, groundwater flow, sedimentation rates). Information on the Millennium Ecosystem Assessment can be found at www.ma-secretariat. org, including a detailed project summary and an overview of the objectives, audience, and conceptual framework of the project.

Science Magazine's Web Supplement, "Bioinformatics for Biodiversity"

Bioinformatics is a collective word for the tools and techniques for managing, analyzing, and disseminating the increasing amount of biological data available around the world (Sugden and Pennisi 2000). In September 2000, Science magazine ran a series of feature articles on "Bioinformatics for Biodiversity." A supplemental Web site to this issue (www.sciencemag. org/feature/data/biodversity/2000/shl) described, listed, and provided links to a number of important Web sites containing information on biodiversity that is relevant to conservation planning. The supplement divides these sites into six categories: global databases; geographical, regional, and national databases; databases for specific taxonomic groups; tools and software; organizations promoting biodiversity study; and other resources. This supplemental Web site is a good starting point for any regional or ecoregional planner who needs information on the biodiversity features of the region.

Although many of the Web sites listed may not provide the information directly, reviewing the Web sites will lead to other linked sites that will contain useful information for a regional planning project. For instance, clicking on the U.S. Government Web site entitled the National Biological Information Infrastructure (NBII), administered by the U.S. Geological Survey (www.nbii.gov), will lead the user to an outstanding collection of links to other Web sites on species and ecosystem diversity.

Socioeconomic Data

A great deal of social and economic data useful in conservation planning can be found on the Internet. For example, the U.S. Census Bureau (http://tiger.census.gov/) maintains an enormous amount of data on such topics as human population estimates (by ethnic group), population trends, housing units, home ownership rates, median household income, percentage of persons below the poverty line, private farms, retail sales, and federal funds and grants. This type of information is available at state and county levels through the above Web site and, depending upon the type of data, in smaller census blocks as well. TIGER (Topologically Integrated Geographic Encoding and Referencing) refers to the digital database developed by the Census Bureau to support its mapping needs for each 10-year census. TIGER line files are also available on the Web site. These are geographic features such as roads, railroads, rivers, lakes, political boundaries, and other administrative boundaries that cover the United States. The files actually contain digital data that are meant to be imported to a GIS, where they can be displayed on maps.

Worldwide, a great deal of socioeconomic data are available for each country. It is beyond the scope of this book to list the various sources. However, for summaries on human population data as well as other social data around the globe, a good source is the Population Reference Bureau (www.prb.org). The Population Reference Bureau maintains links to other Web sites worldwide that provide a variety of population data by country. These links are maintained through a Web site known as POPNET (www.popnet.org). The International Data Base (http://www. census.gov/ipc/www/idbnew.html) maintains information on a variety of social and demographic data for 227 countries in the world, ranging from household, income, and labor force information to population size, life tables, and immigration rates. Probably the best advice to conservation planners is to do a Web search related to the specific area of concern and type of data needed (e.g., human population numbers by administrative region over a specific country). The Web sites noted above maintain summary data and links to other Web sites that may be helpful. The United Nations also maintains numerous Web sites with social and economic data from around the world and links to country-specific Web sites.

Traditional Ecological Knowledge

Traditional ecological knowledge (TEK) is information about the natural world that originates in observations by indigenous peoples (Striplen and DeWeerdt 2002). This information is increasingly being used in ecological management, restoration, and conservation planning in North America, Australia, Latin America, and other areas worldwide. A special features section of the journal *Ecological Applications* contained several articles about TEK in the October 2000 (Volume 10[5]) issue. At the 2002 meetings of the Society for Conservation Biology, ecologist Kim Heinemeyer presented a paper on how TEK of the Taku River Tlingit First Nation is being used to develop a regional land-use plan for the nation's 15,000 km^2 (5792 mi^2) traditional territory in northwestern British Columbia (Canada). Standardized interviews with tribal elders and hunters are helping document the status and distribution of several focal species being used in the planning process (e.g., grizzly bear, salmon, caribou). Conservation planners working in regions with current or former indigenous landholdings can benefit from using TEK in their preparations. In addition, at least in North America, many indigenous groups now have staff fisheries and wildlife biologists employed on reservations and in territories. These indigenous groups may be interested in participating as partners or contributing to a regional planning effort for biodiversity conservation.

Building Block Six: Managing Information

Regional or ecoregional planners should invest considerable time in gathering a great deal of information and data from many different sources to support the planning process. This investment needs to be safeguarded for several reasons. First, the information will be used for years in the future, and it will have to be revised and updated without "reinventing the wheel." Second, in the era of information sharing via e-mail and the Internet, the team will want to share the information and data with external parties. Finally, with staff turnover, data and information can easily get lost. If the data are not properly archived, new team members may find it necessary to start over in collecting certain types of information.

Although it is beyond the scope of this book to provide detailed recommendations about information and data management, the Biodiversity Conservation Information System (BCIS) has developed a series of handbooks for that purpose. BCIS, a consortium of ten international conservation organizations and programs of the IUCN, supports environmentally sound decision making by facilitating access to biodiversity data and information (www.biodiversity.org, BCIS). The eight-volume set, downloadable from BCIS's Web site, includes the following topics: principles, procedures, custodianship, data access, metadata, standards and quality assurance, core datasets, and tools and technologies.

More general recommendations, based on TNC's ecoregional planning experiences, are outlined below:

- One person should be assigned the duties of coordinating information management activities for the duration of the planning project. Sharing these duties among several people usually results in inefficiencies in collecting and managing data and inconsistencies in data formats.
- Much of the data for a regional conservation plan may already reside with governmental natural resource agencies. Similar planning exercises may have already been conducted in adjacent regions, so it is usually worth the time to contact others to determine both the types of information used and its availability.
- Obtaining the required information for a plan can be time consuming. So, too, can getting data into useful formats, analyzing it, and documenting the sources and types of information (metadata). Planning teams should allocate sufficient time from the outset of the project to avoid costly delays. Tasks such as updating older data sets can often take considerable time when discrepancies in the formats and data have to be resolved.
- Some individuals or institutions may be the source of several different types of data. To lessen the burden on these individuals and organizations, planners must anticipate all such data requests in advance and make them at one time, allowing the individual or agency sufficient time to respond to the request. In some cases, it may be necessary to develop a data-sharing agreement with an external party.

- To avoid confusion and difficulties in updating or modifying data, planning teams should manage information in the smallest number of software applications as possible. Archiving some information in spreadsheets, some in databases, and still some in a GIS application is a recipe for disaster and makes the sharing of information in the future considerably more difficult.
- Time spent on quality control of data is usually valuable. Some data records will be outdated, and teams will want to consider cut-off dates for using such records. Other data records may have imprecise or inaccurate information on locations that needs to be checked. In some cases, uncertainties about taxonomy or inconsistencies in taxonomy may occur (e.g., one jurisdiction calls a species or community by a different name than another jurisdiction). There are often varying degrees of confidence in different types of information provided by experts, although this is often not explicitly acknowledged in plans. The Southern Rocky Mountains Ecoregional Assessment Team (2001) developed an innovative method for documenting the degree of field verification that had occurred in the selection of a conservation area. For example, areas with low field verification were those that were selected primarily on the basis of predictive modeling of community and ecosystem distribution and/or on the basis of interpretation and classification of remote-sensing data. Regan et al. (2002) reviewed the different types of uncertainty in ecology and conservation biology. They divided uncertainty into two principal types: (1) epistemic uncertainty or uncertainty associated with the knowledge of a particular system (e.g., measurement or model errors) and (2) linguistic uncertainty arising from ambiguous or vague language scientists use.
- It is critical that planning teams take the time to document important data gaps (see Chapter 5). Again, the Southern Rocky Mountains Team (2001) identified major information gaps and research needs for each different taxonomic type of conservation target (i.e., birds, amphibians, communities, ecosystem types).
- The final step in good information management is the documentation of metadata—information about data, such as source, reliability, scale, and other characteristics. Metadata documentation helps prevent data from being lost and expedites future plan revisions. Some database software applications have built-in data dictionaries to help in documentation. For geospatial data sets, teams may want to use a metadata tool, such as Metalite, a product of the U.S. Geological Survey and United Nations Environment Programme (available from www.edcnts11.cr.usgs.gov/metalite).

A Word on Scale

Recall from Chapter 1 that the elements of biodiversity—species, communities, and ecosystems—occur at a variety of spatial scales. The same applies to ecological processes and natural disturbances that sustain the patterns of biodiversity (Figure 3.6). This multiscalar aspect of biodiversity adds to the challenge of conserving it. Unfortunately, ecology as a discipline has been slow to recognize the influence of both spatial and temporal scaling on the patterns and processes that are observed in nature (Wiens 1989). However, over the last decade, scale has become a core issue in ecology and conservation biology (Wiens 2002), as evidenced by a recent book that focuses on the relevance of scale to predicting species occurrences (Scott et al. 2002). Some ecologists have argued that understanding how different processes operate at different scales of time and space and produce different patterns is the fundamental problem of ecology (Levin 1992).

Data that are available or that can be collected for conservation planning purposes also occur at different scales, and biologists and planners need to be keenly aware of the influence of scale on data and their interpretation.

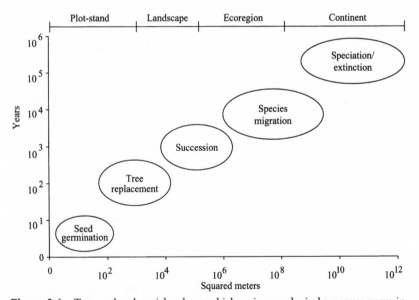

Figure 3.6   Temporal and spatial scales at which various ecological processes occur in subalpine fir forests of the Rocky Mountains. (Modified from Urban et al. 1987. Landscape ecology: a hierarchical perspective can help scientists understand spatial patterns. *BioScience* 37:119–127. Copyright, American Institute of Biological Sciences.)

For example, some types of communities and ecosystems that occur at relatively fine spatial scales (e.g., minimum mapping units of 4 hectares, or 10 acres) will not be resolved on maps that are of coarser resolution (e.g., 40 hectares, or 100 acres, minimum mapping units). Many riparian communities that occur as linear habitat strips in more xeric regions are good examples of communities that will not be resolved on coarser scale maps such as those often used in statewide Gap Analysis projects in the United States. Studies have demonstrated that as habitat maps are generalized (meaning that the minimum mapping unit is increased), the number of habitat types that can be mapped decreases (Stoms 1992) and the accuracy in predictions of vertebrate species distributions from habitat relationship models decreases (Karl et al. 2000). A corollary is that the number of vertebrate species that can be detected on these maps also decreases with map generalization because their distributions are predicted in large part by habitat relationship models. Even within the same vegetation type, maps at coarser scales of resolution will miss smaller remnant patches of habitat that may be important as refugia for rare or endemic species (Stine et al. 1996).

Areas that are identified as "hotspots" of species richness or endemism are also very sensitive to changes in scale. Stoms (1992) showed that as the minimum mapping unit changes, so does the location of areas identified as centers of richness or endemism for vertebrates in Gap Analysis projects. For ecologists familiar with the high diversity of species associated with riparian areas in the arid western United States, this makes perfect sense. In many cases, these riparian corridors will be areas of the greatest species diversity. If these areas cannot be detected because of a relatively coarse scale or minimum mapping unit, other types of habitats and areas will be identified as "hotspot" centers. Conroy and Noon (1996) identified other scale-related problems associated with measures of species richness. Among these are the points that both demographic processes (e.g., birth and death rates) and habitat suitability vary substantially across the distribution of any given species. Coarse-scale maps of species richness tend to "wash out" these finer scale differences.

Issues such as connectivity of habitats that are important in the design of networks of conservation areas are also strongly scale-dependent (see Chapter 8). In a study of two forested habitat types in the American Southwest, Keitt et al. (1997) demonstrated that connectivity not only increased with scale, but did so abruptly. For species to perceive this particular environment as connected clusters of habitat, those species needed to have a capability of dispersing at least 45 km (28 mi). The point is that every species will perceive the environment differently according to its own life history characteristics. What appear to be connected habitats for a beetle or a field mouse will clearly be quite different than the connectivity needs for a wide-ranging carnivore.

Throughout the conservation planning process, biologists and planners need to be evaluating whether the data they are collecting and analyzing are at scales appropriate to the species or ecosystems that are the targets of conservation planning and to the processes and disturbances that sustain them. Distribution patterns observed at one scale may be nonexistent at another scale. By inference, it is both difficult and dangerous to extrapolate results from data collected at one scale to predict what might happen at a different scale (Wiens 2002).

Building Block Seven: Peer Review and Evaluation

Because of the uncertainty involved in both the information used in conservation plans and the science of conservation planning itself, peer review plays a critical role in developing conservation plans for biodiversity. From the selection of conservation targets to completion of the final report, each major step of the planning process can be improved by peer review (see Figure 2.4). Planning teams should establish a peer review process that provides input throughout the planning cycle (target selection, goal setting, assessment of viability/integrity, threats assessment, assembly of a portfolio of conservation areas). Furthermore, the planning methods and process itself should be subject to peer review, especially by external parties who can bring objectivity and fresh ideas to the planning table. Kleiman et al. (2000) provided examples and recommendations for conducting evaluations of conservation programs. Expert workshops are one avenue for obtaining peer review of both the process and the products of conservation planning (see Chapter 5).

A related and critical aspect of program evaluation is the need for planners to be explicit about assumptions and hypotheses they make during the planning process. Stating these assumptions and hypotheses clearly makes the plan more credible and facilitates a more informed understanding of the planning process for partners and stakeholders. For example, if a team assumes that a particular set of conservation targets will serve as a surrogate for particular elements of biodiversity, the plan should explicitly state these assumptions in the methods section. If a team elects to not focus on an entire taxonomic group because of a lack of data, the plan needs to acknowledge this omission and the reasons for it. Throughout the planning process, even teams with abundant sources of information will need to make certain assumptions. The highest quality planning efforts will acknowledge these assumptions, data gaps, and other sorts of uncertainties, and make designs in the future to test assumptions, fill data gaps, and reduce uncertainty where the benefits of so doing outweigh the costs.

Key Points for the Building Blocks of a Regional Conservation Plan

1. Assemble a multidisciplinary team with clear leadership and a strong sense of purpose, commitment, and common goals.
2. Develop and agree to an overarching vision, goals, and objectives for the planning process.
3. Address key strategic questions about the project before it gets started: how much time and money will be invested, who is the audience, what products are expected, should the plan be prepared cooperatively with partners?
4. Develop a detailed project work plan and a schedule with benchmarks. Remember the four components of any project: performance, time, budget, and scope. Of the first three (also known as "good, fast, and cheap"), only two can be realized at any one time.
5. Live by this credo: "The perfect plan never gets done." Treat the plan as a dynamic and iterative process that will be improved over time. Manage the project to closure, staying within time and budget limits if at all possible.
6. Assess who the important players (stakeholders) are in a planning region, and determine whether there are potential partners with whom your institution could profitably collaborate to produce a conservation plan. Consider the pros and cons of forming alliances with other institutions.
7. Assess information needs early in the planning process, and begin investigating major sources of data for the project.
8. Invest in wise stewardship of the information and data the planning team collects. Place one person in charge of all data management. Allow adequate time to collect, analyze, control quality, and archive information and data. Make the extra effort to provide metadata documentation.
9. Build peer review and program evaluation into the planning process at several important stages (target selection, assessment of viability and integrity, threats assessment, assembly of the network of conservation areas).

PART TWO

THE SEVEN HABITS OF HIGHLY EFFECTIVE PLANNING

*Chapters 4–9 expand in
detail on the seven-step planning process
for conserving biological diversity outlined at the
conclusion of Chapter 2. I present the seven
steps in six chapters, incorporating two
of the steps in Chapter 5.*

CHAPTER FOUR

What to Conserve? Selecting Conservation Targets

*The distinction between single species man-
agement and ecosystem management is a
false dichotomy; both are part of a continuum
of steps necessary to protect biodiversity.*

—DAVID WILCOVE (1993)

In its formative years during the 1950s, The Nature Conservancy (TNC)
was staffed almost entirely by volunteers. The first land acquisition by
TNC was in Mianus Gorge, New York, and it was accomplished largely by
volunteers, some of whom even mortgaged their own homes to raise the
funds to secure this important natural area. The term *biodiversity* was still
years from being coined, there were no scientists who worked for the
organization, and there was no clear strategy for conserving nature, other
than one we might call "opportunistic." Such efforts by TNC and many
other organizations to conserve nature are what Australian ecologist Bob
Pressey has labeled ad hoc, meaning that they happened for a variety of
reasons under a variety of circumstances (Pressey 1994). Recall from
Chapter 1 that this ad hoc approach to conservation has resulted in a
biased distribution of conservation lands and waters throughout the world
that falls considerably short of what is needed to conserve the full array of
biological diversity in most regions.

Conservation planners and biologists now recognize that if we are to
prevent vast numbers of species from going extinct, we must adopt more

systematic approaches to conserving biodiversity. One of the most important aspects of such a systematic approach is to be explicit about what features of biodiversity we are trying to conserve and where we are trying to conserve them. We refer to those specific features of biodiversity as *conservation targets* (Redford et al., in press). If Mianus Gorge were a high priority for conservation action today, planners and ecologists would want to know what species, communities, or other biological or environmental features occur there that warrant conservation attention.

Being explicit has several advantages. First, any planning region or ecoregion will be home to many thousands of species and potentially hundreds of natural communities. It is simply impractical to think about planning individual conservation efforts for all of these elements. Therefore, planners and biologists need to focus on a smaller set of features that they believe will have a high likelihood of conserving the full array of biological diversity in a region, including those species known to science and those yet to be discovered. Being explicit and specific about these targets forces planners to go through this winnowing-down process.

Second, because conservation targets will serve as the primary basis for identifying and selecting conservation areas in the region, the targets can be used as a check later to determine how well the final plan performed at representing them (see Chapter 8). Finally, conservation targets will serve as the basis for designing strategies and actions for each conservation area identified during the regional planning process. These strategies will be aimed at abating threats to the conservation targets and improving their overall viability and integrity, through a process referred to as conservation area or site conservation planning (discussed in more detail in Chapter 13).

The Hierarchy of Biodiversity

Scientists have recognized for quite some time that biodiversity occurs at a variety of levels of organization, from genes to ecosystems and landscapes (Noss 1990, Noss et al. 1998; Peck 1998). Theories about the hierarchical organization of biodiversity suggest that the higher levels, such as landscapes and ecosystems, constrain and affect what can happen at the lower levels of genes and populations (O'Neill et al. 1986). For example, in some areas, increased temperatures that are expected to occur under various scenarios of global climate change will cause snowpacks to melt earlier and at faster rates than in the past (watershed or landscape-level changes). In turn, these changes will influence the timing, duration, and magnitude of the hydrological flow regime that will likely affect fish and macroinvertebrate populations (population level changes).

Biodiversity also occurs at a variety of spatial scales. In the simplest of cases, a salamander population might be distributed across only a few square meters around a seep or waterfall, whereas a grizzly bear (*Ursus arctos*) population might range over hundreds of square kilometers. Poiani and colleagues (2000) have developed a practical framework that enables conservation biologists and planners to better account for the influence of spatial or geographic scale in their work. Their framework distinguishes four spatial scales: local, intermediate, coarse, and regional (Figure 4.1). Both the number of scales and the suggested size guidelines accompanying these scales are somewhat subjective. Some planners may find it useful to create slightly different categories, and/or categories with different size guidelines, to better fit the biodiversity and conservation targets found within a particular planning region. The framework is applicable to both species and communities or ecosystems and can be applied across terrestrial, freshwater, and marine systems. The example below refers to species,

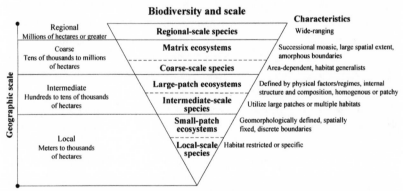

Figure 4.1   Spatial scales and levels of organization for biodiversity. Biological diversity can be described at a variety of spatial scales (local, intermediate, coarse, regional) and levels of biological organization (species, communities, ecosystems). The general range in size for each spatial scale is indicated to the left of the inverted pyramid, and some general characteristics of two types of conservation targets (species and ecosystems) are shown on the right. Some species (e.g., many frogs or salamanders) occur only at local scales, while others (e.g., grizzly or brown bears) are very demanding in terms of space requirements. Similarly, some communities and ecosystems, such as those occurring in caves or along cliffs, are localized in their distributions (patch), while others, such as the shrub steppe of Patagonia, occur over vast areas (matrix). Both the level of biological organization and the spatial scale at which biodiversity occurs are important aspects of conservation planning. (From Poiani et al. 2000. Biodiversity conservation at multiple scales: functional sites, landscapes, and networks. *BioScience* 50:133–146. Copyright, American Institute of Biological Sciences.)

but later in this chapter, I will return to the application of the framework to communities and ecosystems.

The Chaco is a vast plain measuring about 1 million km^2 (366,100 mi^2) in South America. The vegetation consists of a mosaic of grasslands, savannas, open woodlands, and xeric thorn forests. Annual rainfall in the Chaco declines in a gradient from the east, with 1200 mm (47.3 in.) per year, to the extremely arid west, with as little as 200 mm (7.9 in.) per year. The Chacoan fairy armadillo (*Burmeisteria retusus*) is a local-scale species whose individuals range over tens of hectares and are largely confined to sandy soils. It spends most of its time below ground where it feeds predominantly on insects. The southern three-banded armadillo (*Tolypeutes matacus*) is an intermediate-scale species that is also native to the Chaco, occurs in grasses and marshes between forested areas, and ranges over thousands of hectares. The gray brocket deer (*Mazama gouazoubira*), a small deer of the drier vegetation formations of South America, is a coarse-scale species that uses a wide variety of habitats of the Chaco as well as the cerrado and drier portions of tropical forests. Individuals range over an area greater than 1000 km^2 (386 mi^2). Finally, the puma (*Felis concolor*) is a regional-scale species that occurs in the Chaco region but occupies many different habitats, from deserts to tropical forests, thorn scrub, and grasslands. Its home range would typically cover several thousand square kilometers. It is distributed from Alaska to Tierra del Fuego.

The hierarchical ordering of nature, and the variety of spatial scales at which species and ecosystems occur, poses some intriguing questions for conservation planners attempting to select a group of targets for a regional planning exercise. For example, from which levels of biological organization (genes, populations, communities, ecosystems) should targets be selected? This very question has been the subject of considerable debate and discussion in the scientific community. Maddock and Du Plessis (1999) warned against the continued use of species-based approaches to conserving biodiversity, citing poor data coverage, biased biological surveys, and inadequate attention to viability issues as the main reasons. A series of papers published in the journal *Ecological Applications* in 1993, and subsequent responses to those papers, probably best typify the debate. Although the papers were primarily U.S. in scope, most of the debate and discussion had global application. Franklin (1993), for example, argued that efforts to conserve biodiversity must increasingly focus at the ecosystem level because of the impracticality of planning for a large number of species, many of which are not known to science (e.g., below-ground species). Orians (1993a) recommended that the Endangered Species Act should be expanded to include a habitat or ecosystem component that would be more responsive to taking conservation actions before populations reach dangerously low numbers and become endangered, at which time recovery efforts typically are initiated. The primary obstacle to implementing such an approach appeared to be the lack of any systematic and consistent classification of communities or ecosystems.

It is clear that biodiversity occurs—and is threatened—at many levels. In the United States and other countries with endangered species laws, a day hardly passes without a report of a threatened or endangered species. Similarly, we are all familiar with losses being sustained by wetland and tropical forest ecosystems around the world. These stories, and our knowledge of how the natural world is organized, suggest that what we need is a continuum of strategies that seek to conserve biodiversity at all spatial scales and levels of organization, from species to ecosystems (Noss 1996a), a point that is underscored by the quotation from David Wilcove at the beginning of this chapter. In fact, a thorough reading of the "target" papers in *Ecological Applications* indicates that all of these authors support a plurality of approaches; they differ only in what strategies they would emphasize and in how they would implement those strategies. The next section of this chapter describes an approach that attempts to synthesize several different strategies for conserving biodiversity from species to ecosystem levels. It has been referred to as the fine-and-coarse-filter approach.

Fine and Coarse Filters

Not far from where I live in the Pacific Northwest of the United States, four federally designated wilderness areas collectively span several million hectares and arguably represent the largest and most pristine natural area remaining in the continental United States. The heart of this area is the Salmon River and Selway River watersheds in central Idaho, an undammed series of rivers and streams with generally high water quality and a natural flow regime, surrounded by coniferous montane forests where the dominant natural disturbance regime (fire) functions moderately well across parts but not all of the region. There are few roads, even fewer that are paved, and timber harvest in the region has been minimal. Wolves were successfully reintroduced to the area in the early 1990s and are now thriving. The only species that is completely extirpated from the region is the grizzly bear, and habitat analyses suggest that the region could still accommodate grizzlies. At first glance, it would appear that a management focus on the stream and forest ecosystems of central Idaho is an adequate strategy for conserving the region's biodiversity. A closer inspection, however, suggests otherwise. Film footage from the 1950s shows salmon (*Oncorhynchus* sp.) spawning in headwater streams of the Salmon River in such numbers that fishermen claim it was nearly possible to ford streams on their backs. Today, salmon are on the brink of extinction in this region, the principal causes of which can be found far downstream. In some areas, sheep grazing at high-elevation lakes has introduced many non-native plant species, and many of the lakes in the region were naturally barren of fish until the state fish and game agency introduced non-native fish for recreational purposes. Elsewhere, several studies have demonstrated the detrimental effects to native amphibians from introducing fish to aquatic ecosystems that previously harbored no fish species (e.g., Fisher and Shaffer 1996).

The lesson to be learned from the wilderness areas of central Idaho is that focusing on the ecosystem level (i.e., the coarse filter) as the target of conservation is a necessary but insufficient strategy for conserving the region's biodiversity. With some exceptions, the stream and forest ecosystems in these central Idaho wilderness areas are functioning well generally, and the overall patterns of biodiversity remain relatively little affected by humans. Yet few scientists and conservationists in this region would feel that their work is complete until salmon and other endangered fish populations (i.e., the fine filter) are recovered, exotic species populations are brought under control, and a more natural fire regime is returned across the entire region.

The ecological story of central Idaho is illustrative of the importance of taking an approach to conservation that works at spatial scales and levels of biological organization from fine to coarse. Redford's (1992) empty forest metaphor of the otherwise high-quality examples of tropical forest, in which large mammals have been selectively hunted to local extinction, is a similar and equally compelling example of the need to focus on multiple levels of biological organization. Taking a myopic, either–or view (e.g., species or ecosystems) to conserving biodiversity will fall far short of what is really needed. Invoking analogies to Noah's ark remains popular in conservation biology (Scott and Csuti 1997, Shaffer et al. 2002). We now know that Noah's species approach to conservation is only one of the tools in the conservationist's toolbox.

The metaphor of the coarse and fine filters for biodiversity conservation originated with Natural Heritage methodology in TNC (Noss 1987, Hunter et al. 1988). In its original and simplest form, the main idea of the coarse filter was that by conserving representative examples of all the ecological communities of a given region, the vast majority of species would also be conserved. The explicit assumption was that communities could serve as a surrogate or coarse filter for conserving the majority of species of a region. To complete the metaphor, some species would fall through the pores of the coarse filter and would need to be conserved through individual efforts (the fine filter), such as those conducted under state or federal endangered species programs (Figure 4.2). Typical species that were

Figure 4.2   The coarse and fine filters of conservation planning. Conservation efforts directed at ecological communities and ecosystems (coarse filter) may conserve the majority of species in a planning region, but some species, such as rare or wide-ranging ones, are either not predictably associated with communities and ecosystems or range across many ecosystems. These species may pass through the coarse filter and require a more focused fine-filter approach, such as those that commonly occur under endangered species programs.

presumed to fall through this coarse filter were rare and endangered species, such as local-scale plant species confined to specific substrates that by chance alone would not be represented or "captured" by community or ecosystem-based conservation efforts. Indeed, a study of vegetation remnants in agricultural areas of Tasmania demonstrated that there was little overlap between remnants in good ecological condition and the presence of rare and endangered species (Kirkpatrick and Gilfedder 1995). The Nature Conservancy estimated that 85–90% of all species could be conserved by the coarse filter, although the efficiency of the coarse filter has been empirically tested only rarely, and the Conservancy's estimates are probably overly optimistic (Hunter et al. 1988, Hunter 1991, Noss and Cooperrider 1994).

MacNally and colleagues (2002) recently tested the effectiveness of ecological vegetation classes in Victoria, Australia, as surrogates for tree species, birds, mammals, reptiles, terrestrial invertebrates, and nocturnal flying insects. They sampled 80 sites distributed over 14 vegetation classes for these different taxonomic groups. They found that the ecological vegetation classes did a reasonably good job of representing birds, mammals, and trees, but not reptiles and invertebrates. Thus, one of the few field tests of the coarse filter corroborated Margules and Pressey's (2000) statement that there are no perfect surrogates. Their results also supported the notion that a hierarchical approach to selecting conservation targets across different levels of biological organization will likely be most effective in selecting areas for biodiversity conservation.

As it was originally conceived, the coarse filter was thought to offer a number of advantages (Hunter 1991). First, it requires a limited amount of information on the status and distribution of community types, as opposed to the nearly infinite amount of information that a species-based approach would require. Second, it appears to be efficient, focusing conservation action on priority examples of community and ecosystem units, as opposed to much more numerous and widely distributed occurrences of numerous species. Third, by focusing on communities, it begins to incorporate the notion of ecological integrity, paying attention to both the biological and the physical environments in which species occur.

There are a number of limitations as well (Noss 1987, Hunter 1991). The first of these is the manner and scale at which communities are defined. By defining communities based on homogeneous vegetation types, species whose life history requirements include several community types will be missed (e.g., habitat generalists, such as some wide-ranging carnivores). In addition, communities defined in this manner are often too small to maintain viable populations of constituent species, especially vertebrates. A second limitation is that many regions lack a classification of communities or, in some cases, lack any on-the-ground information on the status and distribution of individual units (communities) of these classifications. This is especially true in freshwater and marine systems (see Chapters 10 and 11). Third, there was an underlying assumption that these communities were stable, predictable assemblages of species, an assumption that was easily challenged by the paleoecological literature, thereby demonstrating that species respond to climate change (see Chapter 12) by shifting their distributions in individualistic manners (Hunter et al. 1988, Hunter 1991).

The idea of a fine-and-coarse-filter approach to conservation has evolved considerably since its original conception. Noss (1987, 1996a) recommended expanding the coarse-filter concept to incorporate transitions between communities, functional combinations of communities, and environmental gradients at landscape scales. Noss and colleagues (Noss and Cooperrider 1994, Noss et al. 1997, 1999a) have recommended a three-track approach to conservation planning that is essentially an expanded version of the fine filter and the coarse filter. The three tracks are represented by three different classes of conservation targets: representative habitats, special elements (e.g., imperiled species), and focal species (see the discussion of focal species below). Seymour and Hunter (1999), while discussing the conservation of biodiversity in managed forests, expanded the notion of coarse filters to include managing ecosystems within the limits of natural disturbance regimes established prior to extensive human alteration of landscapes. Haufler et al. (1996) emphasized the importance of developing an appropriate ecological classification system to adequately apply the coarse-filter approach, but also noted the importance of understanding historical disturbance regimes.

The fine-and-coarse-filter approach to conservation represents a practical approach to an otherwise complex problem. Yet in some respects, it can be a confusing and overly simplified concept. Some typical coarse filters, for example, can occur at both fine and coarse scales. Is a wetland community that naturally occurs over just a few hectares a fine filter or a coarse filter? Should we consider a wide-ranging species that is not endangered but might not be represented well by a community or ecosystem approach to be a fine-filter or coarse-filter element? A more useful approach is to recognize that, as articulated by Poiani and colleagues (2000), conservation targets occur over a continuum of spatial scales and levels of biological organization. Limitations on data availability at the species and gene levels necessitate that conservation planners will also focus on targets at higher levels of biological organization.

In reality, every conservation target, at whatever level or scale, is a surrogate to some extent for other features of biodiversity. Margules and Pressey (2000) argued that the best surrogates are determined by available data, the ability to collect new data (see Chapter 5), and the type of data analysis that can be done (e.g., predictive modeling of species distributions). In an extensive evaluation of surrogates in a well-inventoried area of New South Wales, Australia, Ferrier and Watson (1997) found that the poorest-performing surrogates were those derived purely from abiotic environmental data, that vegetation classifications generally outperformed environmental classifications, and that the best overall performance was obtained by modeling of species' distributions. Some parts of the developed world are rich in information about species distributions, but most parts of the world are not. Information about vegetation cover types obtained from remote sensing, as well as various types of environmental or physical data layers (e.g., soil maps, digital elevation models), usually represents the only spatially consistent data available for planners to use as surrogates.

Based on these considerations, planners should consider three general categories of conservation targets: (1) species, (2) communities and ecosystems, and (3) abiotic or environmental units. The following section will discuss the advantages and disadvantages of different types of targets within these categories, provide examples of each, and make some recommendations on how they can be combined into the most effective and efficient set of targets for conservation planning.

Species Targets

There are two general types of species targets. First are those species that would not likely be conserved by efforts that emphasize community or ecosystem approaches (e.g., endangered species). The second type consists of species that by the very nature of their life history requirements infer insights into the conservation needs of other species in the region (e.g., umbrella species). Below I review several types of species that could be considered targets in regional conservation planning exercises.

Threatened, Endangered, and Imperiled Species

The most widely known classification of such species is the World Conservation Union's (IUCN) Red List of Threatened Species (Hilton-Taylor 2000). In the United States, the listing of Threatened and Endangered

Species under the Endangered Species Act (ESA) carries significant legal standing, while the NatureServe/Natural Heritage Program classification of imperiled species is more comprehensive and consistent (Master et al. 2000). Current listings of species under each of these classifications are available at www.redlist.org, www.endangered.fws.gov/wildlife.html/#Species, and www.natureserve.org/explorer, respectively. In addition to these classifications, many countries have endangered species laws and lists that should be consulted, and numerous states in the United States have state endangered species laws and official lists as well.

At a minimum, planning teams should include as targets those species listed as Threatened or Endangered under the ESA; those species classified as critically endangered, endangered, or vulnerable by IUCN Red Lists; and those species classified as globally imperiled or critically imperiled (i.e., species global ranks of G1–G2) by NatureServe. Definitions for these various classifications can be found at the cited references and Web addresses. Although there will be considerable overlap in the species lists from the ESA, IUCN, and NatureServe sources, planning teams should be aware that there are significant differences in how these organizations define the various categories of threat and imperilment as well as other protocols for listing (e.g., the IUCN list does not allow for subspecies, while the ESA list does).

At-Risk Species

For a variety of reasons, additional species that may be at risk in a region are not included on the ESA, IUCN, or NatureServe/Natural Heritage lists. There may already be lists of such species with local, state, or national governments. For example, in the United States, many states maintain lists of "species of special concern." Two federal agencies, the U.S. Forest Service and the U.S. Bureau of Land Management, classify such species as "Sensitive" and maintain lists of Sensitive Species by national forest and state, respectively. Similarly, Natura 2000, an initiative of the European Union, has identified over 200 animal and 500 plant "species of Community interest" to the member states of the Union that are in need of conservation attention (European Commission 2000).

Another extensive effort in this same vein is that of Partners in Flight, a consortium of organizations concerned about declining bird species in the western hemisphere (Carter et al. 2000; see www.cbobirds.org for complete list of priority conservation rankings for U.S. landbirds). In addition, many countries have completed and published analyses of conservation priorities for various taxonomic groups (e.g., Ceballos et al. 1998, Cofré and Marquet 1999) that will be helpful to planners in selecting species targets. Planning teams should review the available lists and publications for potential additions to the species conservation target category for a regional or ecoregional planning project.

One caution is in order, however. Rare species can have a disproportionate influence on the selection of conservation areas especially when using area selection algorithms (see Chapter 8) that employ the principle of complementarity in the selection process (Rodrigues and Gaston 2002). Therefore, planners need to be selective about which types of rare or at-risk species are included as conservation targets, taking care to include only those species that are truly at risk in a substantial portion of their range, and avoiding the inclusion of species that may only occasionally or minimally occur within the planning region (e.g., "vagrant" bird species).

There is also a growing body of papers in the ecological and conservation biology literature on extinction-prone species (e.g., Reid and Miller 1989, Angermeier 1995, Lawton and May 1995, Fagan et al. 2001). In general, species with low population density, low reproductive potential, narrow geographic distributions, and relatively larger body mass within a taxonomic group tend to have a higher likelihood for extinction. Planning teams may find it useful to evaluate whether there are species that fall within these categories that are not included in any threatened, endangered, imperiled, or at-risk classifications.

Endemic Species

Endemic species are those species whose entire distribution is restricted to an ecoregion or small geographic region within an ecoregion. Because of their limited geographic range, they are often, but not always, vulnerable to extinction. In most cases, planning teams will want to include such species on their target lists because populations of these species in the region or ecoregion represent the only opportunity to conserve them.

Flagship Species

Flagship species are charismatic species, often referred to wittily as "charismatic megavertebrates," that are symbols of major conservation projects (Noss 1991, Caro and O'Doherty 1999). They are usually large-bodied species, often endangered, that are used to garner public support for nature conservation at a variety of scales. One of the most visible flagship species is the panda (*Ailuropoda melanoleuca*), the global emblem for the World Wide Fund for Nature (WWF). Although flagship species can be effective public relations tools for conserving particular sites or areas, they have limited scientific value in serving as conservation targets in regional planning and should be used with caution. Simberloff (1997) pointed out that there are real dangers for conservation projects whose success depends on the fate of one population or one species and that there are real problems for areas worthy of conservation that contain no flagship species.

Umbrella Species

Umbrella species are those species whose conservation confers some protective status to numerous co-occurring species (Fleishman et al. 2000). As intuitively attractive as umbrella species may be, until recently there has been little empirical evidence to support the concept and mixed results from various post hoc analyses. Noss et al. (1996) surveyed the literature and found no published empirical studies that documented protection afforded to other species by an umbrella species. Using predicted distribution maps of vertebrate species in Idaho from the Gap Analysis Program, they estimated that a proposed recovery plan for the grizzly bear would also conserve over 10% of the statewide distribution of two-thirds of the vegetation types and 65% of the native vertebrates in Idaho. With a similar retrospective analysis of two regional databases and one national database on imperiled and endangered species, Andelman and Fagan (2000) found that umbrella and flagship species performed no better as surrogates than species randomly selected from these databases.

Sanderson and colleagues (2002a) recently advanced the concept of using a suite of landscape species in conservation planning. Because of the relatively large areas over which they range, the heterogeneous habitats they occupy, and their vulnerability to human activities, these species have the potential to play an umbrella role in the conservation of other species, species assemblages, and ecological processes. Fleishman et al. (2000) applied the umbrella species concept to montane butterfly communities. They demonstrated that a subset of butterfly taxa could serve as umbrellas for a wider group within the same taxon. Yet, even among the umbrella butterfly species, there was considerable variation in life history traits, such as population size and the type of larval host plant, suggesting that there is no easy way to select such species.

In summary, studies to date suggest that umbrella species may be useful conservation targets in regional planning efforts under some circumstances.

Their use, however, should be treated hypothetically and tested as to whether the species actually perform as planners and biologists intend. As Sanderson et al. (2002a) suggested, a suite of umbrella species will probably be more effective than any single species. The best umbrella species are likely to have relatively large geographic ranges, have large home ranges, and be habitat generalists (Hunter 2001). Lambeck (1997) has defined a subtype of umbrella species termed *focal species*. Planners may find it useful to combine the umbrella and focal species types during the selection process of target species.

Focal Species

Lambeck (1997) defined focal species as a suite of species that are consistent with the concept of umbrella species and that can be used to develop explicit guidelines for determining the composition, quantity, and configuration of habitat patches at the landscape scale for restoration purposes. In so doing, these focal species should serve as surrogates or umbrellas for all species that occur in the landscape. Focal species, as considered by Lambeck, fall into four general categories:

- *Area-limited species.* Patches of habitat are usually too small to support breeding pairs or socially functional groups of these species (the northern spotted owl in the northwestern United States is one of the best-known examples).
- *Resource-limited species.* Their populations are limited by the supply of a particular resource, often food (e.g., the dependency of black bears [*Ursus americanus*] on mast and fruit production).
- *Dispersal-limited species.* Suitable habitat patches exist for these species, but the interpatch distance is too great or inhospitable to allow dispersal among patches.
- *Process-limited species.* They depend upon particular disturbance regimes or ecological processes, and their populations may be limited when the process is altered (e.g., plant species dependent upon fire for regeneration, or certain woodpecker species dependent upon fire for creating snags).

Others have defined focal species in different ways. For example, Noss and colleagues (1999a) used focal species to answer two questions: How large do conservation areas need to be, and what should their configuration be? Caro (2000) noted at the 2000 meetings of the Society for Conservation Biology that the term *focal species* was being used inappropriately as a catch-all phrase for umbrella species, flagship species, and indicator species. More recently, Lambeck's concept of the focal species approach has been criticized on the grounds that taxon-based surrogate or umbrella schemes have a number of conceptual, theoretical, and practical problems; the data to actually implement the approach are lacking for most landscapes; and the approach fails to account for the issues of temporal scale and population viability (Lindenmayer et al. 2002). Given the confusion over the use of this terminology, perhaps the best advice is that of Caro (2002): "We should abandon them [surrogate and focal species] altogether and define precisely what we are talking about instead of using insider jargon."

Area-limited species are a good example of being more specific about a type of focal species that is useful in conservation planning. These species are frequently used as conservation targets in ecoregional and regional conservation plans (e.g., Carroll et al. 2001). This is particularly true for wide-ranging species such as many carnivores, migratory ungulates, or anadramous fishes. In these situations, the distribution of individual populations of these species will likely go beyond the boundaries of any single ecoregion or planning region. In these cases, planning teams will find it more useful to work with biologists in adjacent regions and develop a rangewide conservation plan for the species. Recent examples of such plans have been completed for the tiger (*Panthera tigris*) (Wikramanayake et al. 1998) and the jaguar (*Panthera onca*) (Sanderson et al. 2002b).

Keystone Species

The concept of keystone species was first introduced by ecologist Robert Paine to describe the reduction in species diversity that occurred in an intertidal marine community when predatory starfish (*Pisaster* sp.) were experimentally removed (Paine 1966, 1969). As the concept attracted more attention, scientists have advocated that keystone species be "special targets" in programs to protect biodiversity (Mills et al. 1993a). Unfortunately, the term has been poorly defined and broadly applied, thereby limiting its usefulness in conservation applications.

More recently, Power et al. (1996) provided a more specific definition of keystone species as those species whose impact on their community or ecosystems is disproportionately large relative to their abundance. They also developed a mathematical expression to estimate the strength of the effect of a potential keystone species on a particular community or ecosystem trait and provided a summary table of likely keystone species or guilds that have been described in the scientific literature. Kotliar (2000) applied Power's definition of keystone species to prairie dogs (*Cynomys* sp.) and proposed that an additional criterion be added to the definition: Keystone species perform roles not performed by other species or processes.

These recent conceptual advances and applications suggest the following recommendations for use of keystone species as targets in conservation planning:

- Keystone species have utility as conservation targets because of their differential impact on communities and ecosystems, but planners should use caution in their selection and real evidence for the keystone role should be provided.
- For any single community or ecosystem, there will be only a few and possibly no keystone species identified.
- Where keystone species can be identified and used in conservation planning, they may be able to serve as surrogates for some ecological processes or ecosystems of high ecological integrity (Simberloff 1997, Kotliar 2000).

Indicator Species

Indicator species are best defined as surrogate species whose traits (e.g., presence/absence, density, dispersion, reproductive success) are used as an index to the condition of other species for which those characteristics are too difficult or expensive to measure (Landres et al. 1988). Much of the use of indicator species originated with federal natural resource agencies in the United States, specifically the Habitat Evaluation Procedures (HEP) of the U.S. Fish and Wildlife Service and the Management Indicator Species (MIS) in the U.S. Forest Service. Three types of indicator species are recognized: health indicators, population indicators, and biodiversity indicators (Caro and O'Doherty 1999). Health indicators assess changes in habitat, population indicators serve as a barometer for populations of other species, and biodiversity indicators pinpoint areas of high biodiversity. The concept of indicator species has come under considerable criticism (Landres et al. 1988), but until the late 1990s, most of that criticism was directed at them with regard to their use as health and population indicators.

A great deal of research has been conducted on whether well-known taxonomic groups can serve as surrogates or biodiversity indicators for conserving areas of high species richness of sympatric species from more poorly known groups (e.g., Prendergast et al. 1993). The unfortunate news for conservation planners is that the vast majority of studies from several ecosystems around the globe indicate that areas rich in species of one taxonomic group do not overlap with areas rich in another (Pimm and Lawton 1998). Nor do areas of the rarest species generally coincide. Although there have been notable exceptions to these trends (e.g., Dobson et al. 1997, Kerr 1997), in general, conservation planners are likely to find little benefit to using biodiversity indicator taxa as conservation targets. In addition, there are other concerns to using species richness as a metric for biodiversity value. Areas that are high in richness may simply represent the edge of the distributional range of a number of different types of communities or ecosystems. In other cases, species richness may reflect concentrations of widely distributed and abundant species. Finally, patterns of species richness have been shown to be sensitive to the size of sampling units that are being used in the analyses (Stoms 1994). In any event, there seems to be little conservation value in targeting species–rich areas unless those areas represent concentrations of endemic or threatened and endangered species.

Many classifications of species can serve as conservation targets in regional planning exercises. Some of them suffer from imprecise definitions and inconsistent application. To help clarify similarities and differences, Table 4.1 provides the common characteristics of each category of species discussed here. In summary, planners should always consider species classified as threatened, endangered, imperiled, and at-risk; endemic species; and keystone species as conservation targets. For species that may be variously classified as umbrella or focal, planners should designate such species as targets only if they are more precisely defined (e.g., area-limited species), the data exist to support their use, and they are not likely to be conserved by targets at higher levels of biological organization (e.g., communities and ecosystems).

I will address one final issue concerning species as targets—peripheral populations. *Peripheral populations* are those that are on the outer limits or edge of a species' distribution. Conventional wisdom holds that many, if not most, of these populations are ecologically marginal. They tend to have smaller population sizes, occupy suboptimal habitats, and have reduced rates of immigration compared to populations at the core of the range. Lesica and Allendorf (1995) suggested that peripheral populations may be valuable targets for conservation if they are genetically divergent from more central populations. Spatial isolation, poor dispersal ability,

Table 4.1 *Advantages and Disadvantages of Classifications of Species Used for Conservation Planning*

Species Classification	Advantages	Disadvantages	Similarity and Overlap
Threatened and Endangered species under ESA	Strong legal standing, T/E status tends to force collaboration among stakeholders; funding often available for planning and recovery; recovery plans may already exist.	T/E status can be negative stigma with some segments of public; known biases in the listing process; many vulnerable species are not listed as T/E.	Some overlap with IUCN Red List and NatureServe/Natural Heritage list of imperiled and vulnerable species.
NatureServe/Natural Heritage imperiled and vulnerable species (G1–G2)	Compiled in more systematic and comprehensive fashion than ESA lists; ranking process transparent; considerable basic life history and status information available for each species on NatureServe Web site.	Global or G ranks most strongly influenced by number of known populations or occurrences; threat and decline minor influence in ranks.	Considerable overlap with ESA Threatened and Endangered and IUCN Red List of critically endangered, endangered, and vulnerable.
Endemic species	By definition, limited places where these species can be conserved.	Can be defined in different ways in terms of area of endemism; not all endemic species are at risk.	Many endemic species are classified as threatened, endangered, imperiled, or vulnerable.
Flagship species	Easy to garner public support.	Often not most critically at-risk or endangered species; no scientific basis to classification.	Many flagship species are also listed as threatened, endangered, imperiled, etc. (e.g., panda).

98

Umbrella species	Efficient and practical; focus is on one species being able to conserve many species.	No empirical studies have demonstrated validity; retrospective analyses have been equivocal about usefulness of concept.	Well-known examples tend to be wide-ranging carnivores, many of which are listed as T/E, imperiled, vulnerable (e.g., grizzly bear, lynx).
Focal species	A group of species that will often not be conserved by community or ecosystem approaches because of specialized resource use or unique life history.	Have been defined in many different ways by different authors.	Focal species are often classified as T/E, imperiled, vulnerable, etc.
Keystone species	Require planners to think about structure and function (ecological processes).	Only a limited number have been identified; applied and defined in different ways.	Little overlap with other classifications.

short generation time, and habitat differentiation between core and peripheral populations will all contribute to genetic differentiation in peripheral populations. Some recent biogeographic analyses by Channel and Lomolino (2000) have shown that range contractions in declining species around the world result in remnant populations occurring along the periphery of a species range—*not* in the core of the range, as previously believed.

These results, and the recommendations of Lesica and Allendorf (1995), suggest that conservation planners should target species with peripheral populations in the planning area under two sets of circumstances: (1) experts in the region have reason to believe that such populations are likely to be genetically divergent from core populations, and (2) a species range has shrunk by at least 50% over historic times and remaining populations are primarily peripheral ones. For species that are in decline but for which core populations remain extant, conservation efforts should first be concentrated on these core populations, as the analysis of Channel and Lomolino (2000) indicates that they will be among the first to be extirpated.

Community and Ecosystem Targets

The terms *community* and *ecosystem* have been defined and used in many different ways by ecologists. For example, Noss et al. (1995) noted that "an ecosystem can be a vegetation type, a plant association, a natural community, or a habitat defined by floristics, structure, age, geography, condition, or other ecologically relevant factors." From a management or biodiversity conservation perspective, the lack of precision in defining these terms causes considerable confusion in professional discourse. Whittaker (1975) defined a natural community as an assemblage of populations of plants, animals, bacteria, and fungi that live in an environment and interact with one another, forming a distinct living system. He distinguished this community from an ecosystem, which is the sum of the community and its environment treated together as a functional system that transfers and circulates energy and matter. Although most ecologists recognize this distinction between community and ecosystem, as suggested by Noss and colleagues, some do not. As a result, when planners evaluate various community and ecosystem classifications, they should pay particular attention to how those terms are defined and used.

Many different classifications of communities and ecosystems have been described and used in a variety of fashions. Table 4.2 summarizes a number of the more commonly used classifications and some characteristics of each. There is no perfect classification system or one correct way to classify ecosystems and communities (Grossman et al. 1999). Classifications have been developed for three primary purposes: as a heuristic tool for the discipline of biogeography; to identify "representative" ecosystems for management, restoration, research, production, and other uses; and as an aid to programs aimed at conserving biological diversity. It is the third purpose that concerns us here.

Most terrestrial biotic classifications are based on vegetation, although a number have also been described for animals, a process referred to as zoogeographic classification. Although planners may occasionally find that zoogeographic classifications exist and can be used in regional planning (e.g., Angermeier and Winston [1999] describe fish communities), in most cases terrestrial classifications based upon dominant or characteristic vegetation are the only type of spatially consistent information that is likely to be available to planning teams. Classifications of freshwater ecosystems are more limited than those for terrestrial ecosystems, although several have been developed (see Table 4.2). Chapter 10 will focus in more detail on those classifications. In marine systems, there has been an even greater paucity of ecosystem classification work. Those that have been published tend to be at very coarse spatial scales, although there is a growing number of classifications at finer scales of resolution (see Chapter 11). Because of the difficulty in obtaining biological information, especially in pelagic marine systems, these classifications tend to be largely based on environmental or abiotic factors. (See the section on abiotic or environmental units later in the chapter.)

The ability to define a standard set of community or ecosystem units that are comparable across a planning region, to determine their distribution and condition, and to relate those units to other levels of biological organization is critical to using communities and ecosystems as conservation targets in regional planning (Noss et al. 1995). Because of the different classification schemes available, there are five criteria planners should consider when selecting a classification system (Anderson et al. 1999):

1. What criteria are used to define the classification system?
2. Can the classification system be arranged hierarchically for use at multiple scales?
3. Can the units of classification be readily mapped, and to what extent is that mapping complete for the planning region?
4. What is the geographic area for which the classification is most useful?
5. Have successional types of vegetation units, usually resulting from natural disturbances to ecosystems, been incorporated in the classification system?

Table 4.2 *A Summary of Different Types of Ecosystem Classifications*

	Ecological System					Geographic Coverage					Objectives		
	Terrestrial	Freshwater	Marine/Coastal	Estuarine	Wetland	Global	Continental	National	Regional	Local	Research	Management	Conservation
Albert 1994	√				√			√			√		
Allee and Schmidt 1937	√							√			√		
Anderson et al. 1976	√							√				√	
Bailey 1995	√	√	√		√	√	√	√	√		√	√	
Braun-Blanquet 1928	√				√	√					√		
Busch and Sly 1992		√				√		√			√	√	
Cleland et al. 1994	√				√				√	√	√	√	
Cowardin et al. 1979		√		√	√			√			√	√	
Daubenmire 1952	√				√						√		
Dinerstein et al. 1995	√				√		√		√			√	√
Driscoll 1984	√				√			√	√			√	
ECOMAP 1993	√				√			√	√	√	√	√	
Frissell et al. 1986		√									?		
Grossman et al. (TNC) 1998	√							√				√	√
Hayden et al. 1984			√								?		√
Holdridge 1947	√					√					√		
Illies 1961		√				√					√		
Ketchum 1972			√			√					√	?	
Kotar et al. 1988	√									√	√		
Krajina 1965	√					?					√		
Kuchler 1964	√					√	√	√			√	√	
Lammert et al. 1997		√						√				√	√
Maxwell et al. 1995		√					√					√	
Mitsch and Gosselink 1986					√			?				√	
Omernik 1995	√						√	√				√	
Pfister and Amo 1980	√								?		?	?	
Pregitzer and Barnes 1982	√										√	√	
Pritchard 1967, Barnes 1967				√				?		√			
Ray and Hayden 1992			√	√					?				√
Rowe 1984	√							√					
Schupp 1992		√								√			
Scott et al. 1993													
Udvardy 1975	√					√					√	√	
UNESCO 1973	√					√					√	√	
USGS 1982		√						√			√	√	
Wiken 1986	√							√				√	
Zonneveld 1989	√							?			√	√	

Environmental Factors						Biological Factors					Other Factors			
Climate	Soils	Elevation	Geology	Landform/Position	Hydrology/Hydrography	Existing Vegetation Structure	Existing Vegetation Composition	Potential Natural Vegetation	Zoological Guilds	Zoological Composition	Classification Units Are Geographically Referenced	Hierarchical (multilevel)	Multifactored	Extensive Data Requirements
√	√	√	√	√	√			√			√	√	√	
√									√	√	√			
							√					√		√
√		√		√				√			√	√	√	
							√					√		
							√					√		
					√				√	√			√	
√	√	√	√	√	√			√			√	√	√	
	√				√						√	√	√	
	√			√			√						√	
						√	√	√			√	√		
√						√	√							
√	√	√	√	√	√			√			√	√	√	
					√							√		
√	√		√	√		√	√					√		
√					√							√		
				√	√						√		√	
								√					?	
√	√	√	√					√				√	√	
√	√				√			√			√	√	√	
			√	√	√	√	√		√	√	√	√		
	√					√	√				√			√
√	√		√		√	√					√			√
	√										√	√		
√	√			√	√			√			√	√	√	
				√	√						√	√	√	
					√									
√							√				?	?		
					√					√			√	
√	√		√			√					√	√		
√	√		√	√		√	√	√			√	√		
				√	√						√	√		
√	√	√	√		√	√	√	√			√	√		
√	√			?				√			√			

Most of the more widely used vegetation classifications of communities and ecosystems are based on physiognomy (structure), floristics (species composition), or both (Grossman et al. 1999). Physiognomy refers to the life-form (shrub, tree, herb, etc.) of the dominant plant species. Closed-canopy forests and herbaceous grasslands are examples of classification units based on physiognomy. Holdridge's (1967) life zones are an example of a physiognomic-based classification. The advantages of these types of classifications are that they are easy to use by nonexperts and that information about them can often be obtained from remote-sensing imagery.

Floristic classifications emphasize the species composition of dominant or diagnostic species. The best-known classification is that of Braun-Blanquet (1928). Typically, these classifications contain units referred to as associations and alliances, whose naming describes the species composition of the unit (see below for more details). An example of an alliance name might be "American beech–southern Magnolia Forest Alliance"; an example of an association, "Subalpine fir/grouseberry forest" (Maybury 1999). The advantages of floristic-based classifications are that they emphasize biotic patterns and may offer insights into ecological processes (Anderson et al. 1999).

Classifications that combine floristic and physiognomic approaches have also been developed. Probably the best known is the terrestrial International Classification of Ecological Communities developed by TNC, the Association for Biodiversity Information (now known as NatureServe), and other collaborators (Grossman et al. 1998, Maybury 1999). In the United States this system is known as the United States National Vegetation Classification, or USNVC. It uses a physiognomic approach at its upper levels and a floristic one at the lower or finer levels of resolution in the classification (Figure 4.3).

Whether or not a classification of communities and ecosystems is hierarchical is a second consideration when selecting a system for conservation planning. Because information from biological inventories is often incomplete and available at different levels of detail and resolution, a hierarchical system is advantageous because of its flexibility in using information at different levels of a classification. Figure 4.4 provides an example from the Northern Appalachians Ecoregion in the United States of the hierarchical structure of the classification for a coniferous forest alliance. Approximately 1600 alliances and more than 4500 associations have been described for the USNVC (Grossman et al. 1998). Both alliances and associations occur at a variety of spatial scales. For example, a leather leaf bog

SYSTEM: TERRESTRIAL

FORMATION CLASS
    FORMATION SUBCLASS
        FORMATION GROUP
            FORMATION SUBGROUP
physiognomic levels   FORMATION

ALLIANCE *floristic levels*   ASSOCIATION

Figure 4.3 Different levels in the hierarchy of the U.S. National Vegetation Classification (Grossman et al. 1998). The upper levels of this classification are characterized by differences in physiognomy, whereas the lower or finer-scale levels are separated on the basis of floristic (species composition) differences. See Figure 4.4 for application of this classification scheme.

community in Maine may cover hundreds of hectares, while the same community to the south in Pennsylvania might occur in small patches (Anderson et al. 1999). As a result, depending upon the map scale, some community or ecosystem units can be mapped while others cannot. Because alliances generally occur at coarser scales of resolution than associations, it is more likely that they can be mapped. As a generalization, if a classification unit cannot be mapped at a scale appropriate to the goals of the planning project, then that system is of limited use for conservation planning.

The fourth criterion for consideration when selecting a classification system is its geographic coverage. Minimally, a classification must be applicable throughout the planning region or ecoregion, although it is possible to synthesize or "cross-walk" two or more classification systems into one system. Even more desirable are classification systems that span several regions of interest. They allow planners to set goals for individual communities and ecosystems throughout the range of their distribution and to address important questions about how narrowly or widely certain communities or ecosystems are distributed.

Finally, a classification system should either explicitly or implicitly incorporate successional communities resulting from natural disturbance processes. For example, in the eastern United States, the loss of communities dominated by grasses, shrubs, and young trees within the matrix of an overall forest-dominated landscape is of increasing concern to conservationists (Askins 2001). Many species are dependent upon such communities and the processes that maintain them for their persistence (e.g., canopy-gap species that depend on small or large openings in the forest).

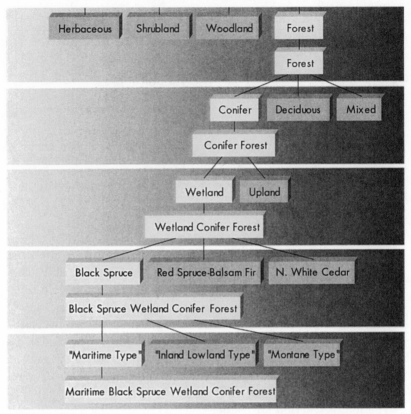

Figure 4.4   An example of the U.S. national Vegetation Classification from the Northern Appalacians Boreal Forest ecoregion. (From Anderson et al. 1999.)

Noss and colleagues (1995) were the first to assess the status of endangered ecosystems across the entire United States, although they did so without any formal classification system. A formal classification system now exists for terrestrial systems (the USNVC) and is being put into practice on both ecoregional and state bases by TNC and the Gap Analysis Program, respectively. Table 4.3 lists a sampling of plant community targets from the USNVC for the Northern Tallgrass Prairie Ecoregion.

Despite these applications, a number of problems remain in using the USNVC for regional-scale conservation planning. First, many Natural Heritage Programs maintain information at the finest level of this classification, the plant association, yet information on these associations is spotty

Table 4.3   *A Sampling of Community Targets for the Northern Tallgrass Praire Ecoregional Plan*

Community Name	Distribution★	Pattern★
Northern Mesic Tallgrass Prairie	Endemic	Matrix
Northern Cordgrass Wet Prairie	Endemic	Matrix
Bur Oak–Aspen Woodland	Endemic	Large patch
Aspen Parkland–Tallgrass Woodland	Endemic	Large patch
Great Plains Calcareous Fen	Endemic	Small patch
Midwest Dry Gravel Prairie	Limited	Large patch
Dogwood–Mixed Willow Shrub Meadow	Limited	Large patch
Saline Wet Meadow	Limited	Small patch
Prairie Transition Wet Fen	Limited	Small patch
Bulrush–Cattail–Bur Reed Shallow Marsh	Widespread	Large patch
Cottonwood–Black Willow Forest	Widespread	Large patch
Reed Marsh	Widespread	Large patch
River Mud Flats	Widespread	Small patch
Water Sedge Wet Meadow	Widespread	Small patch
Central Mesic Tallgrass Prairie	Peripheral	Matrix
Northern Mixed Oak–Hazel Forest	Peripheral	Matrix
Mixed Aspen Swamp	Peripheral	Large patch
Paper Birch–Hazel Forest	Peripheral	Large patch
Black Spruce Bog	Peripheral	Small patch
Bulrush Brackish Marsh	Peripheral	Small patch
Foxtail Barley Meadow	Peripheral	Small patch

★See Chapter 6 for definitions of distribution and pattern.

From The Nature Conservancy 1998.

within states and not comprehensive across them. Second, because of the plant associations' fine scale of resolution, information on their distribution is difficult or, in many cases, impossible to obtain from remote sensing. The relatively large number of plant associations for some planning regions also makes them an impractical target in some cases. Third, as I pointed out in the coarse-filter discussion, classification units like associations will not serve well as coarse filters for species that require a number of different vegetation types to meet their life history needs. Finally, disturbance regimes and processes that sustain the patterns of vegetation often operate at coarser scales of resolution than those at which plant associations typically occur.

In response to these problems, scientists from NatureServe and Gap Analysis Programs are now working toward developing a coarser ecosystem-level of classification units within the USNVC. These units have been referred to as ecological systems (Groves et al. 2000a) and are defined as dynamic, spatial assemblages of ecological communities that occur together on a landscape; are tied together by similar ecological processes (e.g., fire, hydrology), underlying environmental features (e.g., soils, geology), or environmental gradients (e.g., elevation); and form a cohesive, distinguishable unit on the ground. The naming of these units as ecological systems is both confusing and unfortunate. In reality, these units are little different from other classification systems identified in Table 4.2 and more commonly referred to as ecosystems. To avoid further confusion, hereafter I refer to these units as *ecosystem targets*. However, please note that in Chapter 10 the terminology of aquatic ecological systems is retained because it has been previously published and is in wider use. These ecosystem targets are characterized by both abiotic and biotic components and can be developed for terrestrial as well as aquatic systems. Table 4.4 provides a partial list of terrestrial ecosystem targets referred to as vegetation types for the Riverina Bioregion, Australia (Eardley 1999). Mapped information on these terrestrial ecosystem targets is typically obtained from multiple sources, including ground surveys, remote-sensing imagery, and expert opinion.

For many countries, one or more classifications of communities or ecosystems may exist at various spatial scales covering parts or all of the

Table 4.4   *A Sampling of Terrestrial Vegetation Types (Ecosystems) in the Riverina Bioregion, New South Wales, Australia*

Vegetation Type Name	Extent in the Bioregion (hectares)
Riverine Forest	445,837 ha
Riverine Forest–Grey Box Woodland	501 ha
Black Box Woodland	301,928 ha
Dune Crest Mallee	803 ha
Sandplain Mallee	10,095 ha
Belah–Rosewood	33,555 ha
Black Bluebush	204,167 ha
Pearl Bluebush	9390 ha
Bladder Saltbush	586,966 ha
Canegrass	43,246 ha
Callitris Mixed Woodland	86,246 ha
Lignum	365,199 ha
Old Man Saltbush	22,883 ha
Cotton Bush	434,149 ha
Dillon Bush	67,116 ha
Great Cumbung Swamp	4445 ha
Grey Box Woodland	24,433 ha
Boree Woodland	106,718 ha
Grassland	774,409 ha
Casuarina	2955 ha
Water	5611 ha
Cleared or Cropped	4,092,787 ha

Modified from Eardley 1999.

country. In addition, a classification system may have already been used with remote-sensing imagery and other information to develop a map of vegetation cover types or some similar product. Planners will need to use the criteria discussed in this chapter in making decisions about which maps and classification systems will be most useful for biodiversity conservation planning. Lacking either of these products, planning teams can develop vegetation maps from field reconnaissance data and remote-sensing imagery, including satellite imagery, aerial photos, and aerial videography (Jennings 2000). The *Gap Analysis Handbook* (available from www.uidaho.gap.edu) provides detailed information on land-cover mapping for conservation application, and Wilkie and Finn (1996) offer a useful introduction to remote-sensing imagery for natural resource applications. A recent book on Rapid Ecological Assessments (REAs) also provides useful guidance on the development of vegetation classifications and maps (Sayre et al. 2000) (see Chapter 5).

Abiotic or Environmental Units

For many parts of the world, there are no classifications or maps of terrestrial, freshwater, or marine ecosystems, and developing these maps is an expensive endeavor that many countries cannot afford. Fortunately, digital data sets on environmental or physical variables (e.g., terrain, climate, substrate) are increasingly available at regional, national, and global scales. These types of variables have been shown to influence the distribution of species (e.g., Austin et al. 1990, Burnett et al. 1998, Nichols et al. 1998). As a result of the ready availability of environmental data, conservation planners increasingly are attracted to the use of environmentally derived units as conservation targets (e.g., Hunter and Yónzon 1993).

Australians have been the intellectual leaders in this field. Several different approaches have been taken to using environmental variables as targets for conserving biodiversity. These can be grouped into three general categories: land systems, environmental domain analyses, and unique combinations of environmental variables.

Land systems are large areas with recurring patterns of landform, soils, vegetation, and hydrological regimes that can be identified and mapped by interpretations of patterns on aerial photographs. Pressey and Nicholls (1989) used land systems as a surrogate for ecosystems in the western part of New South Wales to identify potential conservation areas. Since that time, land system classifications have been extensively used in Australia and elsewhere as a conservation target in various bioregional planning exercises (e.g., Smart et al. 2000). WWF-Canada has taken a similar approach to biodiversity conservation planning in Canada based on "enduring features"—essentially land systems defined by topography, soils, and geology (Iacobelli et al. 1993).

Wessels et al. (1999) used land facets, the building blocks of land systems, as surrogates to predict assemblages of birds and dung beetles at local scales (1:10,000–1:50,000) in South Africa. Land facets were defined as the simplest terrain unit of uniform slope, parent material, soils, and hydrological condition that could be detected from an aerial pattern. Multivariate analyses demonstrated that land facets adequately represented distinct assemblages of birds and dung beetle species and that these facets could then be used in a variety of area selection algorithms (see Chapter 8) to design a reserve network for these species.

Numerical clustering and ordination techniques can be used to define the second major category of abiotic conservation targets: environmental classes or domains. Kirkpatrick and Brown (1994) defined 68 environmental classes that described the physical environment of Tasmanian forests based on a combination of 12 geology groups, 5 altitudinal zones, and 9 climatic variables. These environmental classes were then used to design a system of nature reserves that represented each of these classes in a reserve. Other planners and biologists have undertaken similar analyses to utilize environmental classes as conservation targets in the design of a nature reserve system (Belbin 1993, 1995, Faith and Walker 1996, Noss et al. 1999b, 2001, Fairbanks and Benn 2000, Fairbanks et al. 2001).

The third basic approach to using environmental data combines several environmental variables in a simple overlay fashion with a Geographic Information System. For example, Pressey et al. (1996a) combined rainfall, temperature, soil fertility, and slope variables to create 81 unique environmental units for the forests of northeastern New South Wales. These units were then used as the basis of a Gap Analysis project to determine how well each unit was represented in different land tenure classes and how vulnerable each unit was to forest clearing. As a result of these analyses, conservation priorities could be established for different forest environments. More recently, Pressey and colleagues (2000) have used abiotic data to describe 1486 landscape units across all of New South Wales and assessed the extent to which each of these units is represented in conservation areas, as well as the vulnerability to clearing of each unit. Ecological Land Units, later used by TNC in ecoregional planning (described below), are an abiotic entity developed by overlaying different environmental data layers in a fashion similar to that described for the Australian examples above.

Several recent investigations in Australia (Kirkpatrick and Brown 1994, Ferrier and Watson 1997, Pressey et al. 2000), Papua New Guinea (Nix et al. 2000), the United States (Kintsch and Urban 2000), and South Africa (Cowling and Heijnis 2001, Lombard et al., in press) suggest that combining biotic targets (species, communities, ecosystems) with abiotic or environmental targets results in the best possible representation of biodiversity in a system of conservation areas. In The Nature Conservancy's ecoregional planning efforts in the United States, most projects are now using a combination of terrestrial ecosystem targets and abiotic targets referred to as Ecological Land Units (ELUs) in the conservation planning process. For example, the Central Appalachians Ecoregion Team (2001) combined data on elevation (digital elevation models), geology, and a combination of topographic features (slope, relative land position, moisture, aspect) to create discrete, mappable ELUs (see Figure 4.5, in color insert). These units were used, in turn, to help predict the presence of certain species and communities known to be associated with specific ELUs and to help ensure that community and ecosystem targets were represented in conservation areas across the range of environmental conditions in which they occur. Similar units, known as Ecological Drainage Units (EDUs), based on physical factors, are used to represent environmental variation in aquatic ecosystem targets (see Chapter 10).

When information is available, conservation planners should use combinations of biotic and abiotic conservation targets. Abiotic data layers help ensure that ecological and genetic variation in biotic-based targets will be conserved. Consequently, such an approach is more likely to result in the selection of a set of conservation areas that best represents a region's biodiversity than if either a biotic or an abiotic approach is used alone. In addition, representing biotic targets in conservation areas across a range of environmental conditions (Smith et al. 2001) is one of the leading recommendations for how to best conserve biodiversity in the face of global climate change (Hunter et al. 1988, Halpin 1998, Noss 2001). Chapter 12, on climate change, discusses in more detail the importance of representing biotic targets across the full spectrum of physical environments in a planning region.

Ecological Processes as Conservation Targets

Recall from Chapter 2 that ecological processes and disturbance regimes have not received as much attention in conservation planning as the composition or patterns of biodiversity have. Some scientists have suggested that ecological processes themselves should be considered conservation targets (Margules and Pressey 2000). However, many scientists have argued strongly against doing so (Tracy and Brussard 1994, Noss et al. 1997, Nott and Pimm 1997, Simberloff 1997). Their major concern is that it is possible to maintain ecosystems in which measurable ecological processes appear to be functioning properly, yet the biotic component of these systems could be substantially impoverished due to losses of native species and introductions of exotic ones. As conservation biologist Michael Soulé (1996) noted, "the processes of ecosystems are universal, but the species are not."

From a more pragmatic perspective, conservation biologists and restoration ecologists can often successfully restore ecological processes, whereas the reintroduction of species or re-creation of communities and ecosystems is a much riskier and more expensive activity, with a lower probability of success (Hargrove et al. 2002). Consequently, it is more logical to first define and locate conservation targets based on the patterns of biological diversity and then determine which locations have intact or restorable ecological processes. Chapter 7 will delve into the details of how planners and biologists can assess the functionality of ecological processes and the viability and integrity of conservation targets.

Key Points in Selecting Conservation Targets

1. Identify conservation targets at a variety of spatial scales and levels of biological organization, as appropriate to the planning region and as available information allows.
2. Review the following categories of species targets and identify target species in each category as appropriate to the planning region: threatened, endangered, imperiled, and at-risk species; endemic species; and keystone species. Species variously classified as focal species and umbrella species should also be considered as conservation targets, provided they are precisely defined, the data exist to support their use, and it is unlikely that higher levels of biological organization (e.g., ecosystem targets) will adequately represent their populations in conservation areas.
3. Determine what classifications of communities and ecosystems may be available for the planning region. Evaluate the usefulness of each classification system based on the criteria used to develop them, whether the system is arranged hierarchically, whether units of the system can be mapped, the geographic area that the classification

encompasses, and the extent to which the system has actually been put into practice.

4. Investigate the availability of digital data sets on various environmental and physical variables for the planning region (i.e., soils and geology maps, digital elevation models, climatic data). Overlay available environmental data layers into a single abiotic data layer. The resulting environmental units can be used alone as a surrogate for ecosystem-level targets or, preferably, in conjunction with biotic targets. The latter approach helps ensure that targets will be represented in conservation areas across the environmental gradients in which they naturally occur.

CHAPTER FIVE

Evaluating Existing Conservation Areas and Filling Information Gaps

*The few biological communities [in Aus-
tralia] which have designated reserves have
no rights outside those reserves. Those reserves
are physically inadequate for the officially
designated purpose. Environmental apartheid
gives the illusion of moral rectitude, but liter-
ally sows the seeds of ecological revolution.*

—EARL SAXON (1983)

Before we begin to identify new conservation areas in a planning region, we need to first determine which conservation targets occur within the existing network of conservation lands and waters. The rationale behind this initial scoping is rather simple. A cost-effective approach to biodiversity conservation necessitates that we build on the programs and projects already in place. Rather than duplicating previous work, the aim of conservation planning efforts should be to fill in gaps. This exercise is more complicated than it may first appear. It is not sufficient to simply determine whether a particular targeted feature of biodiversity occurs within a conservation area. We must also evaluate whether current conservation areas and their surroundings are being adequately managed for the long-term persistence of biodiversity.

In most cases, conservation planners will want to include existing protected areas, reserves, refuges, and other types of conservation areas within any future network. However, there will be cases where the existing conservation area either harbors no targets of interest or lacks the attributes and management necessary for targets to persist over the long term. In both the Northern Tallgrass and Central Tallgrass ecoregional planning efforts (The Nature Conservancy 1998, Central Tallgrass Prairie Eco-regional Planning Team 2000), The Nature Conservancy identified a number of its own preserves that either lacked any conservation targets of interest or lacked sufficient viability to maintain these targets over time. Faced with this situation, a natural resource agency or conservation organization needs to reassess whether these conservation areas should remain within their portfolio or whether they can be feasibly restored to conditions that are more likely to reap conservation benefits over the long haul. These are clearly difficult issues to ponder, but in an era of limited conservation funds and increasing losses of species and habitats at alarming rates, conservation organizations and institutions need to ensure that their own investments are bringing the best possible returns. Once this type of analysis is completed, there will be a certain number of targets not adequately represented in the existing network of conservation areas. For those targets, it will be necessary to identify and establish new conservation areas. In most cases, we will not have adequate biological survey information on the status and distribution of conservation targets. Before we can proceed with completing the design of a network of conservation areas for the region, it will likely be necessary to fill some critical information gaps with data from new biological surveys.

In this chapter, I discuss three important components of the conservation planning process: (1) evaluating existing conservation areas for their biological values and identifying gaps in the network of conservation areas (gap analysis), (2) consulting with experts to help fill important information gaps, and (3) conducting selected biological inventories. I focus on two types of gaps: gaps in the system of conservation areas and gaps in the available information on conservation targets. In reality, there are always information gaps on conservation targets, because few areas of the world have been adequately surveyed. The important question to answer is how critical these gaps are to the design of a network of conservation areas. Whether there are critical information gaps or not, most planning teams will find it worthwhile to consult with experts in the region. Not only does this uncover new information, but it also builds credibility and support for the planning process among scientists, conservation practitioners, and other stakeholders in the planning region.

Evaluating Existing Conservation Areas

Determining the degree to which conservation targets are represented in existing conservation areas is an important step in any systematic conservation planning process for biodiversity and has been recognized as such by

Australian conservation planners for several decades (Margules and Pressey 2000). The identification of gaps in representation for national parks in the United States was first begun in a formal plan published in 1972 (Shafer 1999a). An analysis of existing conservation areas provides critical information on which conservation targets are not adequately conserved and is one of several criteria used to set priorities for future conservation action (see Chapter 9). Gap Analysis Programs in the United States (Scott et al. 1987, 1993, Jennings 2000) and elsewhere (e.g., Fearnside and Ferraz 1995) were designed specifically with this task in mind. The basic steps in Gap Analysis Programs typically involve developing a vegetation map to identify the location of the major vegetation or land cover types of a region, using that map as the basis of habitat relationship models to describe predicted distributions of terrestrial vertebrates, and then overlaying the information on the distribution of vegetation cover types and vertebrate species with mapped information on land ownership and stewardship status (Jennings 2000). By so doing, "gaps" in the network of conservation areas can be identified for vegetation cover types (i.e., terrestrial communities and ecosystems; see Chapter 4) and vertebrate species that are not adequately represented in existing conservation lands and waters. Such an approach can be expanded to include virtually any type of conservation target that can be mapped at a suitable scale and overlaid with data on land ownership and stewardship.

The general approach for evaluating existing conservation areas is to develop a land ownership and stewardship map for the planning region and overlay it (usually with a GIS, although it is possible to do the steps with Mylar map overlay sheets) with a distributional map of the conservation targets. Three types of information are needed for the land ownership and stewardship map: geographic boundaries of existing conservation areas and other types of publicly owned lands, manager/owner information for each conservation area, and the biodiversity management status category for each mapped unit (Crist et al. 1998). Obtaining accurate boundaries for some parks and reserves in developing countries can be a challenging task.

Four broad attributes determine the biodiversity management status:

- Permanence of protection from conversion of natural land cover to unnatural cover (cultivated lands, urban areas, areas dominated by exotic species). Retaining natural land cover is a key ingredient to successfully conserving biodiversity.
- The relative amount of the land unit managed for natural cover. At least 95% of the land cover within a conservation area must be maintained in a natural condition for it to be considered being in a "natural state" (see Chapter 6 for a discussion of naturalness).

- Inclusiveness of management (i.e., management aimed at multiple targets and not just a single species). In general, land managed to retain all of its conservation targets or native elements of biodiversity will maintain biodiversity better than land managed for a single feature.
- Type of management (e.g., suppressing or allowing natural disturbance regimes) and the degree to which that management is mandated through legal and institutional arrangements. Management that allows or mimics natural disturbance regimes will maintain biodiversity better than land units where disturbance regimes are suppressed or not adequately functioning. The degree to which an area can be successfully managed for natural disturbance regimes will depend on the design of that area (Shafer 1999b).

With these site-based attributes in mind, the Gap Analysis Program has developed four categories of biodiversity management status that have been widely applied in the United States (Scott et al. 1993, Crist et al. 1998, Groves et al. 2000a):

Status 1: An area having permanent protection from conversion of natural land cover and a mandated management plan in operation to maintain a natural state within which disturbance events such as fire (of natural type, frequency, intensity, and legacy) are allowed to proceed without interference or are mimicked through management.

Status 2: An area having permanent protection from conversion of natural land cover and a mandated management plan in operation to maintain a primarily natural state, but which may receive uses or management practices that degrade the quality of existing natural communities, including the suppression of natural disturbances.

Status 3: An area having permanent protection from conversion of natural land cover for the majority of the area, but subject to extractive uses of either a broad, low-intensity type (e.g., timber harvest) or a localized, intense type (e.g., mining). It also confers protection on species with legal standing as Endangered and Threatened throughout the area.

Status 4: There are no known public or private institutional mandates or legally recognized easements or deed restrictions held by the managing entity to prevent conversion of natural habitat types to anthropogenic habitat types. Management of the area generally allows conversion to unnatural land cover throughout.

If planning teams do not believe the categories are sufficiently detailed, they can subdivide them. For example, a biodiversity planning effort in Oregon (Defenders of Wildlife 1998) used a ten-point scale to evaluate existing conservation areas on public lands and asked land managers to rate their own lands based on four criteria (Table 5.1):

- *Management objectives:* Lands where biodiversity conservation is the primary objective are rated at the higher end of the scale.
- *Security:* Lands formally designated for long-term protection receive high rating; lands subject to zoning or other land-use regulation to maintain natural values receive a middle-range score.
- *Biodiversity values:* Lands are rated high if they are critical to biodiversity conservation; other lands with quality habitat for fish and wildlife or critical to ecosystem function receive middle-range scores.
- *Size:* Large blocks of land managed as a unit receive higher scores compared to small blocks.

The advantage of this approach is that a planning team will likely garner more support for its plan with managers involved in the rating. The disadvantages are increased inconsistencies in applying criteria and increased subjectivity in the criteria themselves.

The Gap Analysis approach of evaluating land ownership and stewardship has been applied to most states in the United States and many ecoregions as well. Figure 5.1 (in color insert) shows an application of the four-category Gap Analysis status codes to lands of the Columbia Plateau Ecoregion in the northwestern United States. By overlaying this ownership/stewardship map with a map of vegetation types or communities for the same ecoregion (see Figure 5.2, in color insert), conservation planners can assess the extent to which each vegetation or community is adequately conserved in existing conservation areas. In practice, those conservation targets that fall within status 1 and 2 lands are viewed as being adequately managed from a biodiversity standpoint. In the Columbia Plateau example, only a combined 4% of the land cover within the Columbia Plateau

Table 5.1 *Biodiversity Management Ratings for the Oregon Biodiversity Project*

Land managers assigned these scores to lands they managed and were given some flexibility within each of the three categories of scores to assign higher or lower ratings, depending on local information and knowledge.

Score: 8–10	Managed primarily for biodiversity or natural values, substantial in size, committed to long-term conservation, relatively high ecological integrity
Score: 4–7	Farm, forest, and rangelands managed primarily for commodity production, but contributing to biodiversity conservation
Score: 1–3	Urban, industrial, and other lands used intensively by humans

From Defenders of Wildlife 1998.

ecoregion is classified as status 1 or 2 lands. Thus, over 80% of the identified vegetation communities in this region have less than 10% of their distribution in status 1 or 2 lands (Table 5.2).

Making the assumption that status 1 or 2 areas have adequate management for biodiversity conservation is an unacceptable leap of faith in many circumstances. For example, some national parks in developing countries have been legally decreed and their boundaries are mapped, but in reality they are parks only on paper. They may have inadequate management and little to no enforcement of regulations. Rao and colleagues (2002) recently analyzed the officially gazetted protected areas in Myanmar for activities incompatible with protected area status. Among their more important findings were that grazing, hunting, fuelwood extraction, and permanent settlements were found in over 50% of the areas. In 60% of the areas, a management plan existed but had not been implemented.

Even in more developed countries such as the United States, national parks and reserves of various types may be inadequately managed to conserve certain target species or communities. A case in point is the grizzly bear in the Greater Yellowstone area. Biologists and conservationists have debated for years whether grizzly bears are being adequately managed there and whether this area can, by itself, sustain a viable grizzly bear population (e.g., Craighead et al. 1995). In addition, some conservation planning efforts assign gap status ranks on an individual basis to each park, reserve, and other types of conservation area, while others assign these ranks on a more categorical basis (e.g., all national parks are automatically assigned status 1). Although for the sake of efficiency it may be necessary to do the latter, the assumptions in doing so (e.g., all wilderness areas merit status 1 ratings) will be unwarranted in some planning regions.

Determining what is adequate in terms of representation and long-term persistence of targets within conservation areas involves setting specific goals for each target and evaluating whether these areas have the potential to maintain targets at sufficient levels of viability and integrity. These topics are addressed in Chapters 6 and 7. Although we are discussing each step in the conservation planning process in a linear fashion, in reality many of the steps are conducted concurrently by planning teams. For example, most planning teams will find it beneficial to be establishing goals for conservation targets and simultaneously gathering information about the status and distribution of those targets in the planning region, including in existing conservation areas.

More detailed information on the methods for conducting these sorts of gap analyses and examples can be found in Davis et al. (1995), Caicco et al. (1995), Stoms et al. (1998), Groves et al. (2000b), and the *Gap Analysis*

Table 5.2 *Percentage of Land Cover Types by Management Status in the Columbia Plateau Ecoregion, as delineated by The Nature Conservancy*

Formations (see National Vegetation Classification in Figure 4.3) are shown in bold type.

Land Cover Type	Status 1 (%)	Status 2 (%)	Status 3 (%)	Status 4 (%)	Total Mapped Area (km²)
Pinus contorta Forest	0.0	0.0	85.1	14.9	176
Pinus ponderosa Forest	0.0	0.0	14.1	85.9	153
Pinus ponderosa–Pseudotsuga menziesii Forest	0.0	0.0	10.4	89.6	784
Pinus monticola–Thuja plicata Forest	0.0	0.0	1.9	98.1	20
Abies species (A. concolor, A. grandis, or *A. magnifica)* Forest or Woodland	0.5	1.1	10.1	88.2	1,397
Picea engelmannii and/or *Abies lasiocarpa* Forest or Woodland	30.2	0.0	62.5	7.3	83
Pseudotsuga menziesii Forest	0.2	0.5	28.9	70.4	2,149
Populus tremuloides Forest	7.8	5.0	57.6	29.5	740
Seasonally/temporarily flooded cold-deciduous closed tree canopy	0.1	16.6	3.5	79.7	382
Pinyon Woodland (*Pinus edulis* or *P. monophylla*)	0.0	0.1	82.1	17.7	165
Pinyon–juniper Woodland (*Pinus edulis* or *P. monophylla* with *Juniperus osteosperma* or *J. scopulorum*)	22.6	0.0	48.2	29.1	193
Juniper Woodland (*Juniperus osteosperma* or *J. scopulorum*)	0.1	3.4	69.3	27.3	2,101
Juniperus occidentalis Woodland	1.1	2.1	43.3	53.4	18,380
Pinus flexilis or *P. albicaulis* Woodland	71.2	0.0	26.8	2.0	104
Pinus contorta Woodland	0.0	0.0	0.0	100.0	22
Pinus jeffreyi Forest and Woodland	0.0	0.0	25.0	75.0	2
Pinus ponderosa Woodland	0.2	2.6	21.1	76.0	5,804
Pseudotsuga menziesii Woodland	0.0	0.0	29.8	70.2	27
Populus tremuloides Woodland	9.0	0.0	80.6	10.4	184
Quercus garryana Woodland	0.1	0.1	4.9	94.9	463
Artemisia tridentate ssp. *vaseyana* Shrubland	1.1	1.6	51.6	45.7	17,181
Artemisia tridentata–A. arbuscula Shrubland	0.6	5.7	69.7	24.0	45,144
Artemisia tridentata Shrubland	2.1	4.6	65.6	27.8	64,574

Land Cover Type	Status 1 (%)	Status 2 (%)	Status 3 (%)	Status 4 (%)	Total Mapped Area (km²)
Artemisia tripartita Shrubland	0.0	1.5	29.2	69.3	3,696
Purshia tridentata Shrubland	0.3	0.3	35.0	64.4	1,140
Artemisia cana Shrubland	14.2	0.8	59.7	25.3	536
Mountain brush Shrubland	5.4	3.7	51.4	39.5	2,027
Cercocarpus ledifolius or *C. montanus* Shrubland	2.2	2.5	74.5	20.7	516
Seasonally/temporarily flooded cold-deciduous shrubland	2.9	18.7	46.5	31.9	1,279
Sarcobatus vermiculatus Shrubland	0.4	7.9	48.5	43.2	3,576
Mixed salt desert shrub (*Atriplex* spp.)	1.2	1.8	64.2	32.8	11,304
Artemisia nova Dwarf-Shrubland	0.0	0.1	65.7	34.2	164
Artemisia arbuscula–A. nova Dwarf-Shrubland	0.0	8.4	74.9	16.7	1,816
Artemisia rigida Dwarf-Shrubland	0.4	0.3	26.8	72.5	700
Dry grassland— *Pseudoroegneria* (*Agropyron*)–Poa	0.2	5.8	11.4	82.5	15,671
Moist grassland—*Festuca*	0.0	2.2	6.6	91.2	2,671
Agropyron cristatum seedings, *Poa pratensis,* hayfields, and Conservation Reserve Program lands	0.2	0.8	68.4	30.5	8,169
Annual grasses— *Bromus tectorum,* etc.	0.8	0.6	53.4	45.3	10,117
Nontidal temperate or subpolar hydromorphic rooted vegetation wetland	0.2	37.9	7.0	54.9	482
Alpine tundra	0.0	0.0	100.0	0.0	3
Wet or dry meadow	0.0	0.0	54.1	45.9	30
Seasonally/temporarily flooded sand flats	0.1	0.3	47.4	52.2	1,670
Sparsely vegetated sand dunes	1.3	37.1	40.7	20.9	345
Sparsely vegetated boulder, gravel, cobble, talus rock	0.4	0.1	70.3	29.2	69
Urban or human settlements and mining					1,201
Agriculture					69,820
Columbia Plateau Totals (not including water)	**1.0**	**3.4**	**42.9**	**52.8**	**297,230**

Analysis and data courtesy of D. Stoms, Department of Geography, University of California, Santa Barbara.

Handbook (available at www.gap.uidaho.edu). Various examples of applications of the Gap Analysis concept outside the United States can be found in Saetersdal et al. (1993) for western Norway, Bojorquez-Tapia et al. (1995) for portions of Mexico, Fearnside and Ferraz (1995) for Brazil's Amazon region, Pressey et al. (1996a) for New South Wales (AU), Williams et al. (1996) for Great Britain, Rodrigues et al. (1999) for the Scottish Borders region of the United Kindgom, Khan et al. (1997) for northeastern India, Powell et al. (2000) for Costa Rica, Maddock and Benn (2000) for northern Zululand, South Africa, Howard et al. (2000) for Uganda, and the World Wildlife Fund's North Andes Ecoregional Team (2001) for the Northern Andes.

Gap Analysis has a number of limitations (Cassidy et al. 2000, Groves et al. 2000a, Jennings 2000). First, Gap Analysis focuses on the current distribution of conservation targets and typically does not reveal anything about historical distributions and losses of targets or address the current levels of threat and decline. For example, in the Northern Tallgrass Prairie ecoregion, only 4% of the ecoregion has not been converted to agricultural or urban use. When planners are working in regions where large amounts of natural land cover have been converted to other uses, they will need to take historical patterns and uses into consideration (see Chapter 6). Second, Gap Analysis typically reveals little information about the viability and ecological integrity of conservation targets, only about their current representation. Third, some important communities and ecosystems (e.g., riparian communities) may occur in patches that are smaller than the minimum mapping unit of the analyses and may therefore be underrepresented or entirely missing from a gap assessment (see Chapter 4's discussion on scale). Fourth, gap status codes reflect what land managers are legally obligated or allowed to do and will not always be an accurate representation of what is actually happening on the ground, which is much more difficult to ascertain. However, recent analyses in California involving future permitted land uses in a region, projected human population growth, and the spatial extent of road effects suggest that the Gap Analysis management status codes are a good approximation of these more direct indicators of biodiversity threat (Stoms 2000).

The IUCN (1994) has developed a ranking scheme for assessing the management of protected areas that can also be used to evaluate the adequacy of existing conservation lands and waters. Although similar to the Gap Analysis management codes, the IUCN scheme is more restrictive and refers primarily to areas that would fall within Gap Analysis status 1 and 2 lands (Table 5.3). These management categories are used extensively

Table 5.3   *Management Categories for Protected Areas, World Conservation Union*

Ia	Strict Nature Reserve: protected area managed mainly for science
Ib	Wilderness Area: protected area managed mainly for wilderness protection
II	National Park: protected area managed mainly for ecosystem protection and recreation
III	Natural Monument: protected area managed mainly for conservation of specific natural features
IV	Habitat/Species Management Area: protected area managed mainly for conservation through management intervention
V	Protected Landscape/Seascape: protected area managed mainly for landscape/seascape conservation and recreation
VI	Managed Resource Protected Area: protected area managed mainly for the sustainable use of natural ecosystems

From World Conservation Union 1994.

by many countries to evaluate the status of protected areas, especially with regard to the development of National Biodiversity Strategies and Action Plans (NBSAPS) under the auspices of the Convention on Biological Diversity (CBD).

There is a considerable amount of additional information on each of these management categories beyond what is presented in Table 5.3, including a more specific definition, objectives of management, guidance for selection, and organizational responsibility (World Conservation Union 1994). For example, strict nature reserves (Category Ia) are defined as areas of land and/or sea possessing some outstanding or representative ecosystems, geological or physiological features, and/or species and as available primarily for scientific research and/or environmental monitoring. These areas have a number of management objectives, such as preserving habitats, ecosystems, and species in a relatively undisturbed state (by humans) and maintaining ecological processes. Areas selected in this category should be large enough to ensure the integrity of ecosystems, free of direct human intervention, and not require substantial active management. Ownership and management should be by a national or subnational level of governmental agency, or by a university or institution with a research or conservation function. A gap analysis to identify targets that are underrepresented in conservation areas can be carried out using either the U.S. Gap Analysis status codes or the IUCN Protected Area Management Codes. In the development of a biodiversity vision for the Northern Andes, the World Wildlife Fund (Northern Andes Ecoregional Team 2001) evaluated how well targeted "landscape units" were represented within IUCN Categories

I and II protected areas and concluded that approximately 45% of the targets were not represented at all within this system of protected areas.

The World Conservation Union established the World Commission on Protected Areas (WCPA) to provide guidance, support, and expertise on the world's protected areas (www.iucn.org/themes/wcpa/). Together with the World Conservation Monitoring Centre (WCMC), the WCPA maintains a database on the world's protected areas and national parks (www.unep-wcmc.org). This database is an excellent starting point for any conservation planners interested in determining the location and status of protected areas on a country-by-country basis, although it likely has important data gaps as well. Table 5.4 provides a sample printout of information available from this database. For the United States and Canada, there is a more detailed, standardized protected areas database that is spatially referenced with GIS and is available for use by conservation planners at various spatial scales (DellaSala et al. 2001).

More recently, the WCPA has established a Management Effectiveness Task Force, which has developed a framework for evaluating the effectiveness of management of protected areas (Hockings et al. 2000). This framework has six primary components:

1. *Context: Where are we now?* Evaluates the current status of a protected area, based on four criteria: significance, threat, vulnerability, and national context.

2. *Planning: Where do we want to be?* Examines five criteria within a country: protected area legislation and policy, design of protected areas systems, design of individual reserves, tenure and customary issues, and management planning.

Table 5.4   *A Sample Printout from World Commission on Protected Areas Database, IUCN and World Conservation Monitoring Centre*

This example is from the province of Santa Cruz in Argentina.

Name	Type	IUCN Category	Size (ha)	Date Established
Bosques Petrificados	National Monument	IV	80,000	1954
Los Glaciares	National Park	II	539,300	1937
Perito Moreno	National Park	II	85,500	1937
Cabo Blanco	Nature Reserve	IV	1172	1977
Bahia Laura	Provincial Nature Reserve	IV	19	1977

3. *Input: What do we need?* Examines issues such as adequacy of resources, funding, staffing, equipment, and infrastructure.
4. *Processes: How do we go about it?* Evaluates whether the best management processes, policies, and procedures are in place and which areas of management need attention.
5. *Output: What were the results?* Assesses what products and services are being delivered and the extent to which work plans are being carried forth.
6. *Outcome: What did we achieve?* Measures the extent to which specific management objectives are being achieved (e.g., threats abated, species or habitats protected).

Because of its site-based detail, this evaluation framework is most appropriately used as part of site conservation or conservation area planning (see Chapter 13). However, some planning teams should find it helpful in evaluating whether existing conservation areas are likely to conserve the targets of interest within the planning unit. In addition, the World Bank, in alliance with the World Wildlife Fund International (WWF), has developed a simple site-level tracking tool to monitor the progress of a protected area in applying the management effectiveness framework outlined above (Stolton et al. 2002).

For community-level and ecosystem-level targets, planners can get information for evaluating their representation in conservation areas by overlaying the boundaries of existing conservation lands and waters with maps of communities and ecosystems developed from remote-sensing information. Planning teams should also contact the managers of individual conservation areas to obtain additional information, applying portions of the IUCN effectiveness framework as appropriate. In many cases, natural resource management agencies will have conducted biological inventories and will probably have reports that provide more detailed information on the biological features of these areas. For example, the U.S. National Park Service has recently increased its funding for inventory and monitoring programs, and the U.S. Fish and Wildlife Service is now mandated to prepare biodiversity management plans for all of its refuges, many of which will require some levels of inventory (Gergeley et al. 2000).

In many situations, existing information will simply be inadequate for assessing the status and distribution of many conservation targets. In such cases, planning teams have two options: (1) contact individual experts or convene an expert workshop or (2) conduct selected biological inventories. The next two sections of this chapter will address these topics in more detail.

Filling Information Gaps: Convening Expert Workshops

Both governmental and nongovernmental agencies (NGOs) routinely use expert panels and workshops to gather a variety of information used in natural resource planning. The Nature Conservancy (TNC), World Wildlife Fund (WWF), and Conservation International (CI) all rely to some extent on expert workshops in their regional planning efforts. Because most of the regional conservation planning projects that WWF and CI have conducted are located in parts of the world with little published information or systematic databases on biodiversity, expert workshops have formed a central and substantive part of their conservation planning process. The WWF workbook on ecoregion-based conservation provides detailed guidance on holding "biological assessment workshops" (Dinerstein et al. 2000). The primary purpose of any expert workshop in a regional or ecoregional planning process is to identify potential conservation areas for taxonomic groups with which the experts are familiar. In addition, participants may provide feedback on the adequacy of the list of conservation targets and information related to the viability and ecological integrity of occurrences of these targets (see Chapter 7).

There are a number of key steps and considerations in holding an expert workshop (Figure 5.3). The steps below are based on a synthesis of guidance from WWF's workbook, CI (1999), and TNC (Groves et al. 2000a).

1. Develop an invitation list, including natural scientists with expertise that spans the different taxonomic groups (birds, fish, plants, invertebrates, vegetation communities, etc.); social scientists who may have information on threats assessments, population trends, etc.; and natural resource staff from a variety of institutions in the region who will be interested in, and have information to contribute to, developing the plan (e.g., government agencies, indigenous groups, nongovernmental conservation organizations, donor organizations). Having respected scientists from the region participating in the workshop will help build credibility for the planning project.

2. Set aside 3–5 days for actually holding the workshop, several weeks to months for preworkshop preparation, and up to a week of time following the workshop for synthesizing of the information. Make one person responsible for coordinating all the logistical details of the workshop. Appoint or hire an expert facilitator to conduct the meeting. Keep in mind that anything less than a well-conducted meeting could leave participants with a bad impression and not engender the type of support and credibility the planning effort needs.

Expert Workshops for Ecoregional Planning

Purpose of the Workshop
- Review/refine conservation targets
- Identify populations/occurrences of targets
- Gather information on viability/integrity of target occurrences
- Gather information on threats to conservation areas or targets
- Gather opinions on conservation goals for targets
- Identify data gaps and inventory/research needs for targets and conservation areas

Examples of Participants in Workshop	**Products from Workshop**
• The Nature Conservancy or Partner Organization Staff	• Refined conservation targets list
• Federal Agency Biologists, Ecologists, and Managers	• Preliminary conservation areas
• State/Province Agency Biologists, Ecologists, and Managers	• Key information on locations of viable target occurrences/populations
• Academic Ecologists and Biologists	• Key information on threats to target occurrences/populations and conservation areas
• Natural Heritage Program and NatureServe Staff	• List of data gaps for the ecoregion
• Planning and Wildlife Consultants	**Workshop Follow-up**
• NGOs	• GIS staff create spatial attribute databases from all of the data gathered at the workshop for further analysis
• Natural History Museum Staff	

Figure 5.3 Purpose, participants, and products for an expert workshop. Both natural resource agencies and conservation organizations rely extensively on expert opinion, often obtained through expert workshops, to help them develop resource management or conservation plans.

3. Distribute an invitation, workshop agenda, and other pertinent materials well in advance of the meeting. The invitation letter should explicitly state the goals of the workshop, what is expected of participants, why they as individuals are being asked to participate, any advance homework, what sort of information participants should bring to the meeting, and what costs will be covered. Other helpful materials to include in the packet are a map of the planning region, a list of conservation targets, and data sheets with information to be completed on the location and status of conservation targets and potential conservation areas. Examples of an agenda, invitation letter, data sheets, and other hints to holding a successful workshop are provided in Annex 2 and 3 of WWF's ecoregion-based workbook (Dinerstein et al. 2000).

4. In advance of the meeting, gather data layers and prepare maps with a GIS. Some of the more critical data layers required are political and administrative boundaries, rivers and watersheds, vegetation, land use, species' distributions, towns and cities, soils and geology, elevation, threats (e.g., mining and logging concessions, livestock allotments, urban expansion areas, exotic

species distributions), and locations of protected or conservation areas. A more complete list of useful data layers is provided in Dinerstein et al. (2000) and Groves et al. (2000a). Print a variety of large-format maps with these data layers for workshop participants to use. Make decisions about the number and composition of different working groups. The most useful segregation of working groups is usually along major taxonomic lines (e.g., birds, fish, vegetation), although working groups focused on socio-economic data and implementation issues may also be helpful. Depending upon the availability of digital information, the amount of species information to be gathered from museums and other agencies, and whether or not a recent vegetation map of the area has been prepared at an appropriate scale, this step could take several months to a year of preparation. The more information made available to experts at the workshop, the greater the likelihood that the team will receive a higher quality of information in return.

5. The workshop itself should alternate between plenary sessions and breakout sessions of the working groups. Each working group and the overall workshop should be coordinated by designated facilitators who have previous facilitation experience at similar workshops or conferences. The project leader should initiate the workshop with opening comments that reiterate the purpose of the overall project and the goals of the expert workshop.

Standardized data sheets and map overlays should be provided to each working group so the input of experts can be gathered in an efficient, consistent manner (see Annex 2 of Dinerstein et al. [2000] for sample data sheets). Although the most important information to be obtained from the workshop is the identification of candidate conservation areas, other types of important information include threats to conservation areas, assessments of viability/integrity, areas or taxonomic groups for which there is limited information, and information on the capacity of local governmental agencies or NGOs to implement conservation plans. Criteria for identifying conservation areas and for setting priorities among areas should be provided to working group participants. For example, in the Irian Jaya workshop coordinated by CI (Conservation International 1999), participants were asked to establish priorities among areas based on biological importance, degree of human threat, urgency, and the need for additional research. Chapters 8 and 9 will focus in detail on criteria for selecting conservation areas and setting priorities, respectively.

Be sure to note the sources of data that experts bring to the workshop, dates of collection of the information, names of nominated sites or candidate conservation areas, and other useful information deemed appropriate by the planning team. Ideally, GIS staff should digitize areas that working groups nominate as being important for various taxonomic groups during the actual workshop, so that revised maps can be presented to participants as the workshop proceeds. Depending on the size of the planning region, it may be useful toward the end of the workshop to subdivide the region into a few major subunits and bring together the different taxonomic group priorities into one map for that subunit. If groups of scientists from different taxonomic groups can reach consensus over priority conservation areas for different parts of the planning region, this will add to the credibility of the workshop and planning effort. Finally, the planning team should consider a pre- or post-workshop field trip to an important conservation area within the planning region.

6. Immediately following the conference, send the participants a thank-you letter to acknowledge their contributions. The letter should outline the planning process, give the timeline, and name the products that can be anticipated over the next several months. Information gathered at the workshop should be compiled and synthesized into a report that can be provided as feedback to participants. For example, CI typically provides all data gathered prior to and at the workshop, plus a compilation of the workshop findings to workshop participants and other interested parties in the form of a manual report and a CD-ROM. Copies of the final biodiversity plan for the region should be provided to all workshop participants, and the participants should be acknowledged within the planning document. Because a considerable amount of new information will be gathered at the workshop, the planning team needs to ensure that this information will be properly maintained in an information management system along with appropriate metadata documentation (see Chapter 3).

As an alternative to expert workshops, some planning teams may undertake individual interviews of experts. Depending on the circumstances, interviewing may be more cost-efficient and just as effective in the solicitation of expert opinion. It has the advantage of personal one-on-one contact, which often yields more detailed information than can be obtained in a group setting, and interviewing avoids at least some of the problems of expert workshops, such as "group think." However, interviewing precludes the exchange of ideas among professionals, the consensus that can be developed over particular places or issues of importance, and the enthusiasm that is often generated for conservation projects in a group setting. When relatively smaller numbers of experts are involved, individual contact and interviews can be a significant cost savings. Expert workshops take considerable time to plan and hold, and they usually cost several thousand (U.S.) dollars in travel and meeting expenses.

Relying on Expert Judgment

Many aspects of conservation planning outlined in this book rely on the opinions and knowledge of experts. Indeed, it would be very difficult to proceed in natural resource management, planning, and conservation without the knowledge of experts. That said, conservation planning teams need to appreciate some of the assumptions and limitations of expert judgment (Maddock and Samways 2000), as well as some of the best practices in eliciting knowledge and opinions from experts. A great deal of research and literature is available on this topic, although much of it, especially the analytical techniques for analyzing expert opinion, is more applicable to the physical and engineering sciences. However, some points and practices for those in the conservation and natural resource fields are worth considering (excerpted from Meyer and Booker 1991 and Cleaves 1994):

- Experts can provide reliable information on what they know about a given subject, problem solving in their field of expertise, and the degree of confidence or certainty surrounding their answers. Experts are not, however, good predictors of future events or trends in particular variables. In fact, they usually do no better than nonexperts at such predictions.
- Experts are limited in the amount of information they can be expected to process. For example, when experts are asked to use a rating scale (e.g., low, medium, high), many experiments have shown that most experts cannot discriminate more than seven rating levels, after which errors increase substantially.
- Experts make mistakes. However, with regard to ecological or biological data, there is little documentation on the error rates of information provided by experts.
- Experts tend to selectively pay attention to information that confirms their beliefs and discount or dismiss information that conflicts with their beliefs.
- The judgment of experts is subject to biases, the two principal types being motivational and cognitive. *Motivational biases* occur when experts alter their responses to questions based on social pressures or personal motivations. Group think and wishful thinking are two well-known types of motivational bias. Group think occurs when pressures from

others in a group influence an individual to alter his or her answer so it will be more acceptable to the group. Wishful thinking occurs when an expert's hope for a certain outcome influences his or her judgment about what would or should actually happen.

Cognitive biases arise from limitations of the human intellect. Inconsistency, anchoring, and underestimation of uncertainty are examples of cognitive bias. Inconsistency bias occurs when an expert's solution to a problem contradicts earlier statements he or she may have made. A typical example might be that species A is more endangered than B, species B more endangered than C, and species C more endangered than A. Anchoring bias occurs when experts tend to ignore new information that has been provided and give answers to questions based on initial responses or first impressions. Biases related to underestimation of uncertainty, as the phrase suggests, occur when individuals do not accurately acknowledge the uncertainty surrounding their answers. A number of studies have demonstrated that most people substantially underestimate the uncertainty of their judgments.

There are several steps that planning teams can take to minimize the bias in expert judgment, capitalize on the knowledge of experts, and acknowledge the uncertainty of such information. Several recommendations are noted below; more rigorous and detailed approaches can be found in Meyer and Booker (1991), Cleaves (1994), and Morgan and Henrion (1990). In addition, in a recent review of viability assessments conducted by the U.S. Forest Service, a chapter on expert opinion has useful advice for conservation planners and biologists and is applicable to the types of situations encountered in regional conservation planning projects (Andelman et al. 2001). Here are some guidelines:

1. Recruit well-qualified experts who have a reputation for being objective. The expert should have experience and knowledge about one or more conservation targets, have field-based experience in the planning region, and ideally also have information on threats to particular areas or targets, as well as informed opinions or data on the viability and integrity of specific target occurrences. Try not to ask for the participation of experts who are likely to have ulterior motives or preconceived notions for particular outcomes. Balance the participants in expert workshops with a diversity of views and opinions.

2. Ensure that experts clearly understand the overall goals of the project, the data or information the team is trying to gather, and the general question areas (e.g., conservation targets, areas that contain the targets,

the viability of targets, and information gaps). If experts do not clearly understand the project and how their information is to be used, they are not as likely to be as engaged in the overall process and as helpful as they otherwise might be.

3. Ask experts questions with as much specificity as possible. For example, instead of asking whether a particular population of a target species is viable, ask more specific questions about the size of the population, whether or not it is reproducing, and whether or not habitat surrounding the population is suitable for dispersal. This process is known as decomposing a question.

4. Be careful not to ask experts to answer an unreasonably long list of questions in a specified time. In general, interviews of individuals require longer periods of time to answer the same questions that might be presented to a group. Questions that involve problem solving also typically involve longer answers and time periods. Similarly, try not to ask questions that mix "facts" about the effects of a particular threat on a target or area with "value judgments" about what levels of effects may or may not be acceptable.

5. Different types of biases are more likely to occur in certain situations. For instance, in expert workshops, inconsistency, wishful thinking, and group think are the most common. Some participants will come to the meeting with the hope (wishful thinking) that their favorite study site or area will be nominated as a candidate conservation area. The facilitator of the expert workshop should make participants aware of these types of biases and why they happen. Although this awareness will not eliminate biases, it can help participants fight off these tendencies.

6. Use skilled facilitators in conducting interviews and in coordinating and facilitating work groups. Such facilitators can often detect different types of biases and warn group members or individuals about such biases influencing their judgments.

7. For answers to questions that typically involve uncertainty, ask experts to rate the uncertainty levels of their answers and provide a narrative statement that gives some additional explanatory information. For example, if an expert judges a certain occurrence of an aquatic ecosystem to be of high ecological integrity, he or she should note how confident or certain he or she is in that response, using a qualitative rating system (e.g., high = > 90% certain, medium = 50–90% certain, low = < 50% certain), and document the rationale behind the recommendation (e.g., no dams in the system, few exotic species).

Subjective ranks such as high, medium, and low should be defined in specific terms. In addition to providing information about the confidence of their responses, it may also help to ask experts to identify assumptions behind their thinking. For instance, an expert may judge a particular reef system to be capable of sustaining fish populations for a specified period of time, provided that fishing pressure on the reef remains constant or decreases.

8. In expert workshop settings, facilitators should routinely ask participants to explain their answers and provide a rationale. For example, if an expert believes site X to be the most significant population of species Y for conservation purposes, he or she should articulate why that site and population are a high priority. This will help combat the tendencies of group think and wishful thinking. Because fatigue often contributes to inconsistent thinking, groups at expert workshops should not work longer than 2 hours without a break. Reminding participants of questions, assumptions, and definitions also helps combat inconsistency.

9. Carefully document all expert opinions, including assumptions or rationales for opinions and levels of confidence.

As useful as experts and expert workshops can be, planning teams should keep in mind that information from experts is strongly biased and influenced by which experts are involved. There have been few tests of expert opinion in the scientific arena to see how well it holds up against empirical data (see Clevenger et al. 2002 for an example of such a test). In many cases, critical gaps in information will remain despite the best efforts in gathering information from all sources including experts. For instance, in some regions there may have no reliable vegetation map, making it difficult to apply and use a classification of communities or ecosystems. Certain areas may lack information on aquatic biota or ecosystems. Depending on the severity of these gaps and the degree to which the absence of information may affect the selection of conservation areas, planning teams may decide to try plugging the most serious data holes either during the planning process or as an update to an existing assessment.

Filling Data Gaps: Assessments and Inventories

In many parts of the world, biodiversity conservation is a crisis discipline. Species and entire ecosystems, in some cases, are being lost from regions at alarming rates. As a result, there is often not nearly enough time or resources for extensive and expensive biological inventories of entire regions. Over the last decade, conservation biologists have developed stop-gap measures for getting information as rapidly as possible, although for some policy makers or conservation practitioners, "rapid" may still seem like either a euphemism or an eternity. Below, I discuss two different rapid survey techniques, Rapid Ecological Assessments and the Rapid Assessment Program, and conclude with a more general discussion on getting the most benefit out of biological inventories. Both techniques are designed to build on and complement existing information and previous inventories.

Rapid Ecological Assessments

Pioneered by the international arm of TNC, Rapid Ecological Assessments (REAs) combine remote-sensing and GIS technology with selected field inventories to provide a baseline of information on the biodiversity values of a region or country (Sayre et al. 2000). The same approach can also be used to bolster existing information in regional planning efforts. Several TNC ecoregional planning efforts and some Gap Analysis projects have used REAs to improve the quality of existing information or fill important data gaps for a particular ecoregion, planning region, or parts thereof.

The primary objectives of a Rapid Ecological Assessment are characterizing the distribution of vegetation types and selected taxa in a study area; producing baseline maps of biological and physical information (e.g., elevation, soils, geology, climate), land ownership, management, land cover/land use, hydrography, and road networks; conducting analysis of threats to biodiversity in the study area; training scientists in REA methods; and producing reports, maps, classifications of vegetation, lists and locations of species inventoried, and threats identified.

The REA process itself (Figure 5.4) consists of eight steps (Sayre et al. 2000). The eight-step REA process generally takes about year to complete and costs between US$75,000 and US$250,000, depending on available data and the investment in getting more information for various taxonomic groups through field inventories. Most REAs are conducted at a mapping scale of 1:100,000 or 1:250,000, with corresponding typical minimum mapping units of 4 hectares (9.9 acres) or 16 hectares (39.5 acres)—the smallest uniform area that can be delineated on a map during interpretation of remote-sensing imagery. Standard data layers typically included in a GIS for this type of assessment include geology, soils, land use, elevation, existing conservation areas or protected areas, transportation

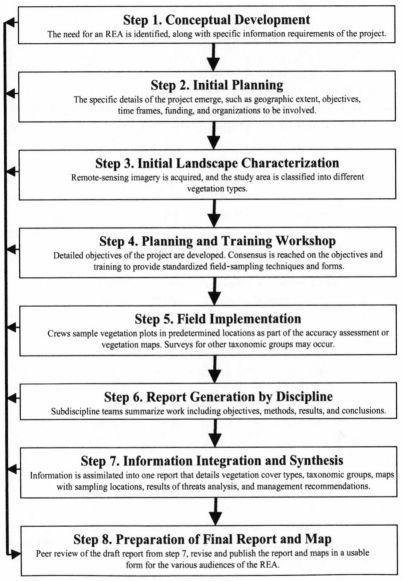

Step 1. Conceptual Development
The need for an REA is identified, along with specific information requirements of the project.

Step 2. Initial Planning
The specific details of the project emerge, such as geographic extent, objectives, time frames, funding, and organizations to be involved.

Step 3. Initial Landscape Characterization
Remote-sensing imagery is acquired, and the study area is classified into different vegetation types.

Step 4. Planning and Training Workshop
Detailed objectives of the project are developed. Consensus is reached on the objectives and training to provide standardized field-sampling techniques and forms.

Step 5. Field Implementation
Crews sample vegetation plots in predetermined locations as part of the accuracy assessment or vegetation maps. Surveys for other taxonomic groups may occur.

Step 6. Report Generation by Discipline
Subdiscipline teams summarize work including objectives, methods, results, and conclusions.

Step 7. Information Integration and Synthesis
Information is assimilated into one report that details vegetation cover types, taxonomic groups, maps with sampling locations, results of threats analysis, and management recommendations.

Step 8. Preparation of Final Report and Map
Peer review of the draft report from step 7, revise and publish the report and maps in a usable form for the various audiences of the REA.

Figure 5.4 The eight-step process for conducting a Rapid Ecological Assessment (REA). This process was pioneered by The Nature Conservancy to obtain biological survey information in a relatively rapid and inexpensive manner. Such information is critical to making decisions about the selection and design of new conservation areas. Conservation International has developed a similar procedure, referred to as Rapid Assessment Program, or RAP. (From Sayre et al. 2000.)

networks, streams and lakes, administrative boundaries, population centers, and land ownership. The primary data source for REAs is remote-sensing imagery from satellites, aerial photography, or aerial videography. Imagery is usually acquired from a commercial vendor and then must be interpreted, a process whereby different spectral signatures from the imagery are assigned to different vegetation types.

Following the classification of imagery, field sampling identifies and verifies vegetation types on the ground. At least one polygon of each vegetation type identified on the map is inventoried, although is preferable for biologists to sample several replicate polygons that are dispersed throughout the study area and represent the same vegetation type. Biologists can inventory other taxonomic groups at the same time as they sample vegetation. Following sampling, the REA team corrects any polygons that may have been misclassified and thereby improve the accuracy of the map. More detailed information on both image processing and field-sampling techniques for vegetation and other taxonomic groups is provided in Sayre et al. (2000). Crist and Deitner (1998) outlined a more detailed accuracy assessment of satellite-derived data for vegetation mapping as part of the national Gap Analysis Program. They recommended that each vegetation type be checked for accuracy with 40 samples, collected either from plots on the ground or from aerial videography. In the New Mexico Gap Analysis project, such a process revealed that ponderosa pine, pinyon pine (*Pinus edulis*), and juniper (*Juniperus* sp.) vegetation types were accurately identified through remote sensing 63% ± 6.3% (S.E.) of the time (Jennings 2000).

In the Central Shortgrass Prairie Ecoregion on the western edge of the Great Plains in the central United States, the ecoregional planning team was faced with a dearth of information on the vegetation communities of the ecoregion (Central Shortgrass Ecoregional Planning Team 1998). In addition, vast areas of native prairie in this region have been converted to agriculture. To fill this important data gap, the ecoregional team used satellite imagery to develop a map of untilled landscapes. Through visual inspection of the imagery, they identified 27 areas that were at least 64.8 km^2 (25 mi^2) in size (the minimum size thought necessary to support populations of target species and important ecological processes) and contained few fields, roads, or towns (Figure 5.5). Ecologists used aerial flights and roadside surveys to identify areas within these untilled landscapes that appeared to be of high quality, were representative of the landscape, or contained rare plant communities. These areas were then the focus of field inventories. An REA working group subsequently collected field infor-

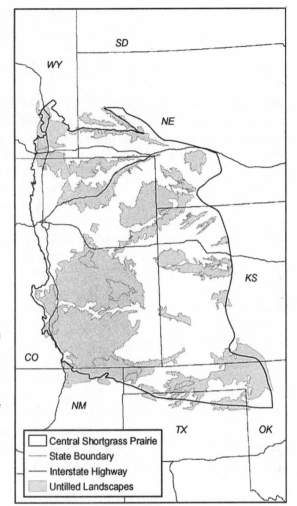

Figure 5.5 Untilled landscapes identified through Thematic Mapper satellite imagery, as part of the Central Shortgrass Ecoregional Plan in the central U.S. The Central Shortgrass team identified 27 untilled areas > 64.8 km^2 (25 mi^2) in size that had the potential to support viable examples of conservation targets. Subsequent fieldwork in these areas resulted in the identification and selection of 50 new conservation areas. (From Central Shortgrass Prairie Ecoregional Planning Team 1998.)

mation on plant communities, their size and distribution, their overall condition or quality, the landscape context within which they were located, and threats to their future existence. As a result of this effort, the Central Shortgrass team collected information on over 75 new occurrences of conservation targets and identified 50 new potential conservation areas. In addition, an analysis of satellite imagery showed that about 40% of the ecoregion remained in large intact parcels of lands that contained only minor amounts of tilled land. In short, the REA process conducted over one field season yielded substantial amounts of new information that significantly improved the ecoregional plan through the selection of many previously unidentified conservation-worthy areas.

Rapid Assessment Program

Somewhat similar to REAs is the Rapid Assessment Program (RAP) of Conservation International (CI), first launched in 1990 to conduct rapid, intensive surveys of ecosystems facing imminent threats. Since the first RAP expedition into northern Bolivia, CI has now completed 32 RAP expeditions and published numerous RAP Working Papers that report results of these expeditions. Expeditions typically involve scientists from several disciplines (ornithology, mammalogy, herpetology, ichthyology, botany, plant ecology, entomology), working as a team, conducting quick biological inventories of approximately 1 month duration. The RAP approach is now being conducted in terrestrial, freshwater, and marine environments. Dozens of new species have been discovered as a result of RAP expeditions, and management recommendations stemming from these expeditions have resulted in the establishment of six new protected areas in five countries. In addition to expeditions, RAP scientists are increasingly providing training to local scientists, thereby leveraging RAP methods to inventory and protect more areas than CI staff could accomplish on their own. Complete reports of RAP methods and results have been published for several expeditions (e.g., Killeen and Schulenberg 1998), a complete list of which is available on Conservation International's Center for Applied Biodiversity Science's Web site: (www. biodiversityscience.org/xp/CABS/research/rap/aboutrap.xml).

Generally, TNC's work in REAs emphasizes the use of remote sensing to develop a vegetation map of a region. In contrast, CI's Rapid Assessment Program puts more emphasis on biological surveys of various taxonomic groups. For example, in Noel Kempff Mercado National Park in Bolivia, CI staff conducted on-the-ground surveys of plants, vegetation communities, birds, mammals, reptiles and amphibians, fish, and dung beetles (Killeen and Schulenberg 1998). Among other findings, they discovered 26 plant species new to science, 21 new records of fish species in Bolivia, and documentation of the most diverse dung scarab beetle assemblage ever recorded in the Neotropics. From these types of floral and faunal surveys, CI RAP reports typically make several recommendations about conservation, management, and scientific research.

Typical RAP expeditions involve leading experts in various taxonomic fields who use standard field methods appropriate to each taxonomic group. As an example of field methods, marine RAP surveys tend to focus on three groups: corals, molluscs, and fishes. Field surveys usually last 15–20 days, with three sites being visited via scuba diving each day. Field workers collect data on the conditions of reefs by placing a line transect across the reef and recording the substrate at intervals of 1 meter (3 feet). These data, along with biodiversity information and information on threats, are combined into a Reef Condition Index that compares survey sites in a region. More details on RAP methods for marine, terrestrial, and aquatic systems are available at the Web site noted above.

Biological Inventories

Survey techniques such as REAs and RAP are efficient tools for filling selective data gaps to help conservation planners make decisions about establishing future conservation areas. Both techniques, however, were designed primarily for use at smaller spatial scales (large sites or conservation areas) than the scale of ecoregions. More systematic and extensive biological inventories, designed in conjunction with regional conservation planning exercises, provide critical information for setting regional conservation priorities, particularly in parts of the world where finer-scale information for such planning is lacking (e.g., Africa; da Fonseca et al. 2000).

Questions that frequently arise with respect to such inventories are: Are they worth it? Does the knowledge they yield improve the selection of conservation areas sufficiently to warrant the investment? Recent work by two British scientists (Balmford and Gaston 1999) suggests that such biodiversity surveys are good value if they can be conducted before many or most conservation opportunities in a country or region are lost. According to their analysis, systematic biological inventories that enable conservation planners to design networks of conservation areas representing all conservation targets in the least amount of areas commonly reduce the amount of land needed for conservation management by at least 5%. By examining the cost of land acquisition and management in several countries, they concluded that even extensive biodiversity surveys such as those carried out for the purpose of establishing nature reserves in Ugandan forests (see below) are usually far cheaper than the additional land acquisition and management that would be necessary if conservation areas were selected without such information. More recent studies support the notion that the completeness or incompleteness of biodiversity surveys has significant implications for the selection of conservation areas and that inventory efforts should be distributed as broadly as possible among sites and taxa (Cabeza and Moilanen 2001).

In Uganda, one of the most biologically diverse countries in Africa, biological inventories were conducted for five indicator groups (woody plants, small mammals, birds, butterflies, selected moth families) to help select and design a forest nature reserve system (Howard et al. 2000). The surveys were conducted over a 5-year period for a total cost of US$1 million. Results of the surveys were used to rank potential forest reserve sites in terms of their species richness, rarity, value for nonconsumptive uses, timber production, and importance to local communities. An iterative area selection procedure (see Chapter 8) was used to select potential forest reserve areas that best represented the conservation targets (in this case, the five indicator groups) and complemented the existing national park system. If the forest reserve system had been designed with just information on major vegetation types (ecosystem targets from Chapter 4), it would have accounted for approximately 82% of the species identified in the biological inventories. By using information from the inventories, planners were able to design a network of reserves that incorporated 96% of the species identified during the inventories. These findings corroborate the results of MacNally and colleagues (2002), who provided one of the first empirical tests of the coarse filter and fine filter mentioned in Chapter 4.

Almost no areas of the world have been comprehensively surveyed for biodiversity, a pattern that is not likely to change in the foreseeable future. The challenge, then, is to make limited biological inventories go further in terms of their conservation application. One way to accomplish this is by using widely available environmental data (e.g., elevation, rainfall, and temperature), combined with information on the range and collection locations of species, to develop predictive models of species distributions. Australian ecologists have pioneered the design of such biological survey techniques (Margules and Austin 1994, Ferrier 1997, Ferrier and Watson 1997) and predictive distributional modeling (Neave et al. 1996, Ferrier and Watson 1997, Nix et al. 2000), although numerous ecologists worldwide are increasingly engaged in species modeling efforts (Corsi et al. 2000, Scott et al. 2002).

Margules and Austin (1994) outlined a conceptual framework for the design of biological surveys and analysis of survey data. The basic idea behind their framework is to identify the major environmental factors thought to influence the distribution of a species, design a set of sampling transects along these environmental gradients that intercept the areas of greatest environmental variation (known as gradsects), and then select replicated sampling sites within the transects using a set of explicit sampling rules (Figure 5.6). Species survey data collected in such a systematic

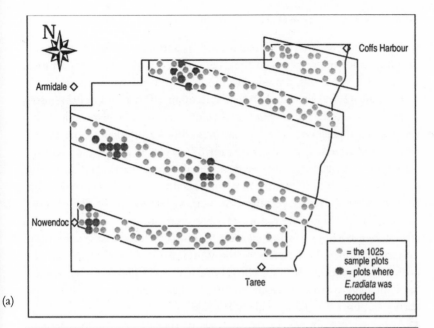

(a)

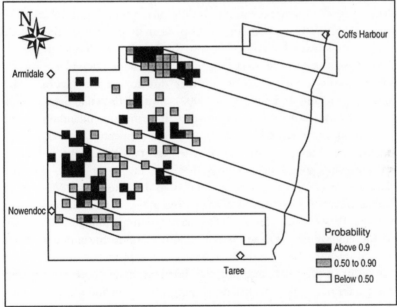

(b)

Figure 5.6 Locations of gradient-oriented sampling transects on the North Coast of New South Wales. (a) Lighter dots represent sampling plots and dark dots represent those plots on which *Eucalyptus radiata* was recorded. (b) Predicted probability of *E. radiata* occurring within 0.1 ha sampling plots within the study area, with original sampling transect boundaries. (Modified from *Biological Conservation*, Volume 50, A. Nicholls, How to make biological surveys go further with generalized linear models, pages 51–75, copyright 1989, with permission from Elsevier Science.)

manner, along with environmental data on altitude, rainfall, temperature, lithology, and topography, can then be analyzed with Generalized Linear Models (GLMs) to build predictive models of species distributions.

Nicholls (1989) provided an example of an effort to model the distribution of one tree species, *Eucalyptus radiata,* from survey data collected in the manner described above. Along similar lines, Australian scientists generated predictive distribution maps for 87 species in Papua New Guinea based on a combination of environmental and climatic data layers and 10,000 species' locations from museum specimen records. These distribution maps were used in combination with vegetation maps and environmental data to help derive a set of conservation areas as part of a comprehensive conservation plan for biodiversity for Papua New Guinea developed under the auspices of the World Bank (Nix et al. 2000). This recent conservation planning effort built upon one previously conducted for Papua New Guinea (Conservation Needs Assessment) by the Biodiversity Support Program (Beehler 1993, Janis 1993).

Although building predictive models of species distributions remains challenging and many parts of the world remain poorly surveyed, there are signs of optimism on both fronts. Gap Analysis projects in the United States have spawned a cottage industry of ecologists developing better models to predict species occurrences. A recent book edited by conservation biologist J. Michael Scott (former director of the National Gap Analysis Program) and colleagues (2002), entitled *Predicting Species Occurrences,* provides a wealth of up-to-date information, including several papers from outside the United States on the conceptual basis of modeling, the role of temporal and spatial scales, accuracy assessment, and different modeling tools and methods. On the inventory front, the U.S. National Park Service has reinvigorated its Inventory and Monitoring Program over the last decade (www.nature.nps.gov/facts/fi&mbase.htm, National Park Service). Since 1992, that program has funded 930 biological and environmental inventories in 250 park units. Inventory work has focused over the last decade on identifying the plant and animal species that occur in the parks, mapping the distributions of endangered species and species of special concern, developing vegetation and soils/geological maps, and gathering basic air quality and water chemistry data.

Several challenges, some scientific and some not, must be overcome in order to make better use of predictive information on species distributions. First, conservation biologists and planning teams need to invest more time and effort into the design of biological surveys so that useful models of species-predicted distributions can be generated with acceptable degrees of accuracy. Too often, accuracy assessments are overlooked or ignored. To be useful, a predictive model of a species distribution should clearly state the rates and sources of error in it. Second, planners should be sure the scale of the modeling effort is appropriate to the needs of the project. A predictive model developed at the scale of a few square kilometers is unlikely to be useful to the needs of a regional-scale planning effort (see Chapter 3).

Third, planners must be careful not to misuse the results of species modeling efforts. The strength of any modeling effort is related to the appropriateness of assumptions behind it and the quality and quantity of information that goes into it. In the age of GIS, it is relatively easy to display model results that are seductively attractive, but may or may not have a sound basis. Fourth, as models get more complicated, the gulf between conservation planners and ecological modelers may widen dangerously. Although there is a real need for planners to use sophisticated models, ecologists who develop these tools must increasingly be able to explain their use in terms that less technical audiences can understand. Without such understanding, the credibility of such modeling efforts will suffer in the hands of conservation practitioners, and the likelihood that the results will be used will decrease.

Finally, planners and biologists need to be more willing to use "predicted" types of information in regional planning exercises. Some conservation biologists are reluctant to use information that is not informed by actual biological inventories. I have even heard some colleagues refer to predicted or modeled species distribution maps as science fiction. At best, such criticisms seem poorly placed, as it will increasingly be impractical to expect conservation practitioners to ever have the detailed sorts of biological survey information for many parts of the world before these areas are seriously degraded or destroyed.

The focus of this section on biological inventories has been on designing surveys with the aim of informing regional planning for biodiversity conservation. The detailed methods of how to conduct biological surveys for specific target species, communities, and ecosystems is a subject that is beyond the scope of this book. However, interested planners and biologists should consult Cooperrider et al. (1986), Heyer et al. (1994), Wilson et al. (1996), Sutherland (2000), and Sayre et al. (2000) for methods and additional references on biological surveys.

There will always be a need for more and better information. As one biologist put it, this need "is almost a bottomless well." Biologists and planners need to distinguish between information that is critical to a planning process and information that is important but not essential. How these two categories are defined will depend on what information is already available, the threats to biodiversity in the planning region, and the staff and financial resources available to the project. At minimum, all projects should have or develop information on ecosystem-level targets or environmentally based targets that can serve as surrogates for representing other features of biodiversity for terrestrial, freshwater, and marine systems.

At this point in the planning process, the conservation targets that will be the focus of planning efforts in the region have been identified. We have also begun to investigate the extent to which some of these targets occur in existing conservation areas and are already being managed for conservation purposes, and to assess where more information may be needed. The next step in the planning process is establishing explicit goals concerning how many, where, and in what condition or quality these target species, communities, and ecosystems need to be conserved in the planning region.

Key Points for Evaluating Existing Conservation Areas and Filling Information Gaps

1. Conduct a gap analysis of existing conservation areas to determine what conservation targets they may contain, the condition of those targets, and the adequacy of management of the conservation area to maintain those targets.

2. Assign a biodiversity management status code to each conservation area using the standard Gap Analysis Program codes, the IUCN protected area codes, or similar codes developed specifically for a conservation planning project.

3. Consider the value of convening an expert workshop to obtain additional information on conservation targets, their quality or condition, threats to these targets, and potential conservation areas in the planning region.

4. Several issues are important to consider when soliciting expert opinion. Experts have biases, are often poor predictors of trends, have limits to how much information they can handle, and make mistakes like everyone else. Planners can take steps to address these issues, as outlined in this chapter.

5. Almost every planning effort will have data gaps, and some may be serious enough to warrant efforts to fill them. Two sets of rapid sur-

vey techniques—Rapid Ecological Assessments and the Rapid Assessment Program—have been developed by The Nature Conservancy and Conservation International, respectively, to help fill important information gaps in a relatively short time frame and a cost-effective manner.

6. Biological inventories have not always been designed as effectively as possible in the past. Because there will never be sufficient funding or time to conduct exhaustive surveys, they need to be designed to allow predictions about species distributions. Predictive modeling of species distributions has made substantial progress over the last decade, and conservation planners will need to rely on this sort of information in the future.

How Much Is Enough?
Setting Goals for Conservation Targets

*The role of science is to give guidance in how
to achieve goals that have been set with
reference to underlying values.*

—LYNN MAGUIRE (1994)

How much is really enough? How many populations of a target species
are needed to sustain that species within the planning region over the long
term? What proportion of an ecosystem needs to be conserved to ensure
that the ecological processes that drive the system remain intact and that
the native species composition and structure are maintained into the
future? These are among the most difficult questions for planners, biologists, and managers to answer and some of the most discomforting. They
are discomforting because, at best, any answer under almost any circumstances is likely to have a high degree of uncertainty. Although principles
of conservation biology and ecology can offer guidance to address these
questions, our knowledge of the life history requirements of species and
how ecosystems function is too incomplete to provide definitive answers.
Moreover, that knowledge base is likely to remain relatively incomplete
within the time frame for making critical decisions—decisions that will
determine the fate of much of the world's biodiversity.

The primary purpose of setting goals is to estimate the effort that will
be necessary to sustain conservation targets well into the future. "Sustain"

146

does not mean keeping populations around at minimal levels that may not be ecologically functional (Groom and Coppolillo 2001) or conserving examples of ecosystems in which ecological processes are no longer intact and now require considerable resources to mimic natural disturbance regimes or compensate for their absence. Instead, the aim in establishing goals should be to provide a setting where target species, communities, and ecosystems are allowed to play out their lives, so to speak, on an ecological and evolutionary stage.

Setting goals for conservation targets serves four useful purposes. First, goals allow an evaluation of how effective a proposed system of conservation areas will be in representing the conservation targets at levels believed necessary to achieve their conservation in the planning region (Figure 6.1). Second, setting explicit goals enables planners and managers to better understand and account for the tradeoffs that often must be made in trying to sustain human communities and ecological communities. Most governmental natural resources agencies are routinely placed in such a position by laws and policies that require them to make these sorts of tradeoffs. As Ruckelshaus and colleagues (2002) observed, conservation planning is often a zero-sum game, even for endangered species, and some populations may be deemed more expendable than others by natural resource managers. Third, goals for targets will ultimately have a strong influence on determining how many conservation areas are needed in a planning region and the extent of area within the region they will occupy. Fourth, goals provide a vision for achieving conservation success.

It is not unusual for biodiversity conservation plans to lack clearly stated goals or contain goals that are not scientifically defensible (Noss et al. 1997). The same could be said for some guidelines and handbooks for conservation planners. For example, a recent handbook prepared by the United Nations Development Programme (UNDP) for countries developing National Biodiversity Strategy and Action Plans (NBSAPs) under the Convention on Biological Diversity offered no guidance for establishing goals or objectives based on actual elements of biodiversity that need to be conserved (Hagen 1999). Recovery plans for Threatened and Endangered species under the U.S. Endangered Species Act (ESA) have been criticized for lacking specific, quantitative goals related to population size and numbers (Tear et al. 1993, 1995; Shemske et al. 1994). Worse yet, nearly 75% of the recovery plans reviewed for vertebrates established population goals so low that the species would remain vulnerable even if recovery goals were achieved. However, this somewhat gloomy picture is improving according to a recent in-depth analysis. Since 1990, more

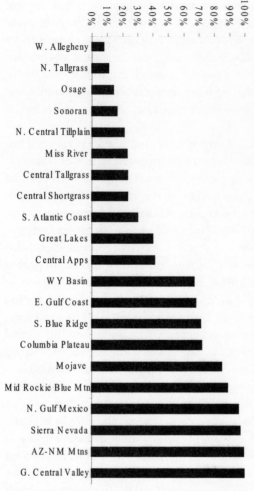

Figure 6.1 The percentage of conservation targets for which goals were met in several Nature Conservancy ecoregional plans in the United States. Meeting goals refers to whether a conservation target is represented in proposed conservation areas a specified number of times across the range of the target within the ecoregion or planning region. This graph indicates a general pattern of lower percentages of goals met for ecoregions where natural vegetative cover has been extensively removed or converted. Where conservation goals are not met, it may be necessary to undertake additional biological inventories or restoration efforts. An assessment of conservation goals is one mechanism for measuring the effectiveness of a proposed system of conservation areas. (From Groves et al. 2002. Planning for biodiversity conservation: putting conservation science into practice. *BioScience* 52:499–512. Copyright, American Institute of Biological Sciences.)

recovery plans have quantitative recovery goals (Gerber and Hatch 2002) and more plans are using population viability analysis (Morris et al. 2002). Finally, a recent review of Habitat Conservation Plans prepared under the ESA recommended that for plans involving species covering large areas or a large proportion of a species' range, more quantitative biological goals and predictions of "take" were needed (James 1999).

Conservation goals need to be clear, explicit, and defensible (Noss et al. 1997, Margules and Pressey 2000). Such goals give credibility to regional conservation plans. Credibility, in turn, is critical if these broader regional plans and goals are to be used and implemented at more local scales. Local governments, for example, are often best positioned and have statutory authority to resolve land-use conflicts and implement conservation programs; however, they often lack staff trained in conservation planning and cannot effectively coordinate regionally with other planning efforts to meet effective conservation goals (Reid and Murphy 1995). In other situations, local agencies responsible for managing natural resources want their lands placed in a broader regional context to determine the most effective ways to manage these lands. The U.S. Department of Defense, for example, has funded several ecoregional planning projects of TNC to assess how to best manage natural areas on military installations from a broader ecoregional standpoint (e.g., Marshall et al. 2000). Finally, as noted in Chapter 1, regional conservation planning is aimed at selecting those lands and waters of highest priority for conservation action. The design and management of these conservation areas is most effectively done through site or project-level conservation planning at relatively smaller spatial scales. Such project-level planning greatly benefits from clearly established goals and conservation targets at the regional level.

Conservation goals in regional planning should have two components: a representation component and a quality component. The *representation component* refers to the question, How much is enough? It is a specific statement about how many or how much of a target species, community, or ecosystem is needed for conservation within the planning region and how the individual occurrences of a target should best be distributed across the region. The *quality component* refers to the level of viability or ecological integrity of individual target occurrences. On the basis of these two components, a goal for a target species might read: "Maintain 10 populations, each containing at least 200 breeding individuals, distributed across all four subsections of the ecoregion in which the species historically occurred."

The assumption behind this goal-setting process is that the conservation of multiple examples of each target across its range of distribution will help "capture" the ecological and genetic variability of the target. In addition, it is a form of ecological bet hedging in that maintaining multiple occurrences is a safety net for the target, if some occurrences are lost as a result of natural processes (e.g., storms, floods) or human-related causes. As a colleague once remarked, "We carry around a set of spare car keys in case we lose ours, and we shouldn't think about doing anything less for nature."

The overarching goal or vision of regional conservation planning, as developed in this book, is to identify a portfolio of priority lands and waters that, if properly managed, would conserve the majority of a region's biodiversity. To achieve that overarching goal, it is necessary to establish target-specific goals. In either case, biologists and planners need to be careful to not confuse goals with values. Scientists often give credence to goals by referring to them as "scientifically defensible" or "justifiable." As Maguire (1994) and others have pointed out, goals and objectives are justified not in reference to science but in reference to values—the things people care about. Scientific thinking helps inform how we go about achieving goals. Why is this distinction important? First, conservation biologists and planners need to be clear about fundamental values if they are going to advocate particular actions, particularly in face of other competing uses and actions. Second, conservation moves forward by people with disparate views and values working together. To work together constructively, people need to explicitly acknowledge and understand the values each person brings to the table (Maguire 1996).

In its report to the U.S. Forest Service on the stewardship of National Forests and Grasslands, the Committee of Scientists (1999) clearly understood that how these lands will be managed in the future is a societal choice about how the lands should be valued. To make such choices, the committee advocated defining desired future conditions. *Desired future conditions* are valued as those that will sustain ecological integrity over the long term (see Chapter 7 on viability and integrity). To be meaningful, these conditions must be defined at the relatively large spatial scales at which integrity is sustained from landscape-scale disturbances (e.g., natural fire regimes). From a social perspective, desired future conditions are those that enable future generations to "support cultural patterns of life and adapt to evolving societal and economic conditions." Goals for biodiversity-based targets, along with assessments of the viability and integrity of these targets, can help define these desired future conditions to a great extent. But by themselves, the goals and desired conditions are statements about what some sectors of society may value, manifested in the above example by laws and policies that govern the U.S. Forest Service. Scientific information can inform planners about how to best achieve these goals and conditions, but not justify them.

Because conservation planners and biologists value biological diversity, it is desirable to articulate future biological and ecological conditions that will conserve as much of a region's biodiversity as possible. To achieve these conditions, it is necessary to focus attention on certain species, communities, and ecosystems, and the areas where these conservation targets occur. For example, sage grouse (*Centrocercus urophasianus*) are a conservation target in the Columbia Plateau ecoregion of the Pacific Northwest of the United States. The fact that sage grouse are a planning target and that the conservation of areas where they occur needs to take precedence over other possible use of those lands is a value judgment. Setting goals for certain sizes and distributions of breeding leks and wintering habitat that sage grouse need to persist in the wild is how conservation science informs planners on how to best achieve goals. Yet, to achieve these conservation goals, scientists and conservation practitioners will need to work with the ranching community and the federal Bureau of Land Management of the Columbia Plateau to improve rangeland conditions for sage grouse while simultaneously ensuring that the needs of private ranchers are also part of the equation that determines the desired future conditions of this ecoregion. To better understand how conservation biologists and planners may best go about this exercise of establishing conservation goals, it is helpful to first review a little history of how such goals have been set in the past.

Historical Context on How Much Is Enough

Various international commissions have recommended percentages of land that should be set aside for conservation purposes. In 1982, the World Parks Congress in Bali recommended that countries set aside 10% of their lands as protected areas but did not provide any rationale for this number (Noss 1996b). Subsequently, in 1987 the Brundtland Commission recommended that all nations place 12% of their land base in protected areas (World Commission on Environment and Development 1987), a figure that was subsequently adopted as a goal by many nations. Several scientists have questioned the reasoning behind these figures (Noss 1996b, Soulé and Sanjayan 1998), with at least some concluding that the 10–12% figures were primarily politically motivated. In addition, the 12% figure refers strictly to protected areas as defined by the IUCN (see Chapter 5) and is strongly biased toward terrestrial biodiversity. Even so, achieving the 12% figure would be a considerable improvement over the most current estimate of over 44,000 protected areas covering 13.6 million km^2 (5.25 million mi^2) and 10.1% of the world's land area (World Commission on Protected Areas 2001).

Noss and Cooperrider (1994) reviewed several conservation plans for estimates of how much or what proportion of a region would need to be conserved to achieve the project's conservation goals (see Table 5.3 in Noss and Cooperrider 1994). Their estimates ranged from as low as 4% to as high as 99%, although most fell between 25% and 75%. In marine ecosystems, scientists and planners have suggested that reserves need to cover approximately 30% of coastal waters to adequately conserve biodiversity (see Chapter 11). The wide range in these estimates is attributable to differences in conservation goals among the plans, as well as other factors that influence the size of conservation areas. Examples of other factors that can influence the total size of the portfolio within the planning region are the type of classification system for describing communities and ecosystems (Chapter 4), whether wide-ranging species were included as conservation targets (Chapter 4), the extent to which population viability and ecological integrity of targets (i.e., the quality component of goals) were considered (Chapter 7), how conservation areas were defined, and the degree to which current protected areas and reserves were automatically included in the portfolio of conservation areas (Chapter 8).

What can we conclude from the work to date in setting conservation goals? First, the variety of methods and approaches makes it difficult to arrive at a rule of thumb for how much land and water need to be placed under conservation management to conserve the biodiversity of any given region. Second, past guidelines for setting goals for the amount of land that should be allocated to strict nature reserves or protected areas have not been firmly grounded in scientific thinking, have often been based on well-intentioned guesses, and have likely been politically constrained. Third, there are differences in how conservation areas are defined. Some institutions, such as the IUCN and the Gap Analysis Program, define them in terms of strict legal protection, whereas other organizations, such as TNC, define them more broadly. Fourth, there is no scientific consensus on how many populations are needed or how large these populations need to be for adequate conservation of target species (Beissinger and Westphal 1998), although most scientists suggest that some minimal level of redundancy is essential for long-term survival (Shaffer and Stein 2000). For communities and ecosystems, there is little empirical or theoretical research that specifically addresses how to best represent these targets in a system of conservation areas. Fifth, in many cases there will be tradeoffs in goals related to the need to conserve multiple examples of targets on the one hand, while conserving areas of sufficient quality (see Chapter 7) to persist over the long term on the other hand. All of these factors confound efforts to develop general guidance for how much area of any given region needs to fall within conservation areas.

Fortunately, a few tools and principles from population and community ecology, outlined below, can help guide the goal-setting process. Although using these tools and principles strongly suggests that the 12% goal of the Brundtland Commission is significantly less than what is needed to avoid species extinctions and maintain biodiversity, there is almost certainly no single figure appropriate for the percentage of land or waters that should be managed for conservation purposes. That number is highly dependent on the needs of the target species, communities, and ecosystems found within any given planning region.

Guiding Principles in Setting Conservation Goals

This following section describes ten general principles for helping planners set conservation goals.

1. Use Recommendations from Recovery Plans and Species Survival Commission Action Plans.

The Endangered Species Act requires that the U.S. Fish and Wildlife Service and National Marine Fisheries Service prepare recovery plans for all species listed as Threatened or Endangered. These recovery plans are the central documents that guide the management of threatened and endangered species. Although several studies have identified weaknesses and failings of recovery plans, used cautiously, they are a valuable source of information for setting goals aimed at recovering threatened and endangered species to viable levels (e.g., Boersma et al. 2001, Gerber and Schultz 2001; see Invited Feature section of *Ecological Applications* 12(3) June 2002 for several papers on recovery plans). An example of a helpful set of conservation goals can be found in the recovery plan for the Oregon silverspot butterfly (*Speyeria zerene hippolyta*) (U.S. Fish and Wildlife Service 2001):

- At least two viable Oregon silverspot butterfly populations exist in protected habitat in each of the following areas: Coastal Mountains, Cascade Head, and Central Coast in Oregon; and Del Norte County in California; and at least one viable Oregon silverspot butterfly population exists in protected habitat in each of the following areas: Long Beach Peninsula, Washington, and Clatsop Plains, Oregon. This includes the development of comprehensive management plans.
- Habitats are managed over the long term to maintain native, early successional grassland communities. Habitat management maintains and enhances early blue violet abundance, provides a minimum of five native nectar species dispersed abundantly throughout the habitat and flowering throughout the entire flight period, and reduces the abundance of invasive non-native plant species.

• Managed habitat at each population site supports a minimum viable population of 200–500 butterflies for at least 10 years.

A complete listing of recovery plans is available at the Web site of the U.S. Fish and Wildlife Service (www.ecos.fws.gov/servelet/TESSWebpage Recovery?sort=1).

Similarly, the Species Survival Commission (SSC) of the World Conservation Union (IUCN), through over 120 volunteer Specialist Groups and Task Forces, publishes Action Plans on many species found on IUCN's Red List of Threatened Species (see Chapter 4). Similar to recovery plans, these Action Plans often contain analyses critical to setting conservation goals for various species or species groups. A complete list of the SSC Specialist Groups and published Action Plans is available at IUCN's Web site (www.iucn.org/themes/ssc/pubs/sscaps.htm for action plans and www.iucn.org/themes/ssc/sgs/sgs.htm for specialist groups). An example of goals found in the action plan for wild dogs (*Lycaon pictus*) in South Africa includes the following (Woodroffe et al. 1997):

• Maintain and, wherever possible, expand the area available to wildlife in Kruger National Park and the reserves that border it. Plans to link Kruger with Gona re Zhou through neighboring Mozambique will have substantial benefits for wild dogs.
• Establish photographic surveys in collaboration with Zimbabwe to assess the contiguity of wild dog populations in Kruger and Gona re Zhou National Parks.
• Maintain links with game and livestock farmers in the areas surrounding Hluhluwe-Umfolozi Park to expand the area available to this population.
• Capitalize on reintroductions carried out in Madikwe and proposed in Pilanesberg by establishing a network of tiny populations in fenced reserves across South Africa, managed together as a metapopulation.
• Consider the reintroduction of wild dogs to the proposed cross-border Limpopo National Park if it is established.

The information provided by these goals can inform the target-based goals of a regional conservation plan, but those goals should be more specific and quantitative. For example, the goals for wild dogs in a regional conservation plan should state how many populations are needed, what their size should be, where they should be located, and where new populations need to be established. Only a portion of this information is provided in the Action Plan for wild dogs.

2. Use the Results of Population Viability Analyses.

Population viability analyses (PVAs) are another critical source of information for setting conservation goals for target species. PVAs can be broadly defined as the use of quantitative methods to predict the future status of a population or collection of populations (Morris et al. 1999, Morris and Doak 2002). Although the total number of species for which PVAs have been performed is relatively small, PVAs are an increasingly valuable tool for conservation planning. For example, in developing a biodiversity conservation plan for the state of Florida, Kautz and Cox (2001) used viability analyses to evaluate the influence of environmental variability on population persistence for 11 focal animal species (birds, reptiles, mammals). Their results suggested that an appropriate goal for all target species in their conservation plan was a minimum of 10 populations of roughly 200 individuals each. Where these sorts of analyses are available, planners should use them to help set goals in regional conservation plans. Unfortunately, because of the paucity of population-level data and the limited resources for biological studies, PVAs will be few and far between in most developing countries.

In a more general sense, the thought processes behind population viability analyses can help conservation planners set conservation goals for target species. There are two important issues to consider related to viability and the distribution of a target species across a planning unit (Morris et al. 1999). The first is the extent to which different populations may be subject to the same extinction pressures due to shared environments (population demographers refer to this relationship as "correlated fates"). For example, if two populations are in close proximity, they are likely to be similarly affected by environmental disturbances such as weather events, disease, or fire. Although there are no simple methods for estimating the degree of correlation of fates, knowledge of weather fluctuations and disturbance events (fire, flood) can be helpful in making these assessments. The second important issue to consider is the extent of movement between populations or occurrences. Movement among populations can mitigate the potential genetic effects of small populations (bottlenecks, inbreeding), "rescue" declining populations from extinction, and, in some cases, reestablish a population that has gone extinct.

When there is little movement among populations and fates are poorly correlated, the probability that all populations will go extinct over a specified time period can be calculated by (Morris et al. 1999)

$$P_{\text{global}} = P_1 \times P_2 \times P_3 \times P_m$$

where P_m is the probability that a population becomes extinct over some specified time period and P_{global} the probability that all populations will go extinct. Using this equation, it is easy to observe how increasing the number of populations helps ensure the overall chances that the target will not go extinct in the planning region (Figure 6.2). On the other hand, it should also be clear that adding populations of low quality (high probability of extinction) provides little insurance for the overall viability of the target species. A useful generalization for any type of conservation target is that the lower the quality of the individual occurrences or populations (i.e., the higher the probability of extinction of that occurrence), the greater will be the need for setting conservation goals relatively higher. This generalization and the above equation offer insights into both the representation component and the quality component of goal setting.

When the fates of populations are correlated and there is little movement among them, the effectiveness of multiple populations is markedly reduced (Figure 6.3). The extent to which gains in reducing the likelihood of extinction from multiple populations are reduced by correlated fates depends upon the degree of correlation. In most cases, this will vary from one population to another and, in any event, will be difficult to estimate. Nevertheless, if populations are thought to be highly correlated, goals (i.e.,

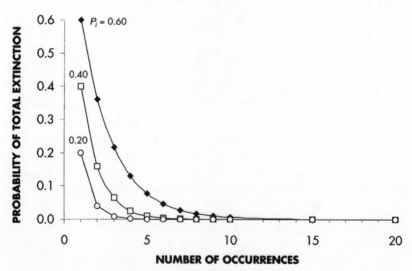

Figure 6.2 The relationship between the number of populations or occurrences of a target and the probability of extinction that all populations die out. These estimates assume that the risk of extinction is uncorrelated among populations and that all populations have the same probability of extinction (P_i). (From Morris et al. 1999.)

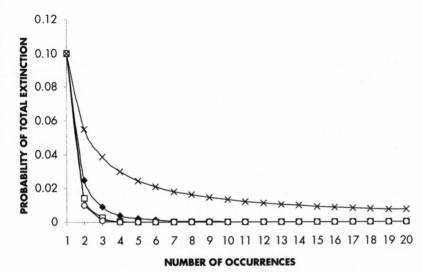

Figure 6.3   A comparison of the probability of extinction and the number of populations or occurrences for no correlation among population and three degrees of correlation among populations. Low: open squares; medium: solid diamonds; high: crosses. As the correlation or relationship between populations grows stronger, there is less added benefit of having more populations to avoid the total extinction of all populations. (From Morris et al. 1999.)

the representation component) should be established to help ensure the conservation of populations that are as widely separated as possible. For example, the conservation of North America's most endangered mammal, the black-footed ferret (*Mustela nigripes*), is highly dependent upon its principal prey item—prairie dogs. Prairie dog populations (*Cynomys* sp.) have been severely reduced in distribution and numbers by several factors, including, in more recent times, the introduction of an exotic disease, the plague. By conserving many widely dispersed populations of prairie dogs, the probability that plague will infect and eliminate these populations is greatly reduced. More detailed information on estimating the probability of extinction of independent and correlated populations is available in Morris et al. (1999).

Finally, in some cases movement occurs among populations with correlated fates. Populations in these situations are often referred to as metapopulations (Hanski 1991, McCullough 1996). A metapopulation is one whose distribution is discontinuous; that is, it is distributed over disjunct patches of suitable habitat with intervening areas of unsuitable habitat. Wiens (1996) summarized the four conditions (based on the work of

Hanski and colleagues) for metapopulation theory and dynamics to have relevance for conservation:

- Local breeding populations occur in discrete "habitat patches."
- No local population is so large that its life expectancy is long relative to that of the metapopulation as a whole.
- The dynamics of local populations are asynchronous, thereby making the simultaneous extinction of all local populations unlikely.
- Habitat patches are isolated, but there is some movement between them so that recolonization is possible.

Some local populations may occur in habitat patches of relatively high quality. These are referred to as *source populations* and *source habitats* (Pulliam 1988, Meffe and Carroll 1997). Other populations may occur in lower-quality habitats and are referred to as *sink populations* and *sink habitats.* Source populations are generally those where reproductive success exceeds mortality, and sink populations are just the opposite—mortality generally is greater than reproduction. Quality of habitat is not the only factor that influences whether local populations persist or go extinct. The size of the habitat patch, its degree of isolation from other habitat patches, the dispersal ability of species within the patches, and the quality of the surrounding habitat (often referred to as the matrix) will all affect the persistence or extinction of metapopulations (Figure 6.4).

The rise of interest in metapopulation theory by conservation biologists can be attributed to one simple fact: Natural, continuous habitats are increasingly becoming fragmented into habitat patches. As a result, species that once occurred over continuous habitats now occur over highly fragmented habitats. The northern spotted owl in old-growth forests of the Pacific Northwest is a frequently cited example of a species for which scientists relied on metapopulation theory to design conservation strategies (Thomas et al. 1990). More detailed discussions on metapopulation theory, ecology, and conservation are available in McCullough (1996) and Hanski (1999).

A number of methods from mathematically simple to complex have been developed for estimating time to extinction of metapopulations (see Morris et al. 1999 for summary). Even the simple models depend primarily on four variables: probability of patch extinction, rate of movement among patches, number of habitat patches, and correlation in extinction probabilities. For conservation planning purposes, these models rarely do an adequate job of mimicking real-world scenarios. In addition, the data they need are rarely available. What can a conservation planner do in this

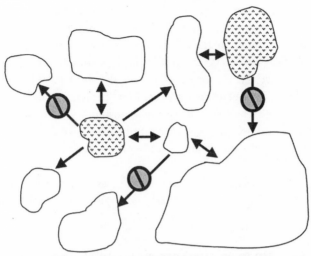

Figure 6.4   Metapopulation theory. Metapopulation structure is characterized by the existence of relatively isolated, local populations (represented here as polygons) that occur in habitat patches surrounded by a habitat matrix. Some of this matrix is suitable for a species to disperse through, while other parts of it is not. Some local populations and their habitat are of higher quality than others. These are often referred to as source habitats and populations, while the lower-quality habitats are referred to as sink habitats (shaded areas) and populations.

situation? Morris et al. (1999) provide some useful advice. Correlated fates of populations and movement between them can be thought of as opposing forces in terms of providing safety from extinction. Populations farther apart are less correlated, thereby providing more safety, while populations closer together have more movement, thereby also generating more safety. Given our limited knowledge on movement and correlated fates for most conservation targets, the best advice is a mixed strategy: Some populations should be clustered close together to facilitate movement, while at the same time, several widely separated clusters of populations should also be maintained.

3. Use Results from Species–Area Relationships to Guide Goal Setting for Communities and Ecosystems.

One of the earliest ecological generalizations established from field studies was the relationship between area and the number of species, commonly expressed as

$$S = cA^z$$

where S is the number of species, A is equal to area, and c and z are constants (Dobson 1996, Meffe and Carroll 1997). A variety of studies from many different areas and many different types of species suggest that values for z usually fall between 0.15 and 0.35. In their studies of island biogeography, MacArthur and Wilson (1967) regularly observed this relationship on islands, but it has been shown to hold for "habitat islands" in terrestrial landscapes as well. The graph of this relationship is shown in Figure 6.5, with the two different curves representing the two different z values obtained from studies on oceanic islands and continental "habitat islands," respectively.

What is the cause of this relationship? The answer to this question has been the subject of great debate among ecologists, and a variety of expla-

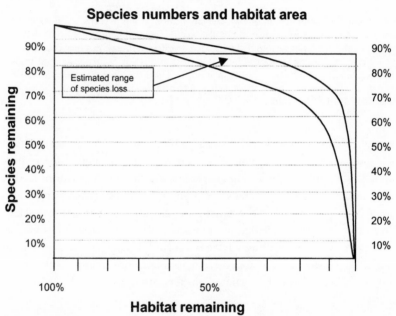

Figure 6.5   Species-area curves. These curves show the relationship between the percentage of habitat loss and the percentage of the number of species likely to be remaining after that habitat loss. Curves like these are used to predict the number of species that will go extinct in a region from habitat loss. The two curves represent results primarily from studies on islands (lower curve) contrasted with studies on terrestrial "habitat islands" (upper curve). Interpreted cautiously, these curves suggest that conservation goals for communities or ecosystems in continental situations that attempt to conserve 30–40% of the area of any given community or ecosystem in a planning region are likely to conserve 80–90% of the species that occur in them. (Modified from Dobson 1996 and Nachlinger et al. 2001.)

nations have been advanced (Meffe and Carroll 1997). The most common explanation is that larger areas, on average, have greater habitat diversity and can therefore support more species. Planners can use the species–area relationship to inform their goals for how much habitat of any given community or ecosystem should be maintained to avoid species losses, while acknowledging that there are criticisms of applying species–area relationships in this manner (Boecklen and Simberloff 1986). These relationships suggest that conservation goals, which range between conserving 30–40% of any given community or ecosystem type, are likely to conserve, on the average, 80–90% of the species in these habitats in continental situations (assuming a z value of 0.15 for "habitat islands" represented by the upper curve in Figure 6.5).

Ecological theory and natural history further suggest that the number of species that are actually lost from a region will depend on the distributional pattern of the species in the planning region (endemic versus widely distributed; see below), the distribution of the remaining habitat, and the life history characteristics of the species (e.g., a good disperser is less likely to go extinct). If the remaining habitat is fragmented into many small pieces, the losses of species are likely to be toward the higher end. Similarly, if most of the species are widespread and not endemic to an isolated area, this will have a mitigating effect on the number of species lost from a region.

The species–area relationship described above assumes that habitat is lost at random from a region. Recent studies of this relationship in California (Seabloom et al. 2002) indicate that habitat loss in most places is not random and that species–area curves may underestimate the potential losses of species as habitat decreases. Given these findings, it seems reasonable to suggest that conservation planners should attempt to conserve at least 30% of any particular community or ecosystem within a planning region to avoid substantial species losses. In areas where there is accelerated development and accompanying habitat losses particularly concentrated on specific community or ecosystem types, the results of the recent California study suggest that planners may want to set this goal even higher.

Based in part on inferences from species–area relationship studies, a conservation planning effort for the Great Basin Ecoregion in the western United States established a goal of 30% of the historic distribution in the ecoregion for endemic large-patch and matrix-forming terrestrial ecosystems (Nachlinger et al. 2001). For terrestrial ecosystems that were more widely distributed, that goal was reduced to 20% (see principle 4 below).

Of course, choosing the time frames and benchmarks for these goals is critical because the current distribution and quality of many habitats have already been substantially influenced and altered by human-related activities in many ecoregions. In a later section of this chapter, I will address this relevant issue of historical ecology.

A cautionary note is appropriate about percentage-type goals. How that percentage is distributed across the planning region is very important from a conservation perspective. As the next chapter shows, conserving 30% of a community or ecosystem in many, small fragments, for example, has its own set of problems. Chapters 7 and 8 will provide additional information about the design of conservation areas that will most effectively implement these percentage-type goals.

Conservation goals set at 30–40% for most communities and ecosystems will not be adequate to conserve all species. Recent studies of habitat loss and fragmentation indicate from both empirical and theoretical standpoints that there is no single threshold value for the amount of habitat needed for species to persist (Fahrig 2001). The minimum amount of habitat that needs to be conserved for the persistence of all species in a region will vary across regions, primarily because of differences in reproductive capability and dispersal capacity of the most sensitive species, but also in relation to the type of natural disturbance regimes and levels of human use. Species with relatively low reproductive outputs and risky dispersal strategies will require larger amounts of habitat to accommodate the larger populations needed to maintain viability. Many wide-ranging carnivores (e.g., grizzly bears) fall into this category (see focal species in Chapter 4). These notions are simply another expression of the coarse-and-fine-filter concept presented in Chapter 4 and are further evidence for the need to identify species, communities, and ecosystems that occur at a variety of spatial scales as conservation targets. Although there appears to be no single habitat threshold value that will allow all species in a region to persist, ecological models suggest about 20%, below which the effects of habitat fragmentation (not habitat loss) on population persistence become negative (Fahrig 1997). The 30–40% goal suggested above for most communities and ecosystems is above this threshold, at which point deleterious effects to populations appear to arise from habitat fragmentation.

4. Account for the Rangewide Distribution of Targets When Setting Goals.

Many conservation targets can be classified according to their rangewide distribution in the following manner:

- *Endemic or restricted:* A target that occurs primarily in one region or ecoregion. For example, the desert pupfish (*Cyprinodon*) of the American Southwest have very narrow range, with some species, such as the Devils Hole pupfish, occurring in a single 21.3 × 9.1 m (70 × 30 ft) pool. Note that certain species, including the pupfish, are naturally restricted in range, whereas others are now restricted but once ranged more widely before habitat and population losses. For instance, in northwestern Yunnan, China, the entire worldwide population of the Yunnan golden monkey (*Rhinopithecus bieti*) is now restricted to a single mountain range.
- *Limited:* A target whose distribution is centered in a few (two to four) adjacent ecoregions. Longleaf pine communities, once widespread throughout the southeastern United States, are now limited in distribution to coastal plain ecoregions. In Australia, the common wombat (*Vombatus irsinus*) is limited in distribution to only a few biogeographic regions in Victoria and New South Wales.
- *Widespread:* A target that occurs across several to many ecoregions. Because of their vagility, many species of neotropical migratory birds have widespread distributions across ecoregions in North America. In Central and South America, paramo vegetation types occur across several ecoregions at higher elevations. Ideally, goals for targets that are widely distributed should be established across the entire distribution of the target, not just a single planning region.
- *Disjunct:* A target that occurs as a distinct population or occurrence of a community in the ecoregion isolated from other populations or occurrences in adjacent ecoregions. A small population of lynx (*Felis lynx*) was recently introduced to Colorado in the western United States. This population is separated from the primary distribution of lynx in North America, which occurs in the boreal forests of Canada and the northern Rocky Mountains.
- *Peripheral:* A target that has a small percentage of its distribution in the ecoregion, with the majority of the distribution occurring in adjacent ecoregions. In the Snake River Birds of Prey National Conservation Area in southwestern Idaho, several species of snakes and lizards (e.g., Mojave black-collared lizard, *Crotaphytus bicinctores*) reach the northern end of their rangewide distribution in the Columbia Plateau ecoregion of the western United States. However, the majority of their distribution is centered in ecoregions to the south.

As a general rule, conservation goals should be set relatively higher for targets with restricted or limited distributions (compared to widespread distributions) because the single or few ecoregions or regions where these targets occur represent the only opportunities to conserve them. For targets that are widespread or limited in distribution, planning teams should consult adjacent regions to determine how goals were established and try to have goals as consistent as possible across regions. Targets that are classified as peripheral may require special attention in goal setting under two sets of circumstances. First, if the target in question is at the northern end of its distribution (or southern end in the Southern Hemisphere), that population or occurrence could become more central and valuable for conservation under various global climate change scenarios (Halpin 1998; see Chapter 12). Second, if a peripheral population belongs to a species that has already suffered declines and habitat losses in a major portion of its range, recent biogeographic analyses of range contractions suggest that these peripheral populations may represent valuable conservation opportunities in the future as the species continues to decline (Channel and Lomolino 2000). From a goal-setting perspective, targets with widespread distributions are of somewhat less concern than targets with limited or restricted distributions.

5. Account for the Pattern of Distribution of Community and Ecosystem Targets.

Community and ecosystem targets vary considerably in the size of areas over which they occur and the environmental conditions across which they are distributed (Anderson et al. 1999). A few communities are often dominant, forming extensive cover across millions of hectares over a broad range of environmental conditions. Northern hardwood forests in the northeastern United States (Northern Appalachian–Boreal Forest ecoregion) are an example of this type of target, referred to as a *matrix-forming community*. Other communities are more localized in their distribution, often nested within matrix-forming types and covering only small portions of land surface. Known as *patch communities*, they are maintained by specific environmental features, in contrast with the regional-scale disturbance factors that characterize the maintenance of matrix-forming communities. Patch communities, such as various wetland/riparian or cave/cliff communities, are often characterized by a distinct flora and by distinct boundaries. The majority of species diversity of a region or ecoregion is often found in these patch communities.

Because of their limited distribution and smaller size in general, patch communities are likely to be more vulnerable to destruction and degradation than matrix communities. Consequently, planners should, as a general rule, set goals relatively higher for patch communities than matrix-forming ones. For matrix communities, the most important factors to consider when setting goals are size and the functionality of the dominant disturbance regime (see Chapter 7). For patches, the number of individual patches and the landscape context around the patches are the most important factors to consider. Table 6.1 provides an example of conservation goals for ecological communities that take into account both the rangewide distribution of these targets (Principle 4) and their different patterns of distribution (patch, matrix).

It is also worth noting that many types of communities and ecosystems that were once matrix-forming have now been reduced by human-related activities to a patchy or fragmented distribution. In North America, tallgrass prairie and longleaf pine communities are two examples. It is possible for a community to be classified as matrix-forming in one part of its range and as a patch community in a different segment of the range. For example, aspen communities in the southern Rocky Mountains of the western United States often occur over extensive, contiguous areas, whereas in the northern Rockies, these same communities naturally occur as small patches. As a final point, planners also need to be aware that the scale at which communities and ecosystems are mapped, and the level of resolution in a community or ecosystem classification, will influence whether a target appears to be matrix-forming or more patchlike on the landscape (Pressey and Logan 1994).

6. Stratify the Ecoregion or Planning Region into Subunits.

Determining how a conservation target should be distributed across a planning region is part of the representation component of goal setting. Planners should ensure, to the extent possible, that targets are distributed across the environmental gradients in which they occur. Doing so helps safeguard against natural catastrophes (storms, disease) that could eliminate

Table 6.1 *Conservation Goals for Species Targets in The Nature Conservancy's Southern Rocky Mountains Ecoregional Plan*

Note that the goals contain both the quantitative and distributional factors of the representation component.

Conservation Targets	Conservation Goal	Goal by Ecoregional Section
Endemic Distribution	20 viable occurrences	At least 3 per section
Limited Distribution	20 viable occurrences	At least 3 per section
Disjunct Distribution	15 viable occurrences	At least 3 per section
Widespread Distribution	10 viable occurrences	At least 2 per section
Peripheral Distribution	5 viable occurrences	At least 2 per section

targets occurring in close proximity, and it also helps conserve the genetic and ecological variation that occurs in target species, communities, and ecosystems across their range. Fortunately, this task is made easier by the fact that many ecoregional classifications are hierarchical and have already been subdivided into smaller units. Recall from Chapter 2 that Bailey's ecoregions, also known as the U.S. Forest Service ECOMAP ecoregions, have been further subdivided into sections and subsections on the basis of similarities in geomorphic process, surficial geology, soils, drainage networks, and regional climate patterns (Figure 6.6). Similarly, freshwater systems (see Chapter 10) and marine systems (see Chapter 11) can be sub-

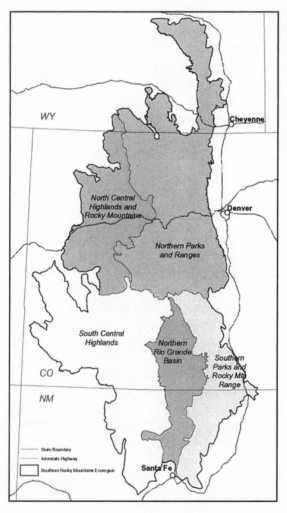

Figure 6.6 Ecoregional sections of the Southern Rocky Mountains ecoregion. Ecoregions are divided into sections on the basis of differences in geomorphic process, surficial geology, soils, drainage networks, and regional climate patterns. These sections, in turn, can be used as stratification units to ensure that conservation targets are represented in conservation areas across the environmental gradients in which they naturally occur within an ecoregion.

divided into smaller units (Zacharias and Roff 1998). Because we almost always lack information on target-specific environmental gradients, we can use these subunits of ecoregions to stratify targets across their range of distribution. For example, the Southern Rocky Mountains Ecoregional Assessment Team (2001) established a goal for endemic species of at least 20 viable occurrences or populations across the ecoregion and at least 3 per ecoregional section in which the species occurred (Table 6.2).

One alternative to the use of subsections and sections of ecoregions as stratifying units is to use data on actual environmental variables to help determine the appropriate distribution of conservation targets. Recall the use of abiotic or environmental variables as conservation targets from Chapter 4. In particular, Ecological Land Units (ELUs), or mapping units defined by two or more environmental variables (e.g., elevation, soils, geology, topography), can be overlaid in a GIS environment with a map of ecological communities or ecosystems to create a combination of vegetation/environmental units, which can then be used to ensure that both the variety of ecosystems and their environmental variation are represented in a portfolio of conservation areas (see Chapter 8). Ecological Drainage Units (EDUs) are used in a similar fashion for planning in freshwater ecosystems (see Chapter 10). The Nature Conservancy's Southern Rocky Mountains Ecoregional Assessment Team (2001) provided a good example of combining vegetation units (ecosystem targets) with ELUs (environmental targets) for a total of 410 vegetation/ELU units, which were then used to capture both the diversity of ecosystem

Table 6.2 *Suggested Conservation Goals for Communities and Ecosystems Based on Pattern of Distribution and Rangewide Distribution*

These goals were based on experience in several ecoregions in the eastern U.S. where there is detailed information on vegetation communities and communities are defined at a relatively fine scale of resolution (see associations in the National Vegetation Classification, Chapter 4).

| | **Pattern of Distribution** | | |
Distribution Type	Matrix	Large Patch	Small Patch
Restricted/Endemic	10	18	25
Limited	5	9	13
Widespread	2–3	4–5	5–6
Disjunct	1*	2*	3*
Peripheral	*	*	*

*Goals should be established case by case.

Modified from Anderson et al. 1999.

targets and the diversity of environments in which they occurred in a portfolio of conservation areas.

7. Pay Special Attention to Wide-Ranging Species.

For some species targets whose rangewide distribution covers multiple ecoregions, planning for those species within any single ecoregion will be inadequate for their conservation needs. There are many examples of such species or species groups. In the Pacific Northwest of the United States, several endangered salmonid fish species range over several adjacent ecoregions. Declining neotropical migratory birds that breed over large areas in North America and winter in the Caribbean and Latin America are a second example. In the case of these species, they are not only wide-ranging, but in contrast to the other example, their distribution is not continuous and the different parts of their distribution represent different phases of their life history. Endangered, wide-ranging carnivores such as the wolverine (*Gulo gulo*) and lynx occupy many different types of habitats over several ecoregions in both North America and Europe. For these types of species, conservation planning in general and goal setting specifically are most appropriately done at the level of multiple ecoregions or planning units (e.g., Wikramanayake et al. 1998 for tiger, Sanderson et al. 2002b for jaguar).

Natural resource agencies, such as the U.S. Fish and Wildlife Service, and federally sponsored programs, such as the Partners in Flight program in the Western Hemisphere, have already undertaken rangewide planning efforts for many of these species. Various specialist groups of the IUCN's Species Survival Commissions have launched similar conservation planning efforts for a number of such species or species groups. The best advice for regional planners is to determine what has already been done and how that work can contribute to the current planning effort, or what additional work needs to be done to set appropriate goals for these species in the planning region.

8. Use History as a Guide to the Past and Future.

To establish meaningful goals for the future conservation of a region's biodiversity, it is helpful to have some understanding of the historic and natural variability in ecological conditions of that region. The historic or natural range of variability (NRV) concept is rooted in a search for a legally defensible strategy for maintaining biological diversity and sustaining the viability of threatened and endangered species (Landres et al. 1999). Today, this concept of range of variability is being used in natural resource plans aimed at sustaining ecological integrity, often in cases where a structural stage of an ecosystem has been significantly altered (e.g., old-growth forests of the Pacific Northwest, tallgrass prairie of the Great Plains) or a key ecological process is no longer operational (natural fire or flooding regimes). The Index of Hydrologic Alteration, a methodology for evaluating the historic flow regime of stream ecosystems, is an excellent application of the NRV concept (Richter et al. 1996; see Chapter 10).

Although conducting an analysis of the historic range of variability is beyond the scope of most regional planning efforts, such analyses are becoming increasingly available via regional or bioregional assessments (see Chapter 3). Natural variability can be defined as "the ecological conditions, and the spatial and temporal variation in these conditions, that are relatively unaffected by people, within a period of time and geographical area appropriate to an expressed goal" (Landres et al. 1999). The use of the concept draws upon many disciplines and is based on the following assumptions (see Landres et al. 1999 for details):

- Landscape-level changes by humans may reduce the viability of many species adapted to past or historical conditions and ecological processes.
- Historical conditions provide a coarse-filter management strategy that is likely to sustain the viability of many species in a region or ecoregion.
- Managing lands and waters within a range of historical conditions is easier and more cost-effective than achieving management goals that are outside the bounds of the systems being considered.
- Natural variability is a useful reference for evaluating the influence of anthropogenic change in ecosystems.
- Analysis of ecosystems over long time frames provides critical insights into the ecological processes that influence the spatial and temporal variability in these systems.

The concept of natural range of variability is only valid when a time period or interval, as well as a specific geographic area, has been specified (Committee of Scientists 1999, Landres et al. 1999). The time period should be one that is similar to the climate, species composition, and disturbance regimes currently found within the planning region. The assumption here is that the NRV analyses will encompass the climate and disturbance regimes under which today's biota evolved and presumably have adapted. Of course, given the rate of climate change that parts of

Earth are currently being subjected to, this assumption is dubious (see Chapter 12). The concept of NRV is probably best understood as a set of frequency distributions of physical and biological conditions. For example, one result might be that the landscape covered by old-growth forest in the coastal range of the Pacific Northwest varied from 25–60% over the past few thousand years (Committee of Scientists 1999). If a planning team wanted to establish goals for old-growth forests that were consistent with conditions on the landscape prior to arrival of Europeans in the region, it would be appropriate to use this range of cover for setting goals of amounts of old-growth forest in conservation areas. In actuality, old-growth forests would represent several different types of terrestrial ecosystem targets in this ecoregion.

The Interior Columbia Basin Ecosystem Management Project (ICBEMP) in the northwestern United States made extensive use of historic or NRV analyses to characterize the native ecosystems of the region (Hann et al. 1997). They used pre-European settlement succession and disturbance models for each vegetation type in the basin and for estimating change over a period of 100–400 years from a historical vegetation map (Figure 6.7). Results of their analyses indicated that native grasslands and shrublands have declined substantially while the presence of exotic species and densities of roads have increased significantly over the last 100 years (Quigley et al. 1996). The ICBEMP team also mapped historical and current distributions of some species groups such as salmonid fish, many of which are currently federally listed as Threatened or Endangered species. The results of these sorts of historical and NRV analyses provide planners with helpful information for setting target-based goals and for conducting assessments of viability and ecological integrity (see Chapter 7).

Having the historical picture in view is especially helpful in regions or ecoregions where much of the land has been converted from native ecosystems to agriculture or urban development. For example, in ecoregions such as the Northern and Central Tallgrass Prairie ecoregions of North America or the Pampas ecoregion of Argentina, only a small percentage of land remains in natural cover. Similarly, the Palouse Prairie region of the Columbia Plateau ecoregion has lost over 99% of its original land cover (Black et al. 1998). Many of the sites where prairie systems still exist may not remain extant over the long term because of small size and poor landscape context (an island of prairie in a sea of agriculture). In their assessment of ecosystems in the United States, Noss and colleagues (1995) reported similar statistics for many different ecosystems in which substantial losses of natural cover had occurred.

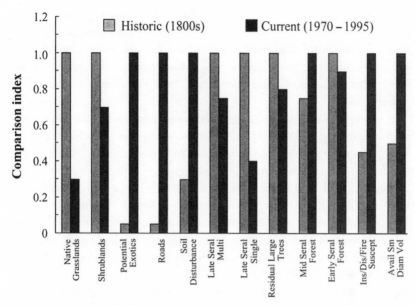

Figure 6.7   A comparison of the historic and current status of various landscape elements in the Interior Columbia Basin in the Pacific Northwest of the United States. (From Quigely et al. 1996.)

In setting goals for communities and ecosystems in these regions and ecoregions, planners must confront difficult issues. Because these communities and ecosystems are greatly reduced from presettlement times and many of the remnants are of poor quality, goals need to be based on conditions other than the current ones, to the greatest extent possible. In some cases, conservationists must face the difficult question head-on of whether a particular community or ecosystem can be realistically conserved or restored within an ecoregion. It may even be necessary to practice conservation triage and conclude that some of these communities and ecosystems are functionally extinct in the planning region and are not feasible to conserve. In the Palouse region of the Columbia Plateau ecoregion, Black and colleagues (1998) compiled current and historical biological, sociological, and environmental information at multiple spatial scales to determine which areas could still be conserved, which areas had potential for restoration, and which areas were likely to be sources of conflict between developmental pressures and conservation.

For most ecoregions or planning regions where detailed NRV analyses are not possible or not available, planning teams may be able to develop historic and current vegetation maps, such as those prepared by the Oregon

Biodiversity Project (Defenders of Wildlife 1998). The maps showed that overall, oak savanna and woodlands, grasslands, prairies, bottomland hardwood forests, and wetlands have suffered the greatest losses. Construction planners should take these losses into account, making every effort to conserve what remains of these ecosystems and restore degraded examples where possible.

Conservationists use a variety of data to reconstruct historic vegetation maps, including historic written records, maps and photos of various kinds, and early land surveys. Written records may include those of early explorations, accounts of travelers, government expeditions, population census data, and local histories (Edmonds 2001). In the United States, the most useful source of information for reconstructing a picture of historic vegetation is the survey notes of the General Land Office (GLO). The GLO established a land survey system consisting of a 9.66 km by 9.66 km (6 mi by 6 mi) township containing thirty-six 2.59 km^2 sections (1 mi^2), each 259 hectares (640 acres) in size (Whitney and DeCant 2001). Stakes were set at the corner of each section, and section numbers were marked on two or more "witness trees" adjacent to the post. Bearing trees—witness trees on which both the bearing and distance from the corner stake were noted—were also marked. Surveyors also recorded types of trees fallen on section lines, general timber types, undergrowth vegetation, and soil type under the section line. The existence of prairies, swamps, tree groves, and various types of disturbances were frequently noted as well. Systematic soil surveys conducted since the late 1800s at the county level in the United States also provide insightful information about plant communities and environmental conditions (Trimble 2001). Taken as a whole, these early land surveys and soils surveys provide critical information from which planners can reconstruct historical vegetation maps of regions.

The emphasis placed on gathering historical ecological data in conservation planning has, not surprisingly, a New World bias. In many parts of the New World, Europeans arrived on the scene only during the last 200–300 years, and in many places in North and South America, for example, the footprints of European humans have been rather light. But what about the Old World, where in most places the imprint of human influence on the landscape has been ongoing for thousands of years? In these locations, the collection of historical ecological information is much more challenging, as is locating areas that have not been significantly altered by humans. Given the expense and difficulty of collecting historical ecological information, conservation planners in these landscapes should develop a conservation vision that focuses on maintaining biodiversity in as many large, unmanaged landscapes as possible that still retain substantial amounts of natural vegetative cover. At the same time, they should promote the restoration of viability and integrity to as many targets as appropriate and feasible.

This same advice may serve conservation planners and practitioners well in many places of the New World, too. Swetnam and colleagues (1999) provide detailed information on methods for constructing historical vegetation maps and the limitations of such data, with some illustrative examples from the southwestern United States. If ecosystems are as dynamic as the nonequilibrium paradigm suggests, then biologists and conservationists need to be cautious in not placing too much weight on one particular time in the past to guide them, and future planners, in establishing desired future conditions or setting goals (Sprugel 1991). As Swetnam and colleagues noted (1999), decisions about what is "natural" and what is desired in the future are value-laden and subjective (see Chapter 7).

The emphasis placed in much of the historical ecology literature on the ecological effects of Europeans on the New World landscape exemplifies the subjectivity and bias involved in defining what is natural, because there is ample evidence that indigenous groups (e.g., Anderson 2001) and aboriginal groups (Vardon et al. 1997) throughout the world also had a substantial influence on the patterns and processes of biodiversity. In a study of disturbance regimes and early successional habitats in eastern North American forests, Lorimer (2001) reached a similar conclusion as Swetnam et al. (1999). Presettlement disturbance regimes were quite variable in space and time, making the determination of what is "natural" a substantial challenge for planners and managers. Although historical ecological information can guide and inform planners in wrestling with these issues, it will not yield unequivocal or unambiguous answers. Its primary value is to provide planners and biologists with insights into the patterns and processes of biodiversity during time periods in which humans were having significantly less influence and impact on the natural world than they are having today.

9. Examine Alternative Goals.

In most situations, planners need the flexibility to meet the needs of many stakeholders in any given region. This need, combined with the scientific uncertainty involved in setting goals, suggests that biologists and planners should set a range of numeric goals for conservation targets (Jennings

2000). In the Cape Floristic region of South Africa, planners established three different goals of 10%, 25%, and 50% of the original extent of each vegetation type within the planning area and then examined alternative portfolios of conservation areas based on these different goals (Heijnis et al. 1999). With the advent of modern GIS and the availability of computer programs to help select conservation areas (see Chapter 8), planner can fairly easily examine these different allocations of land and water to conservation areas based on different sets of conservation goals.

Ease of analysis, however, does little to alleviate the problem that in many ecoregions and planning regions, there is little flexibility left for conservation efforts. As I have noted in earlier sections of this chapter, natural vegetative cover and naturally flowing streams have been so extensively altered in many parts of the world that flexibility in selecting conservation areas, much less the setting of goals, is largely a moot point. Nevertheless, for those planning regions and ecoregions where conservation options remain, it is a practical and worthwhile step to examine a small number of different sets of goals, which, in turn, may provide some much needed flexibility in the implementation of plan results.

10. Observe the Precautionary Principle.

By now it should be abundantly clear that conservation biologists and planners operate in an environment with a great deal of uncertainty. Nowhere is this more obvious than in the setting of conservation goals for target species, communities, and ecosystems. In these uncertain situations, there is no single correct course of action or set of rules to follow, although several decision-making frameworks for dealing with uncertainty exist (Anand 2002). The field of international environmental law has developed a principle to deal with situations that are both large in scale and laced with uncertainty (e.g., global climate change). Known as the *precautionary principle,* this concept shifts the burden of proof of requiring those who would oppose a particular development or action to show it would be harmful to those who propose the action to show it would not be harmful (Kuhlmann 1996, Noss et al. 1997). In fact, the drafters of the Convention on Biological Diversity at the 1992 Earth Summit in Rio de Janeiro recognized the use of the precautionary principle in conserving biological diversity when they wrote in the preamble (Kuhlmann 1996): "Where there is a threat of significant reduction or loss of biological diversity, lack of full scientific certainty should not be used as a reason for postponing measures to avoid or minimize such a threat."

The spirit of the precautionary principle is the need to anticipate the demise and degradation of biodiversity and act accordingly with prudence, especially in the face of uncertainty. Translated to a regional conservation planning context, it suggests the need to set goals for conservation targets with a safety margin and to anticipate activities and development that will cause further losses of biodiversity. Consequently, in ecoregions with rapid rates of human population growth and development, a prudent conservation planner would set goals relatively high for conservation targets that he or she anticipates might suffer substantial losses. This advice is essentially a conservation bet-hedging strategy to compensate for anticipated losses of certain targets. Similarly, for target species and ecosystems for which there is little information about distribution and function, planners should act accordingly by setting relatively higher goals than for targets for which more information is available.

For many biologists, invoking the precautionary principle is a marked reversal from traditional hypothetico–deductive science, in which scientists are primarily concerned with committing what is known as a Type I error—rejecting a null hypothesis when in fact it is true. Conservation planners need to be more concerned with committing a Type II error—failing to reject a false null hypothesis (see Shrader-Frechette and McCoy 1993 for a good discussion of Type I and II errors in ecology). As an example, let's assume the null hypothesis is that a proposed hydroelectric project would have no negative effects on downstream aquatic fauna. If the null hypothesis is, in fact, true, but decision makers reject the project because of concerns over negative environmental impacts, then a Type I error has been committed. If, on the other hand, decision makers allow the project to proceed only to learn later that it had a substantial negative impact on native mussel and fish species, then a Type II error (failing to reject a false null hypothesis) had been committed, and irreparable but preventable losses of biological diversity resulted. Such situations are too often the norm.

Adaptive management (Walters and Holling 1990) calls for designing the management of natural systems as replicable experiments in which participants are constantly learning and improving the management process. Although this approach is possible and desirable in some aspects of conservation planning, it is not very useful for setting goals for conserving biological diversity. Goals that are set too liberally are likely to result in plans that would result in substantial losses of biological diversity. Such losses are often irretrievable, making the learning and adaptive part of management a moot point.

I have reviewed several guidelines and recommendations that planners and biologists should consider in the goal-setting process. Even with the best available information, a degree of uncertainty will always be associated with determining "how much is enough." The guidelines outlined in this chapter will help planners and biologists reduce this uncertainty as much as possible, and the key steps in the process are summarized below. Once goals have been established, the next step in the planning process is to evaluate populations and occurrences of target species and ecosystems to determine whether they are of sufficient viability and integrity to help achieve the target-based conservation goals (see Chapter 7). In Chapter 8, I will revisit the topic of goal setting to evaluate how well portfolios of conservation areas perform in meeting target-based goals.

Key Points for Setting Conservation Goals

1. Use information from recovery plans for endangered species, IUCN Action Plans from the Species Survival Commission (SSC) Specialist Groups, and other expert opinions to help set goals for threatened and endangered species. Determining whether populations share correlated fates, the degree of movement among populations, and whether populations are acting as sources or sinks will also help in the goal-setting process.

2. Use the results of population viability analyses (PVA), when available, to help set species goals.

3. Use the results of species–area relationships to help set goals for communities and ecosystems. Interpreted cautiously and generally, these results suggest that 30–40% of the areal extent of any community or ecosystem within a planning region will need to be conserved in order to also conserve 80–90% of the species occurring in that habitat.

4. Take both the rangewide distribution of conservation targets (i.e., endemic, widespread, disjunct, etc.) and the pattern of distribution (patch, matrix) into account when setting goals for targets, especially for communities and ecosystems.

5. Stratify the ecoregion or planning region into subunits that represent the environmental variation of the planning region. Emsure that conservation goals represent targets across the subregions in which they naturally occur or historically occurred.

6. When available and appropriate, use the historical or natural range of variability (NRV) analyses as a benchmark for settting conservation goals. Such analyses are particularly important in regions that have been extensively altered by human activities.

7. Pay particular attention to wide-ranging species whose goals are most appropriately set at a scale beyond that of a single ecoregion or planning region.

8. Consider setting a range of different goals when substantial opportunities for conservation remain in the planning region and there is considerable uncertainty as to how much or how many of a particular target are needed to ensure its long-term persistence in the region.

9. Invoke the precautionary principle in situations where biodiversity losses are expected to be substantial in the future or a lack of information about targets results in especially high degrees of uncertainty about appropriate levels of conservation goals.

CHAPTER SEVEN

Will Conservation Targets Persist? Assessing Population Viability and Ecological Integrity

The processes of ecosystems are universal, but the species are not.

—MICHAEL SOULÉ (1996)

The worldview held by most ecologists for over half of the twentieth century has been referred to as the *equilibrium paradigm* (Pickett et al. 1992). It focused on end points of ecological processes, such as the climax state of plant succession (e.g., old-growth forests), and viewed ecology through the coarse-scale lens of communities and ecosystems. This classic paradigm also viewed ecosystems as closed and self-regulating, relatively complete systems in and of themselves. The implications of this paradigm for conservation were considerable. Ecosystems left untouched by human hands could be expected to maintain themselves in the state or condition for which they were set aside. If disturbed from their current state, these systems could be expected to return to their original state through natural, self-correcting processes. What happened beyond the boundaries of any particular ecosystem was thought to be of little consequence for that system. These notions about how nature functions, rooted in the Western idea of the "balance of nature," are retained today by some ecologists and certainly by some segments of society at large. Yet, a variety of studies from such diverse fields and topics as plant succession, habitat fragmentation,

178

exotic species invasions, and avian ecology (e.g., nest parasitism and predation) increasingly cast substantial doubt on the validity of the equilibrium paradigm.

A more contemporary view of ecology, termed the *nonequilibrium paradigm,* emphasizes ecological processes themselves rather than their end points and recognizes natural systems as being open and dependent for their maintenance on the context of their surroundings (Pickett et al. 1992). Under this view, ecological processes and disturbances driving ecosystem function occur at a variety of spatial and temporal scales. These new perspectives about how nature operates have a different set of conservation implications. Most ecologists now appreciate that if we are to conserve species, communities, and ecosystems, we must understand and maintain the ecological processes that sustain these conservation targets.

Numerous examples have had a considerable impact on how conservation practitioners now view their work in sustaining species and ecosystems over time. For longleaf pine (*Pinus palustris*) ecosystems and ponderosa pine (*Pinus ponderosa*) ecosystems in the United States, low-intensity understory fires are necessary to help maintain these forests. The scouring action of floodwaters in natural flow regimes is critical to the regeneration of cottonwood (*Populus* sp.) riparian forests of the western United States. Sand corridors and substrates deposited inland by ocean winds are critical to sustaining one of the world's most significant concentrations of endemic plant species in South Africa's Cape Floristic region. Hurricanes and cyclones in tropical environments cause a cycle of changes in forest composition that, in turn, contribute to the high species richness of these forests. And in both temperate and tropical forests, many bird species are important agents of dispersal for tree seeds, thereby helping maintain the structure and composition of forests.

Although this more contemporary view of ecology has been in place to some degree since the 1970s, it has been slow to penetrate the thinking of those concerned with planning for the conservation of biological diversity. Landscape architects and planners, however, starting as early as with the 1969 publication of Ian McHarg's *Design with Nature,* have long recognized the value of planning for the development of urban landscapes while conserving ecological processes that provide critical ecosystem services and other benefits to cities and towns. Research conducted in the 1970s on sand migration on barrier islands of the U.S. National Park system that led to policies and management for dune stabilization was apparently one of the first recognized cases of the importance of ecological process (C. Shafer, personal communication). Landscape ecologist William Baker was one of the first to more formally recognize that ecological processes had received short shrift from conservation planners and that conservation plans needed to consider all levels of biological organization from species to landscapes (Baker 1992). He argued that a successful nature reserve or conservation area must conserve both its targets and their natural disturbance processes. To design a successful reserve system, Baker (1992) recommended that consideration be given to the size of the reserve (i.e., conservation area) in relation to expected sizes of natural disturbances, the location of reserves in relation to expected initiation and export zones of disturbances, and the feasibility of controlling disturbances along reserve boundaries.

More recently, advocates of ecosystem management (e.g., Engstrom et al. 1999, Lugo et al. 1999, White et al. 1999) and scientists focused on biodiversity conservation (e.g., Smith et al. 1993, Balmford et al. 1998, Cowling et al. 1999, Pressey 1999, Margules and Pressey 2000) have recognized the need to explicitly incorporate a consideration of ecological processes, disturbances, and population viability into the planning process. For example, as part of an effort to develop a conservation plan for the Succulent Karoo Desert of South Africa's Cape Floristic region, Richard Cowling and colleagues outlined an approach to conserve both the patterns of biodiversity and the ecological and evolutionary processes that sustain this diversity (Cowling 1999, Cowling et al. 1999). They referred to this dichotomy of pattern and process as planning for retention and persistence, respectively.

The pattern–process dichotomy is also played out in two parallel schools of thought about ecology and conservation. Conservationist and philosopher J. Baird Callicott has referred to these two schools as compositionalism and functionalism (Figure 7.1). The compositionalist view is one of evolutionary ecology with a focus on organisms that are aggregated into populations, which in turn interact in communities (Callicott et al. 1999). Functionalists have an ecosystem orientation with a focus on ecological processes. The vocabulary of compositionalists is one of biological diversity, biological integrity, and ecological restoration, while functionalists speak of ecosystem health, ecological services, ecosystem management, and ecological sustainability. In reality, these schools of thought represent, as Callicott et al. (1999) indicated, two ends of a continuum, and most scientists are not neatly pigeonholed into one school or another. Although this chapter is more focused on function (process) than composition (pattern), the most effective biodiversity conservation plans will have heavy doses of both. A corollary is that the most effective planning teams will

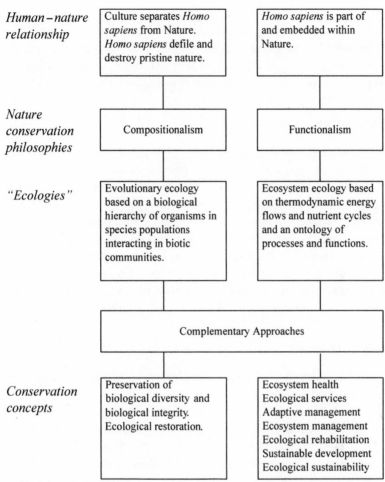

Human–nature relationship	Culture separates *Homo sapiens* from Nature. *Homo sapiens* defile and destroy pristine nature.	*Homo sapiens* is part of and embedded within Nature.
Nature conservation philosophies	Compositionalism	Functionalism
"Ecologies"	Evolutionary ecology based on a biological hierarchy of organisms in species populations interacting in biotic communities.	Ecosystem ecology based on thermodynamic energy flows and nutrient cycles and an ontology of processes and functions.
	Complementary Approaches	
Conservation concepts	Preservation of biological diversity and biological integrity. Ecological restoration.	Ecosystem health Ecological services Adaptive management Ecosystem management Ecological rehabilitation Sustainable development Ecological sustainability

Figure 7.1   The two schools of thought in conservation biology: compositionalism and functionalism. In reality, most conservation biologists and planners are probably somewhere on a continuum between these two philosophical positions. Effective conservation planning for biodiversity necessitates "marrying" these two schools and ensuring that any planning team has representatives from both, or individuals who understand that both composition and function are critical to sustaining biodiversity. (From Callicott et al. 1999. Current normative concepts in conservation. *Conservation Biology* 13:22–35. Copyright, Blackwell Publishing, Inc.)

have members who individually or collectively bring both schools of thought to the planning table.

In this chapter, I review the concepts of population viability and ecological integrity, and discuss several methods for putting these concepts into practice. I also discuss the concept of naturalness as a benchmark for ecological integrity and the role of ecological restoration when the levels of viability and integrity are judged insufficient. Much like the role of quality assurance in a business, using viability and integrity as screening mechanisms in regional planning is designed to help ensure that the targets planners are trying to conserve will persist over the long term.

Definitions

When ecologists talk about populations of species that can persist for long periods of time, they usually speak of *viable* populations. When the conservation targets are communities or ecosystems, the terminology of *ecological integrity* is more appropriate. Any conservation planning effort that utilizes a range of targets over different levels of biological organization will consider, by necessity, both the viability and the ecological integrity of these targets. An important and related concept is that of *naturalness,* or the natural or historical state of a conservation target. Ecologists are concerned with how far removed a species or community can be from a natural or historical state and still retain viability or ecological integrity.

Population Viability

A viable population is one that maintains its vigor and its potential for evolutionary adaptation (Soulé 1987). When biologists undertake analyses to assess the factors that put a population or species at risk of extinction, they are performing *population viability analyses* (PVAs) (e.g., Gilpin and Soulé 1986, Beissinger and McCullough 2002). PVAs help biologists and planners address two critical questions: (1) What is the likelihood that a known population of conservation concern will persist for a given period of time? (2) How many populations must be conserved to achieve a reasonable chance that one or more or them will avoid extinction for a specified period of time?

The first question is the focus of this chapter, and I explored the second in Chapter 6 on conservation goals. In general, both deterministic factors (e.g., human-related habitat destruction) and random or stochastic factors can cause a population to go to extinction. Random factors can be grouped into four categories: demographic, genetic, environmental, and natural catastrophes (Shaffer 1981). Demographic and genetic factors are referred to as intrinsic factors, while environmental and natural catastrophes are referred to as extrinsic factors. Both these intrinsic and extrinsic sets of factors affect variability in the trajectory or behavior of a population over time and therefore also affect the population's risk of becoming extinct. When populations are small, there is greater uncertainty in estimating population dynamics. As a result, population ecologists often refer to these four categories as categories of uncertainty.

Demographic Uncertainty. This type of uncertainty arises in estimating population dynamics due to the effects of random events on the survival and reproduction of individuals in a population. An oft-cited example of demographic uncertainty is when a population decreases to a small size and the only remaining members, by chance, are of one sex; that population is then doomed to extinction. Such was the fate of the dusky seaside sparrow (*Ammodramus maritimus nigrescens*), an Endangered species (under the Endangered Species Act) of the southeastern United States. This species was doomed to extinction in 1980 when its last six known members were all males.

Genetic Uncertainty. As populations decline, the deleterious impact of inbreeding, genetic drift, and founder effects can significantly influence local populations. Although such effects have been difficult to demonstrate in wild populations, a few studies have shown that inbred individuals have lower survival rates and lower reproductive success (Meffe and Carroll 1997). As a general rule, genetic and demographic uncertainty are important factors only in the viability of very small populations, while environmental uncertainty and natural catastrophes can affect the viability of even large populations.

Environmental Uncertainty. This category of uncertainty is due to variability in such factors as climate, weather, food supply, and populations of competitors and predators. For example, the single remaining population of black-footed ferrets (*Mustela nigripes*) in North America following their rediscovery was significantly affected by the introduction of disease, canine distemper, and a decline in its principal prey item, prairie dogs (*Cynomys* sp.), due to the spread of disease (sylvatic plague) through prairie dog colonies (Clark 1997).

Natural Catastrophes. Generally viewed as extreme cases of environmental variability, examples of catastrophes are severe storms, stand-replacing fires in some forest types, and severe flood events. For example, Hurricane Hugo had a devastating impact on the habitat and populations of an Endangered species, the red-cockaded woodpecker (*Picoides borealis*) in the southeastern

United States, when it destroyed many nesting trees and colonies (Jackson 1994).

Ecological Integrity

Like the concept of viability, a great deal has been written about biological and ecological integrity (e.g., Angermeier and Karr 1994, Noss 1995, Pimentel et al. 2000). The concept of integrity is complex, and it is related to many other terms that are in use today in ecology and conservation, such as ecosystem health, resilience, resistance, and stability. A healthy ecosystem, for example, has been defined as one that is stable and sustainable, is active, maintains its organization and autonomy over time, and is resilient to stress (Haskell et al. 1992, Rapport et al. 1998). Understanding this definition hinges on comprehending other terms that have been used variously and often vaguely in ecology, such as resilience and stability. Holling (1986), for instance, defined resilience as a system's ability to maintain structure and patterns of behavior in the face of disturbance.

As these definitions suggest, the literature on ecosystem health, resilience, stability, and integrity is rich and confusing. From a practical standpoint, the principal challenge in using these terms is to define them in an operational manner and then apply them in a conservation planning context. The concept of integrity was first made operational in a natural resource management context under the auspices of the Clean Water Act in the United States, which has the objective "to restore and maintain the chemical, physical, and biological integrity of the Nation's waters." Perhaps because aquatic ecosystems have been so extensively altered and degraded worldwide (as opposed to completely destroyed or eliminated, as is the case with some terrestrial ecosystems), aquatic ecologists are among the leading thinkers about ecological integrity. For example, Karr and Dudley (1981), working primarily with aquatic ecosystems, defined ecological integrity as the sum of physical, chemical, and biological integrity.

Biological integrity, in turn, can be defined as "the capacity to support and maintain a balanced, integrated, adaptive biological system having the full range of elements (genes, species, assemblages) and processes (mutation, demography, biotic interactions, nutrient and energy dynamics, metapopulation processes) expected in the natural habitat of a region" (Karr and Chu 1995). A system possessing integrity can withstand and recover from most natural perturbations as well as human-caused ones. Implicit in these definitions is the notion that integrity is defined as a baseline condition at places where the effects of modern anthropogenic activity are largely absent. This notion will be discussed in greater detail below as I review methods for assessing integrity.

Although there are many other definitions, the approaches outlined in this chapter for assessing integrity will rely on the definitions used by Karr and colleagues, because they have done the most extensive work in this field. To have some understanding or make an evaluation of when a conservation target can be assumed to have sufficient integrity for conservation purposes, it is necessary to have a benchmark. The concept of naturalness provides insight into how this benchmark may be established, at least in some cases.

Naturalness

In Chapter 6 I introduced the topic of the natural range of variability (NRV) in the processes and patterns of biodiversity as a useful concept to consider when setting conservation goals, especially in regions where large amounts of natural land cover have been converted for human-related uses. This notion of what is natural can also be useful in assessing the viability or integrity of conservation targets by serving as a benchmark against which the current conditions of a target can be measured. A great deal has been written about naturalness, how to determine it, and whether it has any utility to conservation.

The most useful articulations of naturalness in the scientific literature come from Anderson (1991) and Angermeier (2000), both of whom developed specific criteria to evaluate naturalness. Their criteria included the degree to which an ecosystem would change if humans were removed, the spatial extent and abruptness of change caused by humans, the cultural energy required to maintain current ecosystem functions, and the complement of native species currently present in an area compared to presettlement conditions.

No single criterion will adequately distinguish between anthropogenic and natural conditions. In addition, the differences between these conditions are not black and white; they are best viewed along a continuum of naturalness. Taken together, the Anderson and Angermeier criteria can help conservation planners assess integrity (and viability as well) by helping define baseline conditions and determining the degree to which the current conditions of any community or ecosystem are different from those baseline conditions. To be useful, however, two assumptions about these criteria need to be clear. First, both the Anderson and the Angermeier criteria primarily relate to the influence of European settlement of the New

World and not to the effects that indigenous people had on the landscape (see below). Second, for parts of the world in which landscapes have been heavily modified by humans for thousands of years (e.g., Europe), the extent of change caused by humans is inordinately more difficult to determine.

There is considerable debate among conservationists as to whether an assessment of naturalness has any relevance to biodiversity conservation (Angermeier 2000, Redford 2000). The argument for relevance is that because humans, along with their technology, have dramatically altered the structure, function, and composition of many ecosystems, it is necessary to examine areas and time periods relatively free of human influence to determine the makeup and function of ecosystems under more natural conditions. The underlying value judgment of this statement is that naturally evolved elements of biodiversity—from genes and individuals to communities and ecosystems—are more valuable than artificial ones, the artifacts of human technology. Conservationists and planners need basic guidance on how to judge the worthiness of conservation targets for conservation action. Areas and time periods relatively free of human influence are the least ambiguous benchmarks for assessing the viability and integrity of these targets (Hunter 1996).

But what does the phrase "relatively free of human influence" really mean? Although many ecologists have traditionally assumed human influence to mean the influence of industrialized societies (Redford 2000), there is a growing recognition that indigenous people and their activities are a "natural" part of the environment. Indeed, there is also a growing body of evidence, as Redford (2000) points out, that most neotropical forests have been modified at some point by human activity prior to the arrival of Europeans. Others have made similar arguments for parts of Europe and the United States. The same arguments can be made that indigenous groups have long had major effects on wildlife populations through hunting.

What constitutes human influence and what can be called "natural" are clearly relative issues, and they make sense only when placed in both temporal and spatial contexts. Furthermore, it can be argued that "natural" is defined primarily in political terms (Redford 2000).

Given this debate and discussion, what is the relevance of the concept of naturalness to conservation planning? I would suggest a couple of guidelines. First, humans have most likely had some influence on most landscapes since their origin. Although influence varies from minimal to substantial, it strongly depends on both spatial and temporal components.

Today, humans as a species are having a greater influence on the natural world than any other species is ever known to have had (Vitousek et al. 1997). For establishing benchmarks, planners should, to the degree possible, look to both time periods and places in their planning regions where the influence of humans is more toward the negligible end of the continuum. Second, in establishing benchmarks about viability or integrity, planners need to be explicit about the time period, conditions, and places being used as reference points.

What is natural, then, is to some extent a matter of choice to be made by planning teams. Their obligation in establishing benchmarks to help assess viability and integrity by the methods outlined below is to make this choice and its rationale explicit.

Assessing Population Viability

There are many different methods for assessing population viability. They range from the quantitative, data-gathering approaches to more expert-driven, qualitative methods.

Quantitative Population Viability Analyses

Many scientific papers have been published that address the different methods for conducting population viability analyses (PVAs), and this body of literature continues to grow. Although a treatment of these different methods is beyond the scope of this book, many good references are available (e.g., Boyce 1992, Groom and Pascual 1998, Morris et al. 1999, Noon et al. 1999, Andelman et al. 2001, Beissinger and McCullough 2002, Brook et al. 2002, Morris and Doak 2002, Reed et al. 2002). Some of the most recent literature on PVAs focuses on whether these models are sufficiently precise and accurate to provide any relevance to conservation biology (e.g., Brook et al. 2000, 2002, Ellner et al. 2002).

In reality, there are relatively few species for which PVAs have been performed, and most of those that have been completed are for species occupying portions of the world that are relatively data-rich. In a large-scale conservation plan for the northern Great Plains in the United States, only 3 of 119 species identified as at-risk species had sufficient data to warrant conducting a PVA (Samson 2002). Because of data limitations, lack of appropriate expertise in PVAs (Shemske et al. 1994), and the time and considerable costs involved in conducting a quantitative PVA relative to the large number of at-risk species identified as conservation targets in most large-scale planning projects, few planning teams will find it cost effective to conduct such analyses. As a result, quantitative PVAs are most applicable for species recovery plans or conservation plans developed at smaller spatial scales (i.e., the site, project, or conservation area scale), where there are more limited numbers of species targets and a need for greater detail of information to guide management actions. Regardless of the scale, Beissinger and Westphal (1998) cautioned that predictions from quantitative PVAs for many endangered species are often unreliable, due to the poor quality of available demographic data, difficulties in estimating variance in demographic rates, and a lack of information on dispersal. They concluded, as did Reed et al. (2002) more recently, that the best use of quantitative PVAs is in a relative sense of comparing different model outcomes based on different possible management scenarios.

Those approaches for estimating viability that are likely to prove most useful for regional-scale conservation planning are rules of thumb and methods based on expert opinion and/or habitat analyses. These approaches are more appropriately referred to as screens or filters for potential viability, rather than actual assessments of population viability.

Rules of Thumb

In two of the first published books on conservation biology, Franklin (1980) and Soulé (1980) recommended that over the short term, populations should be maintained at above an *effective population size* (N_e) of 50 individuals and over the long term at an N_e of 500 to avoid the negative effects of inbreeding, depression, and genetic drift. Effective population size (N_e) is the number of individuals that contribute genes to the succeeding generation. It remains a difficult parameter to estimate in natural populations and difficult to apply to endangered species management and PVAs (Waples 2002). The so-called 50/500 rule has been incorporated into many endangered species programs, but at times it has been misapplied to the actual census population size. Based on mutation research, Lande (1995) later modified the 50/500 rule by recommending that an effective population size of 5000, not 500, was needed for long-term viability. More recent analyses suggest that an effective population size of about 1000 individuals is needed to allow continued evolution and prevent the accumulation of harmful mutations (Allendorf and Ryman 2002). This effective population size translates to a real or census population size of approximately 5000 individuals for many species.

Planners should exercise great caution when using these sorts of rules of thumb. The genetic rules ignore uncertainty arising from a consideration of environmental and demographic factors. As a result, there are cases where viable populations will need to be much larger than these recommendations suggest. On the other hand, these rule-based approaches may encourage conservationists to "write off" small populations, even though there are numerous examples of such populations that have thrived for decades. From a pragmatic standpoint, there are increasingly fewer populations today, especially for vertebrate species, that would meet either the N_e of 1000 or the N of 5000. For many species, the only hope of maintaining such large populations is to increase the connectivity among geographically isolated populations (Allendorf and Ryman 2002). Such a strategy has advantages and disadvantages (see principle 2 in Chapter 6).

Expert Opinion: Natural Heritage Programs

Natural Heritage Programs (NHPs) and their umbrella organization, NatureServe, are one of the primary sources of biodiversity information in North America and, to a more limited degree, Latin America (see Chapter 3; see also Groves et al. 1995). These programs track information on imperiled and endangered species and natural communities (Stein and Davis 2000). The basic data unit NHPs use for tracking and maintaining on-the-ground information on populations of species and occurrences of natural communities is referred to as an element occurrence. An *element occurrence* (EO) is an area of land and/or water in which a species or natural community is, or was, present. For species-level elements, the EO often corresponds to a local population, but it may also be a metapopulation or group of populations.

Element occurrence ranks (EO ranks) provide an estimated viability of individual occurrences of a given species. These ranks are based largely on expert opinion of the relative viability of an occurrence with respect to other occurrences for that same species. Typically, NHPs assign one of the following viability or EO ranks to each occurrence or population that is tracked in their databases: A = excellent estimated viability, B = good estimated viability, C = fair estimated viability, and D = poor estimated viability.

These qualitative EO ranks are derived from a consideration of three criteria (Stein and Davis 2000): size, condition, and landscape context.

- *Size* is a quantitative measure of the areas and/or abundance of an occurrence. For species, it includes components of population abundance, population density, and population fluctuations.
- *Condition* is an integrated measure of the quality of biotic and abiotic factors, structures, and processes *within* the occurrence. Components of

this criterion include reproduction, ecological processes, and physical/chemical factors.

• *Landscape context* is an integrated measure of the quality of structures, processes, and biotic/abiotic factors *surrounding* the occurrence, including the condition of the landscape and connectivity from the occurrence to adjacent habitats.

Size is the primary criterion influencing EO rank for many species, especially vertebrates (e.g., Gurd et al. 2001), with condition and landscape context playing a more secondary role. The rationale is that large size would generally not occur without a favorable rating for condition and landscape context. For species for which there is limited information on size (e.g., many plants and invertebrates), condition and landscape context may be relied upon more heavily. Guerry and Hunter (2002) provided a recent demonstration of the importance of landscape context in explaining the distributions of amphibian species in the northeastern United States. Similarly, Parks and Harcourt (2002) found that extinctions of large mammals in U.S. national parks correlated significantly with human density surrounding the parks. Many conservation biologists have argued that a landscape context of natural land cover surrounding conservation areas, referred to by ecologist Jim Brown as the "semi-natural matrix" (Franklin 1993), is as important to biodiversity conservation as lands that have been set aside for conservation purposes (Hansen et al. 1991). Such lands provide important buffers to conservation areas and are important for connectivity among conservation areas. (I will explore the importance of these so-called off-reserve lands in more detail in Chapter 8 on designing a network of conservation areas.)

In practice, Natural Heritage Program and NatureServe biologists, using the most current scientific literature, develop element occurrence specifications that essentially define what constitutes an occurrence for a given species. These same experts also develop specifications for what constitutes an "A" EO rank, a "B" EO rank, and so on. This information on EO specifications and EO ranks is maintained in a central database available to all NHPs by NatureServe. Natural Heritage Program biologists then apply these specifications and definitions to actual populations and occurrences in the field. Table 7.1 provides an example of EO specifications and EO ranks for a species. These EO specifications are available to the public on an ftp site (a file transfer site) maintained by NatureServe: ftp://ftp.natureserve.org/pub/nhp/animalspecs/outgoing/.

EO ranks of Natural Heritage methodology that are based on the criteria of size, condition, and landscape context can be used to help planners

Table 7.1 *Element Occurrence (EO) Specifications and Ranks for the Harlequin Duck (*Histrionicus histrionicus*)*

The viability or EO ranking is based on only one criterion—size.

EO Class: Breeding

Minimum EO Criteria

EOs are defined by a drainage, or a portion of a drainage, where breeding is known or highly suspected. Minimally, this should be based on three or more independent observations of females or pairs (e.g., one pair in 3 different years, three different pairs in 1 year).

EO Separation

EOs are separated by a barrier (> 2 km over a major divide); or a 10-km separation for completely unsuitable habitat (across land); or a 20-km separation (measured along watercourses) for both rarely used habitat (e.g., lakes, < 1% gradient rivers) and apparently suitable habitat that is not known to be occupied.

Separation Justification

The barrier is based on a lack of movement between streams separated by a 4-km rise over a major divide. Unsuitable habitat separation is based on movements of up to 7 km over a low divide. Movements along watercourses include a 21-km movement across a reservoir, and a few movements up to 31 km have occurred across mixed suitable and unsuitable habitat; all have occurred between years or following a substantial disturbance. Home ranges average 7–10 km of stream length.

A-Rank Specifications

> 100 pairs within a single EO.

B-Rank Specifications

40–99 pairs within the EO.

C-Rank Specifications

3–39 pairs within the EO.

D-Rank Specifications

A yearly average of 1–2 pairs within the EO.

Rank Specifications Justification

A-Rank threshold: The largest currently known breeding class EO was calculated to include 215 adults on the Bow River, Alberta. A-Rank specifications may need to be increased if data from Alaska show substantially larger numbers within single EOs.

C/D-Rank threshold: Given the low productivity and high site fidelity, fewer than three pairs are not likely to be viable over a 100-year period. However, little data are available on numbers of ducks present versus the length of time an occurrence is maintained.

EO Class: Nonbreeding

Minimum EO Criteria

EOs are defined by the presence of > 25 individuals using an area > 1 week in most years on coastal waters, or > 5 individuals for interior staging areas.

EO Separation

EOs are separated by a 20-km separation for both unsuitable habitat and apparently suitable habitat that is not known to be occupied. Of 89 females marked during late summer molt in coastal Alaska, 92% stayed within approximately 20 km of where they were marked through mid-February.

(table continues)

191

Table 7.1   *(Continued)*

A-Rank Specifications

> 3000 birds using an area > 1 month yearly.

B-Rank Specifications

1000–3000 birds using an area > 1 month yearly, or > 3000 birds using an area > 1 week but < 1 month yearly.

C-Rank Specifications

100–999 birds using an area > 1 week in most years, or > 1000 birds using an area > 1 week but < 1 month yearly.

D-Rank Specifications

25–99 birds using an area > 1 week in most years for coastal staging, wintering, and summer nonbreeding areas. 5–99 birds for interior staging areas.

Rank Specification Justification

A-Rank threshold: Reports during the nonbreeding season of up to 5300 are known from Hornby Island, B.C. However, A-Rank specifications may need to be increased if data from Alaska show substantially larger numbers.

From Dr. Larry Master, NatureServe, www.natureserve.org.

make decisions on whether or not to try to conserve a population or occurrence of a conservation target within a regional planning project. Occurrences or populations with ranks of A (very good) or B (good) are usually included within a regional conservation plan, while those with ranks of C (fair) are sometimes incorporated and those with ranks of D (poor) are almost always excluded because they are unlikely to persist over the long term. However, in cases where the only known populations or occurrences of a target species are rated as being fair or poor, they may be candidates for restoration efforts (see the section on ecological restoration later in this chapter).

Planners should be cautious when using EO ranks to make conservation planning decisions, and the same is true for the extensive use of expert opinion. Such opinions are rarely tested as to whether the assumptions on which they are based hold up over a reasonable time period. A research project might attempt to answer the question of whether D-ranked populations are more likely to go extinct than A-ranked populations over a 10-year period (P. Kareiva, personal communication). At the same time, conservation biologists and planners usually have data sets to make more informed decisions for only a small number of species, and expert opinion is likely to be better than decision making in an information vacuum, in most cases.

Habitat-Based Approaches

Species can be predicted to occur in various parts of a planning region by mapping of ecological communities across the region and assigning species presence in or absence from a community according to the quality of habitat in that community. This type of habitat-based approach is referred to as wildlife habitat-relationship modeling (Andelman et al. 2001). It is a method that has been applied primarily to birds and mammals, and usually to those species that occupy a single community or ecosystem type. Wildlife habitat-relationship models used in California (Block et al. 1994) are an example. They provide some indirect information on whether a species may be present in quality habitat and, by inference, in viable numbers. This approach does not directly measure viability and is, as noted at the beginning of this section, more of a screening technique for potential viability. It is unlikely that this sort of detailed mapping of ecological communities will be available for anything greater than a portion of a planning region.

Gap Analysis Programs (see Chapter 3) also develop predicted maps of vertebrate species distributions across entire states in the United States, based on known associations with broad vegetation types (Scott et al. 1993) and additional variables (e.g., slope, soil type, climate) in some cases. However, they lack the habitat-quality specificity of the habitat-relationship modeling described above. Recall from Chapter 5 that conservation planners and biologists (e.g., Ferrier and Watson 1997, Scott et al. 2002) have made considerable advances in such predicted species distributions. Although these predicted distribution maps are good screening techniques for whether the appropriate types of habitat for a species have been represented within a conservation area, they provide little information on the quality of that habitat and little inference about whether it can support viable populations of target species. Redford's (1992) reference to the empty forest—one in which quality habitat remains, but many of its carnivorous and herbivorous animals have been hunted to extinction—is a pointed example of how distributions of species predicted on the basis of appropriate habitat could be very misleading to conservation planners.

Because of the relatively large areas over which individuals range and the variety of habitats they often use, wide-ranging species typified by many larger carnivores present some unique challenges in conservation planning in general and viability assessments in particular. In addition, many of these species are thought to serve as umbrella or focal species for which conservation action is likely to benefit an array of other species as well (see Chapter 4). Carroll et al. (2001) developed empirical habitat suitability models using resource selection functions (see Boyce and Mac-donald 1999 for details) for several carnivore species in the Northern Rocky Mountains of the United States, based on trapping and sighting records combined with predictor variables of habitat derived from satellite imagery. Examples of data layers used to develop the predictor variables included vegetation (e.g., tree canopy closure), wetness and greenness indices, topographic variables (elevation), climatic variables (e.g., mean annual precipitation), and human-impact variables (e.g., road density). Results of modeling efforts for these species allow map-based conserva-tion planning at a scale that is relevant to the population processes of wide-ranging carnivores (Carroll et al. 2001), including the incorporation of important linkages among different habitats.

Noss and colleagues (2002) used the same focal species approach of Carroll et al. (2001) for conservation planning for several carnivore and ungulate species in the Greater Yellowstone area, but also specifically incorporated a PVA component using the program PATCH (Schumaker 1998). PATCH links fecundity and survival rates of individual animals to various GIS variables that are related to habitat productivity and mortality risk. Similar modeling efforts are now being incorporated into TNC's plans in the Canadian Rockies ecoregion and the Utah-Wyoming Moun-tains ecoregion (Greater Yellowstone area), into the conservation plans of several Wildlands Projects across the United States, and into plans for wolf (*Canis lupus*) recovery (Carroll et al. 2003). This combination of static habitat modeling using GIS and dynamic population viability modeling holds great promise for better incorporating the needs of many at-risk, wide-ranging animal species in regional conservation plans.

In many situations, what were once large and widely distributed popu-lations of wide-ranging species may now be reduced to populations that have a much smaller range. They may remain viable in that they are likely to persist for long time periods, but may no longer be playing the ecolog-ically functional role in an ecosystem they once did.

Ecologically Functioning Populations

Conner (1988) was the first to suggest that it may not be desirable to man-age for viable or minimum viable populations; instead managers should focus on maintaining species populations that are ecologically functional within their communities. An *ecologically functional population* (EFP) is a conservation planning objective that specifies the conditions necessary for the population of a species to fulfill its specified ecological role at a particular site or conservation area (Groom and Coppolillo 2001). Ecological roles may be defined by a species' effects on the abundance, distribution, or behavior of other species; by an abiotic component of its ecosystem; or by unique ecological phenomena it represents. Work in both marine and terrestrial systems has shown that it is possible to maintain species populations at viable levels from a genetic or demographic standpoint, while at the same time reducing these populations to what is functionally a level of "ecological extinction." In these situations, the population numbers are sufficiently low that the species no longer has significant interactions with other species in its community (Estes et al. 1989). For example, in the Patagonia region of South America, the dominant native prey in the region, guanacos (*Lama guanicoe*) and lesser rheas (*Rhea pennata*), has been essentially replaced by introduced livestock and European hares. This change is thought to be having significant effects on nutrient dynamics, plant–herbivore interactions, and disturbance regimes (Novaro et al. 2000).

There are numerous examples of potentially viable populations that have been reduced to smaller sizes or that are nonfunctional from an ecological perspective (e.g., Redford and Feinsinger 2001). Such functions are difficult to describe quantitatively and, therefore, challenging for goal-setting purposes. Most of the insight about what constitutes an ecologically functioning population is likely to come from historical information on the distribution and abundance of target species or natural history observations of the interactions of target species with co-occurring species or their environment. These historical perspectives and observations may provide conservation planners with some idea about how to better set conservation goals and how to judge whether specific populations are at ecologically viable levels and not just demographic or genetically acceptable sizes.

In some parts of the world, the notion of EFPs is a fleeting fantasy, but in other parts, maintaining EFPs is a daunting but doable task. For example, in the short-grass prairie region of the central United States, black-tailed prairie dogs (*Cynomys ludovicianus*) were reduced in their total distribution and population size by over 90% in the last century. The activities of prairie dogs (*Cynomys* sp.) are known to change the structure of prairie ecosystems and provide habitat for many other species. Many biologists consider them to be a keystone species (Kotliar 2000). In many ecoregions, the habitat for this species remains, despite the fact that prairie dogs were locally driven to extinction by poisoning, shooting, and disease. Thus, the forward-looking conservation planner would attempt to maintain short-grass prairie habitats that might someday allow for the reestablishment and expansion of prairie dog populations to ecologically functioning levels.

At this point, we have considered a number of methods and concepts related to population viability. As previously suggested, when ecologists talk about the "viability" or "functionality" of a community or ecosystem, it is more appropriate to refer to the concept of ecological integrity. In the next section, I address methods for assessing ecological integrity in more detail.

Assessing Ecological Integrity

The challenge in using the concepts of population viability and ecological integrity is to make them operational in both conservation planning and conservation action contexts. From a regional perspective, planners need to identify generalized conservation areas that are likely to have high ecological integrity over the long term. For communities and ecosystems, these areas largely have their native species composition intact, ecological processes and disturbances are operating within natural ranges of variability, and the areas are presently disturbed to the least extent possible by humans, although they may have been significantly altered in the past. I review several possible methods for assessing ecological integrity during regional conservation planning below.

The Disturbance-Based Approach

Work on the ecology of natural disturbances to communities and ecosystems provides one useful approach for assessing ecological integrity. Natural disturbances do not occur evenly across a landscape. Disturbances tend to break homogeneous areas into patches of heterogeneous habitat, creating what has been referred to as a shifting mosaic (Bormann and Likens 1979) of various vegetation successional stages resulting from the spatial variability in intensity and extent of disturbance. The fires of Yellowstone National Park in the summer of 1988 are a well-documented example of such a disturbance pattern, burning hot and intensely in some areas but not at all in others, leaving a forested mosaic of burned patches across the park. How a particular patch of habitat recovers from a disturbance is a function of the availability of organisms for recolonization and the biological legacies (Franklin et al. 2000) remaining at the site (e.g., seed bank, soil organic matter, root systems). The area necessary to ensure survival or recolonization has been referred to as the *minimum dynamic area* (Pickett and Thompson

1978). A parallel line of thinking in aquatic ecosystems has examined natural flow regimes (Poff et al. 1997) and their relationship to ecological integrity (Richter et al., in press; see Chapter 10).

How to estimate the size of the minimum dynamic area has been the subject of some debate among ecologists. Shugart and West (1981) recommended that it be 50 times the mean patch size of disturbance. Baker (1992) thought it more appropriate to scale it to the maximum disturbance size. Similarly, Peters et al. (1997) suggested that the minimum dynamic area be assessed as the largest disturbance event expected over a period of 500–1000 years. In the Northern Appalachians ecoregion in the northeastern United States, Anderson (1999) scaled the minimum size of the matrix forest to the grain and extent of several disturbance patches over long periods of time. Based on information from several studies, Anderson (1999) adopted a general guideline that occurrences of matrix forests to be identified as potential conservation areas should be about four times the size of the largest, most severe disturbance patch to replicate the natural pattern of disturbed and undisturbed forests in New England (Figure 7.2). Although disturbance sizes and patterns are a promising approach

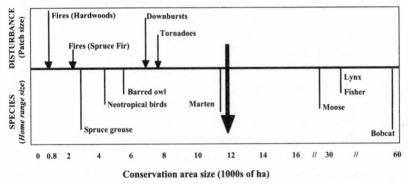

Figure 7.2   Factors used to assess the adequacy of size for proposed conservation areas of forested ecosystems in the Northern Appalachians ecoregion. Two principal factors can be used to assess size: the home range of wide-ranging or area-sensitive animal species and historical patch sizes from natural disturbances. In this figure, the area necessary to ensure the long-term persistence of an example of a forested ecosystem is defined as four times the patch size of the most severely disturbed patch. This estimate, referred to as the minimum dynamic area, is based on historic data suggesting that about 25% of the forested areas of New England were expected to be severely disturbed at any one time. The home range estimate is based on the area needed to accommodate a viable population of each species. In the Northern Appalachians Ecoregional Plan, the minimum size for forested conservation areas (large vertical down arrow) was set at approximately 12,000 hectares. (From Anderson 1999 and Groves et al. 2002. Planning for biodiversity conservation: Putting conservation science into practice. *Bioscience* 52:499–512. Copyright American Institute of Biological Sciences.)

for assessing the integrity of potential conservation areas, there remains a considerable divide in opinion among experts on the most appropriate guidance relative to disturbance size. Conservation planning would clearly profit from more research into the disturbance ecology of different communities and ecosystems and greater consensus among experts on recommendations for minimum dynamic areas based on natural disturbance regimes.

Area-Sensitive Species

A second approach to estimating the size of areas needed for the ecological integrity of communities and ecosystems is to assess the home range requirements of area-sensitive species that are endemic to or typical of particular communities or ecosystems. For example, Anderson (1999) examined the home range needs and habitat requirements of a community of neotropical migratory birds and several wide-ranging mammalian species in the matrix forests of the Northern Appalachians ecoregion (see Figure 7.2). By combining this type of data with information on natural disturbances, he could better assess the size requirements of potential conservation areas necessary to maintain the ecological integrity of communities and ecosystems.

Many planning efforts of TNC in the United States have used this approach for screening the potential integrity of occurrences for some types of target communities and ecosystems (e.g., matrix forest). It is instructive to note that this approach requires some estimate of viable population size for whichever species are used in the analysis. The PVAs for wide-ranging species discussed in the earlier section on habitat-based approaches would obviously be useful here, effectively serving dual purposes: informing the planning team with regard to viability estimates for target species and assessing ecological integrity for target communities and ecosystems.

NatureServe/Natural Heritage Methodology

Natural Heritage Programs and NatureServe staff members have developed element occurrence or ecological integrity specifications for many natural communities and ecosystems similar to those described for species earlier in this chapter and based on the same criteria of size, condition, and landscape context (Table 7.2). Although these specifications and ranks are referred to as "viability" specifications by NatureServe and NHPs, they are

Table 7.2   *Ecological Integrity Specifications for a Mesic Tallgrass Prairie Ecosystem, Central Tallgrass Prairie Ecoregion*

Size Specifications

A-Rank Size: > 4046.9 ha (10,000 ac), B-Rank Size: 809.4–4046.9 ha (2,000–10,000 ac), C-Rank Size: 161.9–809.4 ha (400–2,000 ac), D-Rank Size: < 161.9 ha (400 ac)

Justification for minimum A-rated criteria: This matrix community should occupy extensive areas on the landscape to provide habitat for large fauna, including bison. The A-rated size should, ideally, be set at > 10,000 ac. However, tallgrass prairie has been reduced to less than 1% of its former extent throughout most its range, and few large examples remain. Justification for C/D threshold: Edge effects become increasingly problematic for EOs below the threshold, particularly in fragmented landscapes. Edge effects include dust and salts from roadsides, pesticide sprays, and the presence of exotic-dominated communities.

Condition Specifications

A-Rank Condition: Typical native composition with indicator species present, as these relate to natural disturbances. Key disturbances, including human disturbances that mimic natural ones, include fire and grazing. Typical structure is dominated by graminoids and forbs. Few to no exotics present. Lack of negative human impacts, such as gravel roads.

B-Rank Condition: Lack of some typical native indicators, particularly as these relate to absence of some natural disturbances. Structure not always typical, with native forbs or graminoids overly dominant or shrub encroachment. Some exotics present. Some negative human impacts.

C-Rank Condition: Many native indicator species absent. Structure not typical with native forbs or graminoids excessively dominant, and shrubby encroachment high. Exotic may be extensive, but rarely dominate over native component. Extensive negative human impacts, including pesticide spraying, some dirt or gravel roads, or heavy cattle grazing.

D-Rank Condition: Most, if not all, native indicator species absent. Weedy native dominants are still present with many exotics. Structure is not typical. Exotic species dominate over native species component, as listed in C-rated condition. Extensive negative human impacts evident as listed in C-rated condition.

Landscape Specifications

A-Rank Landscape Context: Highly connected, the element occurrence (EO) is surrounded by intact natural vegetation, with species interactions and natural processes occurring between the EO and all adjacent communities. The area around the EO is > 2000 ha (5,000 ac) with at least 50% natural vegetation, and the rest some mix of permanent cultural grassland.

B-Rank Landscape Context: Moderately connected, the EO is surrounded by moderately intact natural vegetation, with species interactions and natural processes occurring between the EO and most adjacent communities. The area around the EO is between > 800 and 2000 ha (2,000 and 5,000 ac) with between 20% and 50% natural vegetation, and the rest some mix of permanent cultural grassland and tilled fields.

C-Rank Landscape Context: Moderately fragmented, the EO is surrounded by a combination of cultural and natural vegetation, with barriers to species interactions and natural processes between the EO and many adjacent natural communities. Surrounding landscape area is undefined, but EO is surrounded by between 10% and 20% natural vegetation.

D-Rank Landscape Context: Highly fragmented, the EO is entirely or almost entirely surrounded by cultural vegetation or other urban/suburban/rural land uses. Surrounding landscape area is undefined, but EO is surrounded by < 10% natural vegetation.

From Central Tallgrass Prairie Ecoregion Planning Team 2000.

consistent with the terminology of ecological integrity as used in this chapter. For communities and ecosystems, their persistence at a specific location depends upon the ability of habitat patches to survive stochastic events or for propagules to recolonize a site following a disturbance (Anderson et al. 1999). The amount of area (size) needed to ensure survival or recolonization was discussed above under the topic of minimum dynamic area.

For the criterion of condition, planners should consider two factors related to communities and ecosystems: anthropogenic impacts and biological legacies or historic continuity (Anderson et al. 1999). Fragmentation, altered disturbance regimes, exotic species introductions, and pollution are among the more significant human-related impacts. Biological legacies (Franklin et al. 2000) refer to critical features that take generations to develop, such as fallen logs and rotting wood in forest ecosystems, a well-developed herbaceous understory, and structural complexity in the canopy.

The status of the surrounding landscape is critical to the integrity of all communities and ecosystems, but especially for some patch communities (see Chapter 6) that are sustained by ecological processes that occur at scales larger than those of the community. Natural fires that leave a mosaic of burned patches in coniferous forest ecosystems and groundwater flow that sustains a wetland community are examples of such processes. Other types of patch communities such as raised bogs, perched wetlands, or cliff communities are less dependent on the surrounding landscape. An extreme example of the importance of landscape context has recently emerged in studies of coral reef communities where scientists demonstrated that the presence of reefs up to 600 km (373 mi) away were the most important determinant of local reef diversity as measured by fish and coral species (Knowlton 2001).

Outside North America, there may be little or no EO-type data to inform planners about the integrity of occurrences of conservation targets. Even within North America, this type of data is spotty and uneven across the network of Natural Heritage Programs (Groves et al. 1995). In these situations, it may still be helpful to use the criteria of size, condition, and landscape context and to rate the potential integrity of target occurrences based on expert opinion. For matrix-type communities or ecosystems, size is likely to be the criterion that is easiest to assess and the most meaningful in terms of integrity. For communities or ecosystems that are naturally more patchlike in distribution, information on landscape context and condition may be more critical. For ecosystem-level targets, it may be feasible for a planning team to develop ecological integrity guidelines for a limited number of ecosystem types as part of the regional planning process. For example, in the Southern Rocky Mountains ecoregion, the Ecoregional Assessment Team (2001) developed and applied integrity guidelines based on size, condition, and landscape context for approximately 35 different broad terrestrial ecosystem types as part of TNC's ecoregional planning process (see Table 7.3 for sample guidelines). These guidelines were then used to evaluate the potential long-term integrity of occurrences of these ecosystem targets within the ecoregion.

Suitability Indices and Spatial Surrogates

For many parts of the world, conservationists lack site-specific data on the ecological integrity of conservation targets. However, planners can use a variety of surrogate data that can be mapped with a Geographic Information System (GIS) to provide an indirect assessment of an area's potential as a biodiversity conservation area. In fact, landscape architects and planners were among the first to overlay a variety of social and biophysical data layers to determine areas suitable for development and areas where development should be constrained. In a book entitled *The Living Landscape* (2000), landscape architect Frederick Steiner devotes an entire chapter to these "suitability analyses." Several different projects are detailed below as examples of suitability analyses in conservation planning for biodiversity.

In the Columbia Plateau ecoregion, Davis et al. (1999) used available digital information on road density, human population density, aquatic integrity, percentage of land converted to human use, distance from existing conservation areas, and other factors to develop an index of how suitable an area would be for conservation purposes (Figure 7.3). This suitability index was computed as a weighted sum of all the individual factors and was calculated for each of the 4674 watershed units that make up the ecoregion. Areas with relatively higher road density, higher human densities, lower aquatic integrity, higher percentages of private land, and higher percentages of converted lands were assumed to be less suitable for conservation. This suitability index was subsequently used in a multiobjective model to select conservation areas for the Columbia Plateau ecoregion. This model selected conservation areas based on predefined goals for targets and the need to balance the dual objectives of minimizing area while maximizing suitability. A similar approach has been taken subsequently in many Nature Conservancy ecoregional plans throughout the western United States. Chapter 8 will explore the use of models and algorithms to select conservation areas in more detail.

Table 7.3 *Ecological Integrity Guidelines for a Montane Wet Meadow Ecosystem in the Southern Rocky Mountains Ecoregion*

Typical Species

Calamagrostis canadensis, Carex aquatilis, Carex lanuginosa, Carex lasiocarpa, Carex praegracilis, Carex saxatilis, Carex simulata, Carex utriculata, Eleocharis palustris, Eleocharis rostellata, Juncus balticus, Spartina gracilis, Triglochin maritimum

Scale and Range: Small Patch and Widespread

Montane wet meadow ecological system is a small patch system in the western U.S. montane ecoregions. Within the Southern Rocky Mountains ecoregion, this system is widely distributed in both elevation and latitude and confined to specific environments defined primarily by hydrology. Water levels in this system are often at or near the ground surface for much (or all) of the growing season, but also may fluctuate considerably through the year. Surface inundation may occur, but it typically does not last for long. Physical disturbance during inundation (e.g., during flood events) may be significant for the structure and composition of these systems. Wet meadows occur on mineral soils that have typical hydric soil characteristics, including relatively high organic content and redoximorphic features. This system usually occurs as a mosaic of several plant associations. The surrounding landscape often contains other wetland systems, e.g., riparian shrublands, or a variety of upland systems from grasslands to forest. Although this system usually occurs in small patches, the large intermountain valleys (San Luis Valley, South Park, and North Park) have some large examples of montane wet meadows.

Minimum Size: 0.4 ha (1 ac)

Separation Distances: (1) Substantial barriers to natural processes or species movement, including cultural vegetation greater than 0.4 km wide, major highways, urban development, or large bodies of water; (2) natural community from a different ecological system wider than 0.8 km wide; (3) major break in topography, soils, geology, etc., especially one resulting in a hydrologic break.

Rank Procedure: (1) Condition, (2) landscape context, (3) size. Condition and landscape context are the primary ranking factors, with size secondary because even small examples of this system can have high biological significance.

Condition Specifications

A-Rank Condition: Natural hydrologic regime intact. No or little evidence of wetland alteration due to increased or decreased drainage, clearing, livestock grazing, or anthropogenic nutrient inputs. No or very few exotic species present with no potential for expansion. Native species that increase with disturbance or changes in hydrology/nutrients (e.g., nitrogen and phosphorus) are absent or low in abundance.

B-Rank Condition: Natural hydrologic regime nearly intact. Alteration from local drainage, clearing, livestock grazing, or nutrient inputs is easily restorable by ceasing such activities. Few exotic species with little potential for expansion if restoration occurs. Native species that increase with disturbance or changes in hydrology/nutrients are absent, low in abundance, or restricted to high-nutrient microsites that represent less than 5% of the total wetland area.

C-Rank Condition: Natural hydrologic regime altered by local drainage. Alteration from local drainage, clearing, or livestock grazing is extensive, but potentially restorable over several decades. Exotic species may be widespread but potentially manageable with restoration of most natural processes. Native species that increase with disturbance or changes in hydrology/nutrients may be prominent.

D-Rank Condition: Natural hydrologic regime or disturbance regimes not restorable. System remains fundamentally compromised despite restoration of some processes. Exotic species may be dominant. Native species that increase with disturbance or changes in hydrology/nutrients are prominent to dominant.

Size Specifications

A-Rank Size: Very large > 30.3 ha (75 ac)

B-Rank Size: Large 8.1–30.3 ha (20–75 ac)

C-Rank Size: Moderate 0.4–8.1 ha (1–20 ac)

D-Rank Size: Small 0.4 ha (1 ac)

Landscape Context Specifications

A-Rank Landscape Context: Uplands within 1.6 km of the occurrence are largely unaltered (> 90% natural) by urban or agricultural uses such as clearcuts, crop cultivation, land development, or heavy livestock grazing. There are no unnatural barriers present between adjacent lands and the occurrence, allowing free flow of organisms and materials across the wetland/upland boundary. Connectivity of habitats allows natural processes and species migration to occur.

B-Rank Landscape Context: Uplands with 0.4 km of the occurrence with moderate urban or agricultural alteration (60–90% natural), but with no major barriers to water or species movement within the occurrence. There are few unnatural barriers present between this occurrence and nearby occurrences that would inhibit species movement among occurrences. Some natural processes such as flooding and fire may have altered frequencies or intensities.

C-Rank Landscape Context: Uplands surrounding occurrence are fragmented by urban or agricultural alteration (20–60% natural), with limited connectivity between this occurrence and the next nearest occurrence. Some barriers are present, and natural processes few. Activities (development, clearcuts, heavy grazing, etc.) in surrounding uplands alter the hydrologic regime. Restoration of the hydrologic regime and species composition resembling the historic composition is feasible.

D-Rank Landscape Context: Uplands surrounding occurrence are mostly converted to agricultural or urban uses. Connectivity and natural processes are almost nonexistent. Restoration is not feasible within reason.

From Southern Rockies Ecoregional Assessment Team 2001.

A biodiversity plan for the Redwoods ecosystem of northern California employed a similar approach (Strittholt et al. 1999, Noss et al. 2000). Planners there used a model with nine criteria to evaluate focal areas for biodiversity conservation—relatively intact and unaltered areas that still retain native biodiversity. Although the objective of their modeling effort was not to develop an index of ecological integrity or suitability per se, the net effect of combining the nine criteria into an overall ordinal index (ranked 1–5) was to create an index that assessed the overall potential of each focal area to maintain ecological integrity. The nine criteria could be subdivided into three broad functions: patches—discrete mappable units of defined composition, neighborhoods—immediate surroundings of a patch, and watershed—the landscape or hydrological unit in which a group of patches exists (Figure 7.4).

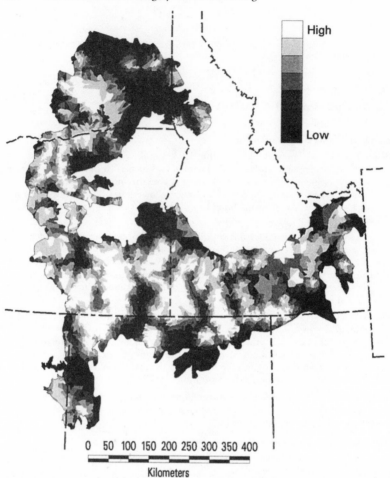

Figure 7.3   A suitability index for the Columbia Plateau ecoregion. Suitability indices are used to help select conservation areas in regional planning efforts. They provide an indirect assessment or screening of the long-term potential of a conservation area to support occurrences of conservation targets at acceptable levels of viability and integrity. The index is developed by overlaying several data layers, such as road density, human population density, and percentage of natural land cover, with a GIS in a composite fashion. (From Davis et al. 1999. Reprinted by permission of World Conservation Union [IUCN] and *PARKS*—the international journal for protected area managers.)

The nine criteria used in the model are large late-successional forest patches (patch), concentrations of late successional patches (neighborhood), road density (neighborhood), locations of imperiled species (patch and neighborhood), forest neighborhood age (neighborhood), forest fragmentation (neighborhood), connectivity to existing conservation areas

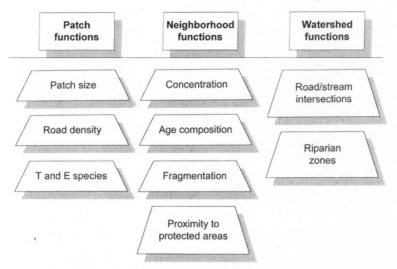

Figure 7.4   Criteria used for identifying potential conservation areas in a biodiversity plan for the Redwoods Ecosystem in the western United States. The nine criteria could be subdivided into the three broad functional categories of patches, neighborhoods, and watersheds. Each of the criteria was assigned an ordinal rank of 1–5. Used together, these nine criteria represented an index to the suitability of particular areas for conservation and an indirect evaluation of the ecological integrity of these areas. (*The Redwood Forest: History, Ecology, and Conservation of the Coast Redwoods,* Reed F. Noss, ed., Copyright 2000 by Save-the-Redwoods League. Reprinted by permission of Island Press, Washington, D.C.)

(neighborhood), road-stream intersections (watershed), and forest riparian zones (watershed). Each criterion was scored individually from 1 to 5 (lowest to highest integrity). For example, the rationale for the road/stream intersection metric (Figure 7.5) was that an increasing number of road intersections with streams would contribute more sediment to streams, thereby lowering water quality and integrity. Individual criteria were then combined into one overall composite score for integrity (Figure 7.6).

One of the most extensive evaluations of ecological integrity was conducted as part of the U.S. government's assessment of ecosystem management in the Interior Columbia Basin in the Pacific Northwest (see Chapter 3). Scientists in this assessment evaluated ecological integrity separately for forested lands, rangelands, forestland hydrology, rangeland hydrology, and aquatic systems (Quigley et al. 1996). They used descriptive data layers, expert judgment, empirical models of ecological processes, and trend analyses in the integrity evaluations. For example, the factors affecting forestland integrity included amount and distribution of exotic species,

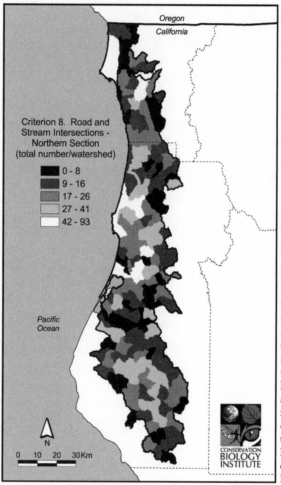

Criterion 8. Road and
Stream Intersections -
Northern Section
(total number/watershed)

■ 0 - 8
■ 9 - 16
■ 17 - 26
□ 27 - 41
□ 42 - 93

Pacific
Ocean

N

0   10   20   30 Km

Oregon
California

Figure 7.5   Road and
stream intersections is
one criterion used in an
overall evaluation of areas
suitable for conservation
in the Redwoods ecosys-
tem. The assumption
behind this criterion is
that increasing numbers
of road intersections with
streams result in increasing levels of sedimenta-
tion in streams, thereby
lowering water quality
and overall ecological
integrity of the stream
ecosystem. (From Strit-
tholt et al. 1999.
Reprinted by permission
of Conservation Biology
Institute.)

amount of snags and down woody material, disruption to hydrological
regimes, tree stocking levels, absence or presence of wildfires, and changes
in fire severity and frequency. To actually assess the integrity of forested
lands, scientists in the Columbia River Basin used proxies for these factors,
including the proportion of area in dry and moist forest potential vegetation groups, the proportion of area exceeding a certain threshold of road
density, the proportion of area in wilderness or nonroaded areas, and the
proportion of area where fire frequency declined between historical and
current conditions by a specified amount.

As a result of assessing these proxies, they judged 17% of the forested
sub-basins in the region to have high integrity (Figure 7.7). Individual

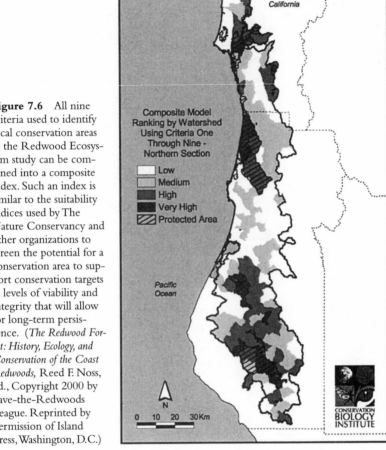

Figure 7.6   All nine criteria used to identify focal conservation areas in the Redwood Ecosystem study can be combined into a composite index. Such an index is similar to the suitability indices used by The Nature Conservancy and other organizations to screen the potential for a conservation area to support conservation targets at levels of viability and integrity that will allow for long-term persistence. (*The Redwood Forest: History, Ecology, and Conservation of the Coast Redwoods,* Reed F. Noss, ed., Copyright 2000 by Save-the-Redwoods League. Reprinted by permission of Island Press, Washington, D.C.)

assessments of integrity for all five systems were combined into a composite integrity rating for the entire Columbia River Basin. The overall ratings for the basin indicated that 16% of the area had high ecological integrity, 24% had moderate integrity, and 60% had low integrity. Many of these low-integrity lands were dominated by agricultural uses. Because the rating system emphasized ecological processes, it tended to rate human-dominated systems lower than those where more natural ecological processes were functioning.

The U.S. federal government has conducted several bioregional assessments similar to the one described here for the Columbia River Basin. Other assessment areas include the Sierra Nevada (www.ceres.ca.gov/

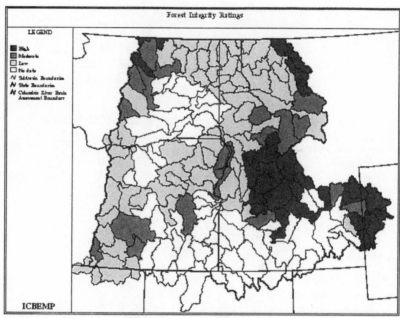

Figure 7.7 Qualitative ratings of forest integrity as developed in the Interior Columbia Basin Ecosystem Management Plan in the northwestern United States. To develop these ratings, scientists from the federal government used data on factors such as proportion of areas in a non-roaded condition and proportion of area where current fire frequencies were significantly reduced from historic conditions. (From Quigley et al. 1996. Reprinted by permission of U.S. Forest Service.)

snep/pubs), Southern California Mountains and Foothills (Stephenson and Calcarone 1999), Northern Great Plains (Samson et al. 2000), and Southern Appalachians (Southern Appalachians Man and the Biosphere Cooperative 1996). Similar data and information may be available on viability and integrity in these bioregional assessments.

Margules and Pressey (2000), in their seminal review paper on systematic conservation planning, suggested that the most practical way to account for ecological processes in conservation planning is to map the locations of spatial surrogates for these processes—watershed boundaries, migration routes, areas with a lack of roads, and so on. In planning for the conservation of southern Africa's Succulent Karoo Desert, Cowling and colleagues (1999) put this advice into action. For example, they identified and mapped entire sand movement corridors that they determined were critical to the conservation of some soil-specific plant assemblages. River catchments were also identified and mapped that were important to distinct species assemblages, that provided movement corridors between uplands and coastal regions for plant species, and that also provided dry-season refuges for large ungulates. Finally, areas incorporating steep climatic gradients that could help facilitate changes in distribution patterns in response to climate changes were also included in the planning process.

Although individually different, most of the examples discussed in this section on using suitability indices and spatial surrogates for assessing integrity have several points in common. First, the planners examined a set of factors that are primarily anthropogenic in origin (e.g., road density) and that are presumed to have negative effects on the pattern and processes that sustain biological diversity. Second, they used several different GIS data layers to arrive at some overall score or assessment of integrity. Third, they focused on factors that can be assessed at landscape scales, most often from remote-sensing data. Depending on what data are available, each of the above methods offers a mechanism for screening the potential integrity of areas. In turn, the results of such screening can be used to select conservation areas that are likely to have relatively higher ecological integrity, to identify other areas that could benefit from ecological restoration (see below), and to pinpoint existing conservation areas that may be suffering now or in the future from inadequate attention to the role of ecological integrity in the persistence of conservation targets.

The use of suitability indices and other similar metrics for assessing the potential ecological integrity of areas has some drawbacks. For example, little research has been done to examine the effects of placing greater weight or emphasis on one or more metrics compared to another. Similarly, the effects of eliminating or adding particular data layers on the overall composite "integrity" scores are unknown, as are the consequences of changing the arbitrary boundaries of certain categories of data such as the road stream intersection metric of Figure 7.5. Finally, few, if any, of these types of indices or spatial surrogates are ever "ground-truthed" for their accuracy in the field. In other words, does an area that is identified as having a low integrity score from overlaying these various digital data layers actually turn out to be in poor ecological condition when one visits or samples sites on the ground? These questions and criticisms are not meant to dissuade conservation planners from using such approaches. Instead, they are simply meant to be a recognition that GIS is an attractive tool, the products of which (maps and analyses) are only as good as the data behind them. In most cases, these products would be greatly improved by a greater (or in some cases, any) investment in field checking the accuracy of the analytical results and maps.

All of these approaches for screening or assessing viability and integrity can be helpful to conservation planning teams, but what should a team do when efforts to identify high-quality examples of conservation targets fall substantially short of target-based goals? In these situations, planners need to consider whether it may be possible to identify degraded or remnant examples of targets that could be feasibly restored to acceptable levels of viability and integrity. In the Columbia River Basin example discussed above, 24% of the basin was rated as having moderate ecological integrity, suggesting that some conservation areas located in this part of the basin may be candidates for restoration. The subdiscipline of ecology that focuses on this topical area is known as restoration ecology.

Ecological Restoration

In many regions and ecoregions of the world, native habitats have been converted and fragmented, often extensively. For example, in the Northern Tallgrass Prairie ecoregion of the central United States (The Nature Conservancy 1998), only 4% of the ecoregion remains in native vegetation cover; most of it was converted to agriculture or urban areas. In these situations, many of the communities and ecosystems in the region are likely to lack the ecological integrity necessary to sustain them over the long term. Conservation goals that were established for these targets are unlikely to be met (see Chapter 6). As a consequence, conservationists face hard decisions. There appear to be three plausible courses of action. First, conducting more biological inventories might result in uncovering new examples or occurrences of target communities that are worthy of conservation action. Second, some communities and ecosystems may be so degraded and converted that it is no longer practical or feasible to continue efforts to conserve them in that ecoregion. A third, and more promising alternative, is that it may be possible to restore the ecological integrity of some degraded communities and ecosystems.

The Society for Ecological Restoration (2002) defines *ecological restoration* as "the process of assisting the recovery of an ecosystem that has been degraded, damaged, or destroyed." As Figure 7.8 shows, ecological restoration is best viewed as a continuum of activities, from reclamation to the reestablishment of predisturbance conditions, including the reintroduction of locally extinct species (Hargrove et al. 2002). To date, most restoration efforts have taken place at relatively small spatial scales, although the need to restore ecosystems at much larger scales has clearly been identified (Simberloff et al. 1999). In particular, the Wildlands Projects in the United

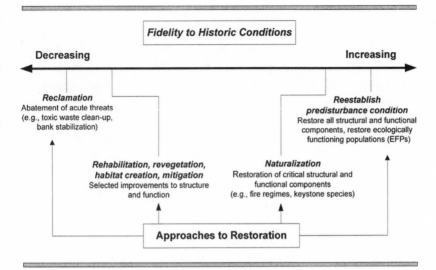

Figure 7.8   Ecological restoration activities can be viewed as occurring on a continuum from reclamation and revegetation of areas that have been heavily damaged to reestablishment of disturbance regimes and ecologically functioning populations. Costs increase in going from right to left on the continuum while probabilities for successful restoration mirror the cost trends. (From Hargrove et al. 2002.)

States have included detailed prescriptions for large-scale restoration efforts in their conservation plans. For example, the Sky Islands conservation plan for the southwestern United States contains goals and objectives for "healing the wounds" of wildlife extirpation, watershed damage, fire suppression, habitat fragmentation, exotic species introductions, and forest degradation (Foreman et al. 2000).

From a regional planning perspective, planners should consider identifying *potential restoration areas* that are likely to increase the size or improve the condition or landscape context of occurrences of communities and ecosystems identified as conservation targets. In short, restoration should focus on how to improve the viability and integrity of existing conservation targets (Hargrove et al. 2002), not re-create target communities and ecosystems from scratch. As a number of authors have indicated, ecological restoration is expensive, and as a science, it is in its infancy. The results of restoration are enveloped with uncertainty, and at least one restoration ecologist has noted that it is "always a poor second choice to preservation of original habitats" (Young 2000). That said, in a number of situations in regional planning, planners and biologists can identify potential restoration areas that are cost-effective, are imperative for conserving one or more conservation targets in the region, and have an acceptable degree of success or failure associated with them.

The restoration of dry tropical forest in Guanacaste National Park in northwestern Costa Rica is, in many respects, the "poster child" for successful large-scale ecological restoration. This ecosystem type once extended from southwestern Mexico to Panama, but is now restricted largely to small remnant patches amidst a sea of introduced grasses and agricultural crops (Meffe and Carroll 1997). The keys to successful restoration of dry tropical forests are threefold: (1) eliminate the conditions that caused forest loss (e.g., human-ignited fires); (2) ensure that forest remnants are of sufficient size to serve as a seed source; and (3) maintain animal species critical to the dispersal of these seeds. In Guanacaste National Park, Costa Rican biologists and American colleagues have been successful at restricting fire by policy and regulation changes, as well as mowing fast-growing African grasses in places where fires could easily start. To augment seed dispersal, horses are fed meal containing seeds of some native forest trees; the seeds are subsequently deposited in grasslands in rich piles of manure. For some tree species, seedlings are planted manually. Through these relatively simple techniques, the dry tropical forest of Guanacaste National Park is rapidly being restored, in some places as quickly as over an 8-year period. In essence, this project was designed to reconnect fragments or remnants of dry forest by "healing the intervening landscape," a strategy that TNC is similarly using in the restoration of the Kankakee Sands (Indiana, U.S.) conservation area, a 14,000-ha (34,580-ac) project containing remnants of sand prairie and oak barren natural communities (Hargrove et al. 2002).

As these examples demonstrate, degraded samples of conservation targets can be elevated to acceptable levels of ecological integrity through restoration efforts. Other examples of such projects that have received far more publicity are efforts to reestablish sand beaches along the Colorado River in the Grand Canyon (U.S.) through the simulation of pre-dam spring flooding conditions and a massive restoration effort to reestablish seminatural flow regimes in Everglades National Park (U.S.) and the adjacent Kissimmee River watershed. Reaching a decision to undertake such projects is never easy. The decision tree shown in Figure 7.9 can help planners make these decisions. It first revisits the topic of goal setting (see Chapter 6), then assesses whether additional inventory is a plausible avenue for meeting conservation goals, and finally considers restoration as a strategy once other alternatives have been exhausted.

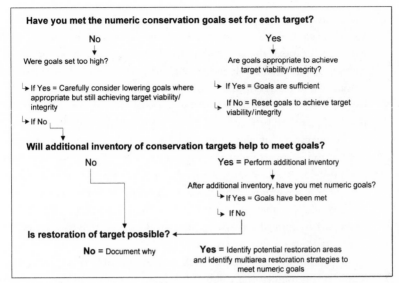

Figure 7.9 Decision trees can help conservation planners evaluate the feasibility of restoration activities as part of regional conservation planning. (From Hargrove et al. 2002.)

Throughout this chapter, I have considered a number of alternative methods for assessing the population viability and ecological integrity of conservation targets. Which approaches to take will depend on data availability and the staff and financial resources that any given planning effort has at its disposal.

Key Points in Assessing Population Viability and Ecological Integrity

1. For species-level targets, use available information from population viability analyses (PVAs) and recovery plans of endangered species to assess the viability of individual populations and occurrences. Be careful, however, not to place too much emphasis on a single number resulting from a quantitative PVA. Overall, quantitative PVAs will be available only for a handful of species.

2. In North America and parts of Latin America, use viability and integrity information on species and a limited number of natural

communities available from Natural Heritage Programs and Conservation Data Centers. Based on the criteria of size, condition, and landscape context, the qualitative viability/integrity ranks that are an output of this methodology are derived from syntheses of scientific literature and expert opinion. As a general rule, occurrences or populations with poor viability ratings should be excluded from the regional planning process, unless they represent the only opportunity for conservation of the species in the planning region and there appears to be some potential for restoration. NHP methodology does not provide an assessment of viability per se, but instead should be viewed as a screening technique for populations or occurrences that are potentially viable. In general, any methodology that relies heavily on expert opinion should be applied cautiously as such data are rarely tested to assess validity or accuracy.

3. Apply cookbook approaches such as the "50/500 rule" and more recent recommendations calling for effective population sizes of 1000 individuals and census population sizes of 5000 individuals with considerable caution, and only when other more useful information is lacking. Based primarily on genetic concerns, these "rules" may underestimate viable population sizes because they do not account for demographic and environmental uncertainty.

4. For community and ecosystem targets, the development of ecological integrity specifications based on expert opinion but utilizing the criteria of size, condition, and landscape context holds considerable promise for regional planning. Once in place, field reconnaissance efforts or expert opinion must be used in combination with these specifications to judge the integrity of individual occurrences of community and ecosystem targets. An especially useful concept is that of minimum dynamic area—the minimum size of an occurrence of a community or ecosystem that is likely to recover and maintain itself in the face of natural disturbances.

5. Recent methods for assessing the viability of wide-ranging species that combine static habitat models with dynamic viability analyses are a promising approach for incorporating the needs of wide-ranging species and other focal species in regional conservation plans.

6. Suitability indices and other approaches based on spatial surrogates should be developed and used for indirect assessments or screening of the ecological integrity of communities and ecosystems. These approaches can easily be custom-designed to match the needs of a particular ecoregion or region, depending upon what digital data are

readily available. In general, these approaches consider landscape-scale metrics that effectively steer planners away from areas that have been significantly altered or degraded by anthropogenic activities. These approaches are particularly effective when on-the-ground information on the condition and quality of community and ecosystem targets is either not available or infeasible to obtain. A variety of such approaches have been used. There is little to recommend one approach over another, because no studies have been undertaken to compare various GIS-based indices and metrics aimed at assessing integrity. The weaknesses of these approaches are that they are rarely field-verified for accuracy and that sensitivity analyses aimed at evaluating the effects of weighting or eliminating certain data layers have not been conducted.

7. When conservation goals cannot be met due to degradation and alteration of conservation targets, consider the feasibility of ecological restoration. Restoration efforts should focus on improving the viability or integrity of existing occurrences of targets, not re-creating targets. A decision tree that considers alternatives other than restoration can help determine whether restoration may be a feasible endeavor to pursue in any given situation.

Drafting Nature's Blueprint: Selecting and Designing a Network of Conservation Areas

One of the early lessons of research on systematic reserve selection is that the development of what seems a promising and important idea does not cause the world to beat a path to one's door.

—ROBERT PRESSEY (1998)

Just as an architect's rendering of the blueprint of a house provides structural details for a builder, a conservation blueprint provides detailed information for conservation scientists and practitioners on the areas and targets that require attention for conserving the biological diversity of a particular region. A conservation blueprint—a map that identifies conservation areas of the planning region and associated information on the conservation targets contained in these areas—is the most significant product of a regional biodiversity planning project. It represents the culmination of a great deal of information gathering and analysis, and is the core of any comprehensive regional plan to conserve biological diversity.

This chapter will focus on two concepts: assembling or selecting a *portfolio* of conservation areas and, from that portfolio, designing a *network* of conservation areas. Like a financial portfolio, a conservation portfolio represents an investment in a variety of conservation areas. Ideally, there will be some areas that conserve freshwater and marine ecosystems, some areas that conserve endangered species, and still other areas that will afford attention to outstanding examples of various terrestrial ecosystems. In essence, a portfolio is a group of conservation areas that represent, to the degree possible, the full range of conservation targets in a distribution and amount as set by the target-based goals. Just as in financial investing, conservation planners, biologists, and practitioners should be investing in a well-balanced portfolio. The problem, of course, is that the current conservation portfolio in most parts of the world is out of balance. As the result of ad hoc conservation efforts first mentioned in Chapter 1, most of the world's conservation areas are situated on poor soils, steep slopes, and/or at relatively high elevations. In addition, much more attention has been paid to terrestrial species and ecosystems than their counterparts in the freshwater and marine realms (see Chapters 10 and 11).

Why is the conservation portfolio so out of kilter? Australian conservation biologist Bob Pressey (1997) outlined three possible explanations. First, there is a tendency to designate conservation areas where it is financially easiest and politically expedient to do so, and these are often the areas where there will be the least competition with socially and politically powerful user groups. Second, conservationists like to champion areas that are well known and charismatic. For example, prior to ecoregional planning, The Nature Conservancy's (TNC) programs in many Rocky Mountain states of the United States rarely focused on conservation areas outside of scenic mountainous areas of the region. Third, the various government agencies and NGOs usually involved in land and water conservation have conflicting agendas and mandates, making a coordinated and strategic approach to the selection of conservation areas exceedingly difficult. Primary goals of this book, and this chapter in particular, are to ensure that regional plans for biological diversity emphasize a balanced conservation portfolio and to encourage all the entities on whom conservation success will depend to work together in a more coordinated fashion to achieve this goal.

Selecting a well-balanced portfolio of conservation areas is not, in itself, sufficient for conservation success. The portfolio needs to be more than a collection of independent conservation areas to which little thought has been given to important linkages among areas, the location of one area relative to another, and the intervening matrix of lands and waters between any two conservation areas. This line of thinking leads to the idea of a network of conservation areas, a concept that will be more fully developed in the second half of this chapter.

Conservation Areas

Although not explicitly used by many organizations, the term *conservation area* has many connotations and it is easily confused with other commonly used terms (Figure 8.1). For example, scientific literature commonly uses

Figure 8.1 Conservation areas are geographic areas of land and water specifically managed for the targets of biodiversity found within them. Many other terms are used around the world that either have the same or a similar meaning. Some of these terms imply areas that are legally protected, while others do not.

the term *reserve,* as in the field of reserve selection and design, with the implication that reserves are areas that are strictly managed for their natural values while other activities are largely prohibited. Conservation organizations such as The Nature Conservancy and the National Audubon Society often refer to lands they manage as *preserves, sites,* or *sanctuaries.* The Wildlands Project in the United States focuses on *core areas* as areas where the conservation of biodiversity, ecological integrity, wilderness, and other similar values take precedence over other values (Noss et al. 1999a).

There are also a number of terms related to governmental or United Nations' conventions or programs that imply conservation. For example, much of the international conservation community, especially under the auspices of the Convention on Biological Diversity (CBD; see Chapter 1), uses the term *protected areas* to denote areas that have been set aside for their natural and social values with legally demarcated boundaries and management objectives (e.g., national parks). Chapter 5 presented widely accepted definitions of different types of protected areas that have been promulgated by the World Conservation Union (IUCN) and the World Commission on Protected Areas (WCPA). In the United States, the Gap Analysis Program classifies lands according to their biodiversity management status, with those lands classified as status 1 or 2 denoting some sort of legal protection and implicitly regarded as conservation areas (see Chapter 5).

Biosphere reserves are areas of terrestrial and coastal ecosystems that are recognized under the United Nations Educational, Scientific, and Cultural

Organization's UNESCO's Man and the Biosphere (MAB) program as having an international biodiversity conservation function, a sustainable development function, and a logistic function (research, education, and monitoring) (see www.unesco.org/mab for details on this program). Finally, World Heritage Sites are those areas of outstanding cultural or natural significance that are designated under the auspices of UNESCO's World Heritage Convention. By signing this agreement, over 150 countries have pledged to conserve World Heritage Sites within their boundaries (see www.unesco.org/whc for details on the program and sites).

All of these terms can fit under the rubric of *conservation areas.* These are geographic areas of land or water (freshwater and marine) that harbor and are specifically managed for conserving features of biodiversity (target species, communities, or ecosystems) and the ecological processes that support them. Some of these areas are privately held, but many are in the public domain. Some have legal restrictions on uses, others do not. Some allow various types of human activity, including resource extraction, while others limit such activities. The key point here is that whether the area in question is privately held with a conservation easement that prohibits development or a strictly managed, legally designated wilderness area, the primary emphasis of a conservation area should be management that promotes the long-term viability and ecological integrity of specific conservation targets. Just as importantly, each conservation area should be viewed and managed in the context of the contribution it makes to a broader regional network of conservation lands and waters.

The degree to which conservation areas should preclude most human activities has been the subject of great debate among conservation biologists, conservation practitioners, and the environmental community at large. For example, a series of articles published in the *Wildlife Society Bulletin* concerning the management of old-growth forests in the northwestern United States (Dellasalla et al. 1996, Everett and Lehmkuhl 1996, 1997, Noss 1996c) and, more recently, several articles published in *Conservation Biology* on tropical forest conservation (Redford and Sanderson 2000, Schwartzman et al. 2000, Terborgh 2000) typify these debates. As I write this book, a great national debate in the United States is under way about whether to allow oil exploration in the Arctic National Wildlife Refuge. This debate includes an element of both values (e.g., human use of the land) and science (e.g., the impact of oil exploration on arctic ecosystems), but the two are commonly intertwined and confused. Nothing that will be written here will resolve these larger debates. What can be said here is that conservation areas need to be managed to meet the requirements of the target species and communities if they are to be sustained over the long haul. Some targets will demand greater restrictions on use than others. Conservation areas selected to sustain endemic plant species of the Karoo Succulent Desert region of South Africa will not have the same management requirements as those needed to perpetuate jaguars on the Osa Peninsula of Costa Rica.

All natural areas occur along a continuum—from limited human use and impact on one end, to substantial use and impact on the other. Although debatable, it is likely that most human activities have some measurable, negative impact on the composition, structure, or function of biological diversity (Redford and Richter 1999). Naturalness, a topic discussed in Chapter 7, can be used to assess the degree of human activity and impact on a conservation area. Povilitis (2002) recently proposed a set of indices that can help distinguish when a conservation area is more or less natural. Indices in the human behavior category that tended to make an area less natural included the use of fossil fuels, the presence of materials and structures that harm wildlife, air and water pollution, nonrecycled products, and the suppression of ecological processes. As a general rule, those areas that place some restrictions on various human uses, particularly ones that degrade natural resources, will have a substantially greater chance of providing a lasting legacy for biodiversity conservation. Conservation planners and biologists should provide natural resource managers and policy makers with the best possible information on how conservation area should be managed to best conserve their biodiversity. At the same time, planners and biologists should recognize the obvious: The degree to which any conservation area remains more or less natural will usually rest upon values held by a society and its governing institutions.

Noah Worked Two Jobs

This catchy phrase was the title of an important notion, advanced by biologists J. Michael Scott and Blair Csuti (1997) of the U.S. National Gap Analysis Program, that the selection of areas representative of a region's biological diversity should largely be a separate process from the design and management of those individual conservation areas. As Scott and Csuti noted, "The modern Noah has two jobs: a conscientious travel agent who ensures that all creatures have tickets and a master shipbuilder whose staterooms afford a comfortable passage for a long and difficult journey." The focus of this book is the former job—ensuring that the full array of biological diversity is represented in conservation areas. How those individual conservation areas are designed and managed involves a detailed assessment of threats, development of strategies to abate and mitigate those threats, and a consideration of other important issues, such as land ownership, boundaries of conservation areas, and mandates and policies for other nonconservation uses of lands and waters in the region.

Although Scott and Csuti (1997) were among the first to clearly articulate the difference between selecting conservation areas and designing their management, they were hardly alone in this view. In *Saving Nature's Legacy,* Noss and Cooperrider (1994) treated these subjects separately. Other conservation planning projects have found the selection/design dichotomy to be a useful one (Maddock and Benn 2000, Howard et al. 2000). Two major conservation organizations, The Nature Conservancy (2001) and the World Conservation Union (Davey 1998, Hockings et al. 2000), have elaborated on separate planning processes for the selection of conservation areas and their design and management. Although the emphasis of this book, and this chapter in particular, is on the selection of lands and waters for biodiversity conservation purposes, the topic of how those areas are designed and managed will be discussed in Chapter 13.

Even if Noah really only had one job—selecting conservation areas— that would be more than enough. Many criteria and principles have been advanced for selecting conservation areas. In the next section, I review and synthesize some of the more exemplary work in this field, with an eye toward reaching some conclusions on a few overarching principles.

Principles for Selecting Conservation Areas

Australian and British scientists have long been the recognized leaders in systematic conservation planning for biodiversity. Margules and Usher (1981) reviewed criteria for evaluating wildlife conservation values and determined that five were commonly used: diversity, rarity, naturalness, area, and threat of human interference. They observed the difficulty of comparing among conservation areas because of the factors used to evaluate and designate these areas, different discrepancies in definitions of factors, and a lack of consensus over which criteria were most important. They suggested the need for an overall index that combined several criteria, but noted the problem of combining factors that were both quantitative and qualitative. Subsequent reviews demonstrated the value of using multicriteria scoring procedures to evaluate sites for their conservation potential (Smith and Theberge 1986, Usher 1986). For example, Bolton and Specht (cited in Pressey and Nicholls 1989) used a "priority × diversity index" for reserve selection in Queensland, Australia. Until the initiation of ecoregional planning, TNC used a scoring system called the B Rank (Biodiversity Ranking) system to evaluate potential conservation areas (Table 8.1).

Pressey and Nicholls (1989) were among the first to demonstrate that these scoring systems have a serious limitation if the goal is to conserve examples of the full range of diversity of any given type of target in a region (e.g., conserve one or more examples of all ecosystem types in ecoregion X). If sites are conserved based on their scores for a particular criterion or multicriteria index, large numbers of sites will be required to achieve the goal of full representation of all target types. In essence, scoring procedures are an inefficient way to achieve representation because of duplication of targets among high-ranking sites. For example, several conservation areas that contained an endangered species might be ranked similarly high by a scoring procedure, and all of them might be included in a proposed system of conservation areas without regard to what other targets were included.

Table 8.1 *The Biodiversity Ranking System Formerly Used by The Nature Conservancy to Assess the Relative Biodiversity Value of Potential Conservation Areas*

Biodiversity Rank	Definition
B1	Outstanding significance, such as the only known occurrence of any target, the best or an excellent (A-ranked; see Chapter 7 for definition of ranks) occurrence of a G1 Target (see Chapter 4 for definition of G ranks), or a concentration of high-ranked (A- or B-ranked) occurrences of G1 or G2 Targets. Site should be viable and defensible for targets and ecological processes contained.
B2	Very high significance, such as one of the most outstanding occurrences of any community Target. Also includes areas containing any other (B-, C-, or D-ranked) occurrence of a G1 Target, a good (A- or B-ranked) occurrence of a G2 Target, an excellent (A-ranked) occurrence of a G3 Target, or a concentration of B-ranked G3 or C-ranked G2 Targets.
B3	High significance, such as any other (C- or D-ranked) occurrence of a G2 Target, a B-ranked occurrence of a G3 Target, an A-ranked occurrence of any community, or a concentration of A- or B-ranked occurrences of (G4 or G5) S1 (state-ranked) Targets.
B4	Moderate significance, such as a C-ranked occurrence of a G3 Target, a B-ranked occurrence of any community, an A- or B-ranked or only state (but at least C-ranked) occurrence of a (G4 or G5) S1 Target, an A-ranked occurrence of an S2 Target, or a concentration (4+) of good (B-ranked) S2 or excellent (A-ranked) S3 Targets.
B5	Of general biodiversity interest or open space.

Alternative approaches to reserve selection that overcame this ineffi-ciency in scoring procedures were first developed in Australia. Known as stepwise or iterative procedures (Kirkpatrick 1983, Margules et al. 1988), this type of approach is specifically directed at sampling the full range of attributes of some conservation target (e.g., different types of wetlands). It does so by first taking into account which attributes have already been represented in conservation areas and then proceeds in a stepwise fashion to select those sites or conservation areas with the greatest number of attributes not already represented. Pressey and Nicholls (1989) showed that these iterative approaches to reserve selection were more efficient than scoring procedures. They were able to represent all the desired con-servation targets or features in fewer reserves, covering less land, than scor-ing procedures using the same data.

Based on extensive subsequent work in conservation evaluation and reserve selection, Pressey and colleagues (1993) elaborated on three prin-ciples for systematic reserve selection for biodiversity conservation: com-plementarity, flexibility, and irreplaceability. Employing the principle of *complementarity* is essentially an application of the iterative or stepwise pro-cedures described above. It involves examining the existing conservation areas to determine which targets are already conserved within them, then proceeding in a stepwise fashion to select the next conservation area that contains the greatest number of targets not already represented in an exist-ing one. These steps are subsequently repeated until all targets are repre-sented in conservation areas at the desired levels. In this manner, each new conservation area that is selected is the most complementary one to the existing areas. A complementarity approach to selecting conservation areas results in an efficient portfolio, meaning that the portfolio represents the conservation targets at the required goal levels in the least number of sites or the smallest possible total area. Given the limited amount of resources that can be devoted to conservation areas, efficient designs are highly desirable as long as they do not compromise other goals.

The second principle of *flexibility* refers to the fact that in many regions there are several ways to combine potential conservation areas to best meet the desired conservation goals. For example, in an environmental province in western New South Wales, Australia, there were 24 possible conservation areas (referred to as sites or pastoral holdings in Table 8.2), and the goal was to represent each conservation target (17 different land systems) at least once in these conservation areas or sites (Pressey et al. 1994). As Table 8.2 demonstrates, there is a minimum number of sites nec-essary to represent all 17 land systems (five), but there are seven different

Table 8.2   *The Possible Number of Combinations of Sites (Potential Conservation Areas) and Combinations of Sites That Represent the Conservation Targets (17 Different Land Systems) in an Environmental Province in Western New South Wales, Australia*

Number of Sites	Possible Combinations	Representative Combinations
1	24	
2	276	
3	2,024	
4	10,626	
5	42,504	7
6	134,596	134
7	346,104	1,150
8	735,471	5,980
9	1,307,504	21,457
10	1,961,256	57,043

Reprinted from *Trends in Ecology and Evolution*, Volume 8, R. L. Pressey et al., Beyond opportunism: key principles for systematic reserve selection, pages 124–128, copyright 1993, with permission from Elsevier Science.

combinations of sites that will achieve that representation. As the number of sites increases, so does the number of potential representative combinations. In the real world, many of these sites will likely be unsuitable for conservation for a whole host of reasons. Among them is the admonishment from Chapter 7 that many plans don't give sufficient attention to integrity and viability, and to the factors that will allow long-term persistence of conservation targets. The important point from this example, however, is that this sort of flexibility is important for planners in the real world who must juggle competing demands for land use to achieve conservation goals, while trying to maximize the design of networks of conservation areas and minimizing costs. Yet, we must also recognize that many regions and ecoregions have been so extensively altered that there is little flexibility in what remains for conservation action.

The third principle for reserve selection is irreplaceability. Once goals have been set for conservation targets, *irreplaceability* can be defined in one of two ways (Pressey et al. 1994, Pressey 1999): The probability that a potential conservation area will be required as part of a portfolio or network of conservation areas that represents the conservation targets at the desired goal levels, or the extent to which options for achieving the conservation goals for the targets are reduced if a particular area is not available for conservation purposes.

Values for irreplaceability can range from totally irreplaceable (100%) to zero (Figure 8.2). An area is totally irreplaceable if it contains the only known occurrence of a conservation target. For example, the Devils Hole pupfish (*Cyprinodon diabolis*) is known from a single spring system in Ash Meadows, Nevada, in the southwestern United States (Chaplin et al. 2000), and therefore Ash Meadows should be included in a comprehensive

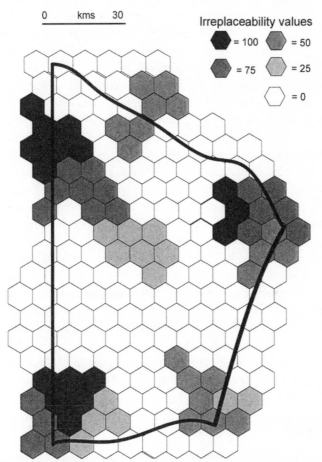

Figure 8.2 A hypothetical planning region with varying degrees of irreplaceability of conservation targets. Irreplaceability can be defined as the probability that it will be necessary to select a particular area to conserve a target species or ecosystem at a desired level. Alternatively, it can be defined as the extent to which opportunities for conservation are foreclosed if a particular area is not available. Values for irreplaceability can range from zero to 100, with 100 being an area that harbors, for example, the only known population of a species.

portfolio. Another way of making the point about irreplaceability would be to consider that if Ash Meadows were not available for conservation purposes, the options for conserving the Devils Hole pupfish would be nonexistent. At the other end of the spectrum are areas with lower irreplaceability values for which there are more options to meet a specified conservation goal and less impact on achieving of goals if that area is destroyed, degraded, or otherwise unavailable for conservation.

Irreplaceability can be estimated for a single type of conservation target or collectively for several conservation targets. For small numbers of targets and relatively small numbers of potential conservation areas, calculating irreplaceability is relatively straightforward. However, for the more usual situation in which there are large numbers of targets and areas, estimating irreplaceability becomes increasingly complex, even for the largest computers. For these situations, predictive statistical approaches to estimating irreplaceability have now been developed (Ferrier et al. 2000). The concept of irreplaceability is increasingly being used in regional biodiversity planning projects in Australia, South America, the United States, and South Africa. In New South Wales, Australia, it has been used as a tool in an interactive computer system with a GIS interface in government-facilitated negotiations concerning which forest parcels would be either reserved for logging concessions or set aside as conservation areas (Pressey 1998). In Chapter 9, I will explore how irreplaceability can be combined with information on threats to conservation areas to help set priorities for conservation action.

In more recent writings, Margules and Pressey (2000) and others have implied that a fourth principle should be considered in reserve selection, a concept I introduce in Chapter 9 on viability and integrity: persistence. *Persistence* refers to the notion that conservation areas should promote the long-term existence of the conservation targets that occur within them through a consideration of the ecological processes and habitat conditions that sustain population viability and ecological integrity.

The World Commission on Protected Areas has advanced five key characteristics or sets of criteria for the selection of conservation areas (Davey 1998):

1. *Representativeness, comprehensiveness, and balance.* Protected areas should represent the full range of biogeographic diversity within a region or country in as balanced a manner as possible (not underrepresenting some targets, while overrepresenting others).
2. *Adequacy.* Conservation targets within protected areas should have sufficiently high levels of viability or ecological integrity to persist over long time periods.

3. *Coherence and complementarity.* Each protected area needs to add value to a system of areas (i.e., it needs to represent conservation targets that are not conserved elsewhere).
4. *Consistency.* Management objectives and policies need to be consistently applied across a system of protected or conservation areas.
5. *Cost effectiveness, efficiency, and equity.* A minimum number of protected areas achieves overall conservation goals and distributes the benefits of these areas to local communities in an equitable fashion.

Compared to the principles of Pressey and colleagues, these criteria also focus on the concepts of complementarity and persistence (adequacy), while at the same time articulating some additional criteria on representation, consistency of management, and equity to local communities. Although this first criterion is relevant to the selection of conservation areas, the last two are more related to the management of such areas.

The Nature Conservancy uses six criteria in the selection of conservation areas in its ecoregional planning work (Groves et al. 2000a):

1. *Coarse-scale focus.* Represent coarse-scale targets (ecosystems, wide-ranging species) in conservation areas as a first step in systematic conservation planning, and determine the extent to which these targets "sweep" along other conservation targets.
2. *Representativeness.* Represent the full spectrum of communities and ecosystems that occur in a planning region within conservation areas.
3. *Efficiency.* Give priority to those areas that contain occurrences of several different conservation targets. (Note that this definition of efficiency differs somewhat from how the term has traditionally been used in the scientific literature on reserve selection and design.)
4. *Integration.* Give priority to areas containing high-quality occurrences of both terrestrial and aquatic (freshwater and/or coastal marine) targets.
5. *Functionality.* Ensure that all conservation areas contain targets that are viable or have sufficient integrity to persist over long time periods.
6. *Completeness.* Ensure that all conservation targets are represented in conservation areas at the desired goal levels to the greatest extent possible.

Notice that the criteria of representativeness, persistence (i.e., functionality), and efficiency emerge again as being important, along with some new criteria related to selecting coarse-scale targets first in the selection process and identifying areas that contain both terrestrial and aquatic targets.

The state and national governments in Australia have used three principles in the design of a national reserve system: comprehensiveness, adequacy, and representativeness (Commonwealth of Australia 1999). *Comprehensiveness* refers to the inclusion of samples of all ecosystem types within reserves at the bioregional scale. *Adequacy* refers to how much or how many of different ecosystems and species need to be included in the reserve system. *Representativeness* is defined as being comprehensive at a finer scale by incorporating with a reserve system the environmental variability within an ecosystem on a bioregional basis.

The Wildlands Project is a scientific program for conserving biological diversity at regional and continental scales in North America (Soulé and Terborgh 1999a, 1999b). The conceptual basis of these projects is the argument that keystone species, especially large carnivores, are essential to stabilizing prey and smaller predator populations and to helping sustain ecological diversity in many ecosystems. Large carnivore populations are maintained through a network of large and well-connected protected areas, referred to as core areas or wildlands. These areas are selected on the basis of three criteria (Noss et al. 1999a): representation, special elements, and focal species. Representation refers to conserving intact examples of each vegetation or habitat type across the environmental gradients in which they occur. Special elements are rare species and communities, pristine sites (e.g., roadless areas), and other unique features to a region (e.g., artesian springs) thought to have high conservation value. Focal species are conservation targets whose needs define answers to two questions: How large do conservation areas need to be? What should their configuration be? With the exception of some special elements (e.g., artesian springs), the three types of conservation targets used by Noss et al. (1999a) are consistent with those identified in Chapter 4 on targets.

Although these examples of principles and criteria for selecting conservation areas from Pressey and colleagues, the WCPA, the Australian national and state governments, The Nature Conservancy, and The Wildlands Project are by no means exhaustive, they certainly illustrate the current thinking in conservation planning for biodiversity. A synthesis of these ideas suggests three overarching principles of importance in selecting conservation areas:

1. *Representation or representativeness:* The need to represent occurrences of each community or ecosystem across the environmental gradients in which they occur in a system or portfolio of conservation areas.
2. *Persistence, functionality, or adequacy:* Synonyms for a system of conservation areas to promote the long-term persistence of conservation targets by maintaining high levels of viability and ecological integrity.

3. *Efficiency:* The ability to achieve the regional goals for biodiversity conservation in the least total area and/or the least number of conservation areas.

In many conservation planning situations, computerized algorithms are used as a tool to aid conservation planners in making these principles operational. After reviewing the use and utility of such algorithms below, I focus on the challenges of putting these overarching principles, as well as other important planning principles, into practice.

Site or Area Selection Algorithms

If the number of conservation targets in a region is small or the total area potentially available for conservation is not great, area selection is a relatively straightforward task. Often it is best accomplished by a set of experts in a room with maps and accompanying information regarding the locations of conservation targets, land ownership and management, and existing conservation areas. For example, in the Northern Tallgrass Prairie ecoregion in the Great Plains of the central United States, less than 4% of the ecoregion has not been converted to agriculture or urban areas. As a result, the remaining options for conservation are relatively limited. However, in situations where the number of conservation targets is larger, the goals for these targets are variable, and/or substantial areas remain that are potentially available for conservation (i.e., greater flexibility and less irreplaceability), selection becomes computationally challenging without the use of computer programs. Using computerized algorithms with a GIS as a tool can help, and fortunately there are many site or area selection algorithms available.

In a review of site selection approaches, Williams (1998) argued that the value of area selection methods is that they provide "sets of rules designed to achieve particular goals efficiently and with a transparency that aids accountability." The vehicle for applying these methods is computerized algorithms. Algorithms also enable planners to rapidly assess alternative portfolios of conservation areas (see flexibility above) through changes in conservation targets, goals, assessments of viability or integrity, or revisions to data sets. In addition, algorithms can deal effectively with incomplete information, and they can help in communicating alternatives or consequences of certain selections to planning team members, stakeholders, and the general public (Pressey 1998). Area selection methods can be and have been applied to hotspots of rarity, hotspots of diversity, sites identified on the basis of scoring systems, and complementary areas (Williams 1998). By far, the most work with area selection algorithms has been done in relation to identifying complementary sets of conservation areas. Scores of scientific papers have been published using area selection algorithms to identify a portfolio of complementary conservation areas (see Williams 1998 for citations).

Conservation planners use two general types of algorithms (Possingham et al. 1999, Scott and Sullivan 2000, Cabeza and Moilanen 2001): iterative (stepwise) or heuristic algorithms (rule-based approaches), and optimization algorithms (more formal mathematical approaches). Possingham et al. (1999) provided good examples of the differences, largely mathematical in nature, between iterative and optimizing algorithms. For relatively large data sets, optimizing algorithms provide more efficient solutions than heuristic approaches. In other words, optimizing algorithms will select a set of conservation areas that satisfy conservation objectives to the greatest degree possible in the least area or least number of conservation areas.

Several scientific papers compare the relative advantages and disadvantages of iterative and optimization algorithms (e.g., Pressey et al. 1996b, 1997, Possingham et al. 1999, Scott and Sullivan 2000). A detailed discussion of these differences is beyond the scope of this book. Suffice it to say that in general, heuristics are relatively fast and easy to understand. They basically operate in a stepwise manner, as described in the previous section under the principle of complementarity. Their primary disadvantage is that they do not guarantee an optimal solution to the efficiency problem, although in many cases they will provide near optimal solutions. In addition, a recent study by Araújo and Williams (2001) suggested that heuristic algorithms employing a complementary approach may preferentially select conservation areas on the margins or periphery of species ranges, the implications of which may or may not be negative for conservation (see Chapter 4).

Optimization algorithms, on the other hand, have the disadvantage of producing only a single solution, may fail with very large data sets, can take hours or days of computer processing time, and are mathematically complex, relying on Integer Linear Programming and mathematical programming techniques such as branch and bound methods. This complexity may lead to a "black box" effect on practitioners and planners who cannot understand the box and are therefore less likely to support solutions stemming from using these types of algorithms.

Some iterative algorithms emphasize sequential selection of new conservation areas that are richest in targets or features not already represented within conservation areas. These are referred to as "greedy" algorithms

(Williams 1998). Others preferentially select areas that contain rare species first and are known as rarity-based algorithms. Some of the better-known algorithms are WORLDMAP (Williams 1996), CODA (Bedward et al. 1992), C-Plan (Pressey et al. 1995), and Sites (Ball and Possingham 2000, Groves et al. 2000a). The first three of these are iterative algorithms, whereas the Sites algorithm uses optimizing methods. The choice of which algorithm to use will depend on the quality and quantity of data, the representation goal, the importance of an optimal solution, and the amount of time that can be devoted to analyses.

A related but more comprehensive set of tools is BioRap, a methodology and tools for rapid biodiversity appraisals and selection of priority conservation areas. The BioRap project was conducted by a consortium of Australian scientists from the Commonwealth Scientific and Industrial Research Organization (CSIRO), the Environmental Resources Information Network, the Australian National University, and the Great Barrier Reef Marine Park Authority. Its tools and methods are detailed in a four-volume handbook (Belbin et al. 1996, Boston 1996, Faith and Nicholls 1996, Noble 1996).

The ability to analyze large sets of biodiversity information with area selection algorithms has provided conservation scientists and planners with new insights into the selection of areas for conserving biodiversity. Beginning with a seminal paper published by British biologist John Prendergast and colleagues (1993), scientists have asked a number of provocative questions concerning the selection of conservation areas and strategies in general for conserving biodiversity:

- Are habitats that are species-rich for one taxonomic group also species-rich for other groups (or can one taxonomic group serve as a surrogate for another)?
- Does the distribution of rare species coincide with the distribution of species-rich areas?
- Do areas that are hotspots of richness for one taxonomic group overlap with "coldspots" (areas with few species) for other taxonomic groups?
- Can complementary sets of conservation areas for one taxonomic group "sweep" along another group?
- Do areas high in endemism for one taxonomic group overlap with species-rich areas of other taxonomic groups?

If the answer to any of these questions were yes, the job of conservation biologists would be made easier because they could focus on fewer conservation targets, knowing that conserving these targets would likely conserve other taxonomic groups. Unfortunately, the answer from numerous studies addressing these questions has largely been a resounding no (e.g., Kershaw et al. 1995, Williams et al. 1996, Kerr 1997, Pimm and Lawton 1998, van Jaarsveld et al. 1998, Pharo et al. 1999). These sobering results strongly imply that there are few shortcuts in conservation planning for biological diversity and that all available information on various taxonomic groups should be used in the development of regional conservation plans. As Margules and Pressey (2000) suggested, there are no best surrogates, and these studies support that finding.

Beyond area selection algorithms, multiobjective models are being developed and implemented for integrating political, economic, biological, land-use, and other sorts of information into a biodiversity conservation planning framework (e.g., Rothley 1999, California Resources Agency 2001, 2002). For example, the state of California is undertaking a project termed the California Legacy Project that promotes the conservation of five different types of natural resources: aquatic and terrestrial biodiversity, working landscapes (i.e., agricultural and ranch lands), watersheds, natural areas for education and recreation, and open space near urban areas. Information and tools developed by the project are intended to help decision makers throughout the state make the most sound and strategic investments in conservation. The multiple conservation goals make this project particularly challenging. Scientists associated with the project will use a wide range of models to develop the most optimal allocation of conservation resources, including ecosystem models, urban growth models, climate change models, and, ultimately, multicriteria decision models.

I cannot reiterate enough times that area selection algorithms are only a tool to help planners select and design a network of conservation areas. In some cases where there are very limited amounts of data or very limited options for conservation, these tools may not be helpful or could even be counterproductive. No team should accept at face value the results of algorithms as the final word in the selection of conservation areas. The reasoned knowledge and opinion of planning team members and experts in the region should always be used in conjuction with algorithms in selecting the most efficient and effective portfolio of conservation areas.

Data Issues in Selecting Conservation Areas

Planners need to consider several important data issues when selecting conservation areas. The first of these is the degree of classification or subdivision of coarse-scale targets. Pressey and Bedward (1991) analyzed 14 Australian studies and determined that as relatively coarse-scale targets such as land systems were defined at finer spatial scales, the percentage of these targets occurring in conservation areas declined. In an expanded study on the same topic, Pressey and Logan (1994) concluded that the degree to which conservation areas represent coarse-scale targets is related to both the level of subdivision and the conservation goals. Depending on the type of goals (e.g., presence/absence of a land class in a conservation area versus a percentage area of the land class occurring within conservation areas), the effect of subdivision could be to either increase or decrease the representation of targets within conservation areas. These findings suggest that assessments of how well a system of conservation areas purport to represent targets must be interpreted with caution. At the very least, such data will only be comparable across planning regions if the level of subdivision is consistent.

These findings also beg the question: What is the most appropriate level of classification or subdivision to be using with coarse-filter targets? Pressey and Logan (1994) suggested a very pragmatic answer: The most appropriate level is probably whatever is available at the time the conservation planning takes place. Ecoregional planning projects conducted by The Nature Conservancy in the United States suggested a somewhat different answer (Groves et al. 2000a). Initial planning efforts focused on using a relatively fine classification of vegetation communities referred to as plant associations in the United States National Vegetation Classification (Grossman et al. 1998; see Chapter 4). Subsequent efforts have placed emphasis on developing and using a coarser level of vegetation classification referred to by TNC as ecological systems (Groves et al. 2000a).

Pressey and colleagues (1999) also examined the effects of other data characteristics on the results of area selection algorithms, including the degree of nestedness of conservation targets, the rarity of targets, and the size variation in potential conservation areas. They found that increasing the nestedness of targets (i.e., more targets co-occurred within the same conservation area) reduced the number and total area of conservation areas needed to achieve a particular representation goal. Conversely, as the proportion of rare targets in the data sets increased, the number and total acreage of conservation areas increased. The implication for conservation planners is that representation of a given number of targets in a portfolio of conservation areas is likely to be more expensive (and less efficient) if the targets are distributed in a random fashion and if a significant proportion of the targets are rare.

Perhaps the most important data consideration in regional conservation planning is depicting potential conservation areas in terms of spatial data units on a map. Conservation planners have done this in many different ways. For example, TNC's Central Appalachian Forest Ecoregional Plan defined the boundaries of conservation areas, in part, on the basis of road-less blocks of forested habitat (see Figure 8.3, in color insert). A conservation plan for the Congo Basin used square grid cells (Figure 8.4). Still another regional conservation plan from the western United States repre-

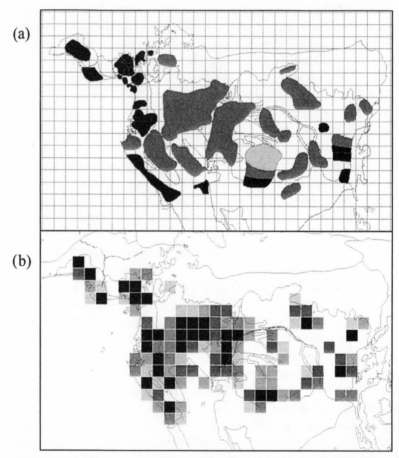

Figure 8.4 Proposed conservation areas in WWF's ecoregional vision for the Congo Basin. (a) A map of conservation opportunities in the Congo Basin superimposed on a one-degree grid cell and WWF ecoregional boundaries. (b) A map of the same data converted to grid-cell scores by sampling the (a) map 16 points per grid cell. Black represents high opportunity, dark gray is medium, and light gray is low. (From Blom et al. 2001. Reprinted by permission of WWF International.)

sented conservation areas as hexagons in a contiguous grid system developed by the Environmental Protection Agency (Figure 8.5). All of these are legitimate ways to depict conservation areas. Some have more detail than others, and some are consistent in size while others are not.

When conservation planners subdivide a regional planning area a priori into smaller units to which data and information on conservation targets,

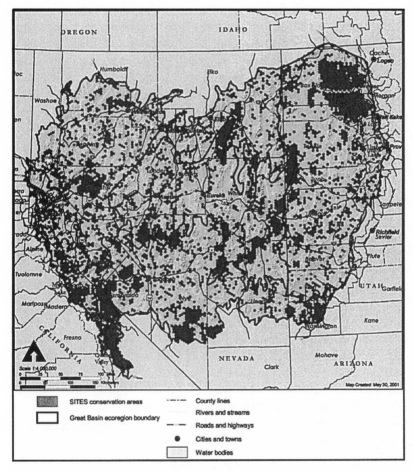

Figure 8.5   An example of using EPA's hexagonal grid system as selection units in conservation planning. A hexagonal grid system, originally developed by the U. S. Environmental Protection Agency, can also be used as selection units for conservation areas. This map portrays conservation areas selected on the basis of hexagons. Each hexagonal units is 2000 hectares in size (approximately 5000 acres). (From Nachlinger et al. 2001)

viability and ecological integrity, and other ancillary data sets (e.g., elevation, human population density) can be attributed, these units are referred to as *selection units* (Pressey and Logan 1998). These selection units allow the identification of generalized conservation areas or areas of biodiversity significance (Groves et al. 2000a). Detailed delineation of boundaries for these areas usually occurs at a later stage in the planning process when biologists and planners consider the design and management of individual conservation areas (see Chapter 13).

A wide variety of selection units have been used in regional conservation planning, including, but not limited to, watersheds of various sizes (Davis et al. 1999), rectangular grid cells (Lombard et al. 1997), hexagons (Csuti et al. 1997), cadastral or farm parcels (Heijnis et al. 1999), and forestry management compartments (Pressey 1998). From an analytical standpoint, these selection units facilitate comparisons among alternative portfolios of conservation areas, support information management, ensure that the selection process is repeatable and transparent, and can effectively "hide" sensitive locational information on endangered species and their habitats. On the other hand, the arbitrary nature of these selection units, the lack of detailed boundary information, and the implied analytical approach behind their use have elicited negative responses from some conservation practitioners and natural resource managers. In addition, because the actual targets occur only over a portion of the selection unit, the units tend to exaggerate the amount of area that is actually needed for conservation purposes. Pressey and Logan (1998) discussed 12 issues in detail that conservation planners should consider when choosing a selection unit. Some of the more important issues are the number of units that can be handled by the analysis, the size of the selection unit relative to the conservation targets and ecological processes, the size of selection units in relation to their mappability, and considerations for public presentations.

Whether or not selection units are used in the process for identifying conservation areas depends on several factors. In regions where a great deal of information is known about the distribution of biodiversity, or where a portfolio of conservation areas is identified largely through an experts workshop, planners may choose to select and delineate boundaries for conservation areas largely on the basis of natural features and the occurrences of conservation targets themselves (e.g., World Wildlife Fund 1997, The Nature Conservancy 1998, Kautz and Cox 2001). However, for many parts of the world where the status and distribution of biodiversity is poorly known or where the complexity of the planning situation requires a computer-based approach, the use of selection units to identify generalized areas of biodiversity significance is a sound and practical approach.

Putting the Principles into Practice

Lessons from conservation biology suggest a number of potentially useful principles, such as complementarity (and efficiency that results from taking a complementary approach), representation, persistence, and flexibility. Algorithms are a useful tool for putting these principles into practice. A critical question, of course, is: How useful are these principles and algorithms in the real world? After nearly 20 years of development, the answer, according to British ecologist John Prendergast and colleagues (1999), is that the gaps between conservation science and conservation practice in the selection of nature reserves (conservation areas) remain substantial. However, there have been changes since Prendergast and colleagues published these observations. The principles of representation and complementarity are seeing wider application in both the governmental and nongovernmental organizations in Australia, South Africa, and the United States. In the continental United States alone, TNC will complete biodiversity conservation plans for every ecoregion by 2003. These plans will emphasize the principle of representation of community and ecosystem targets in conservation areas and will use the principle of complementarity to select many of these conservation areas. Figure 8.6 (in the color insert), taken from a TNC ecoregional plan, demonstrates how data on the distribution of terrestrial ecosystems can be combined with data on environmental attributes (see Figure 4.6) to produce a portfolio that emphasizes the principle of representation. The portfolio of conservation areas, assembled with the aid of an area selection algorithm, is shown in Figure 8.7. The World Wildlife Fund is developing biodiversity visions for numerous ecoregions around the world (i.e., The Global 200), and these visions are also emphasizing the principle of ecosystem representation, along with the persistence of species and ecological processes, and the conservation of functional landscapes.

The principle of flexibility is of more limited value, for several reasons. First, in many regions of the world, there is little or no flexibility on where new conservation areas can be established. The opportunities for conservation have been limited by the conversion of natural habitats to urbanization, industrialization, and agriculture. Second, as I discussed in Chapter 7 on viability and integrity, much of the early work in reserve selection and

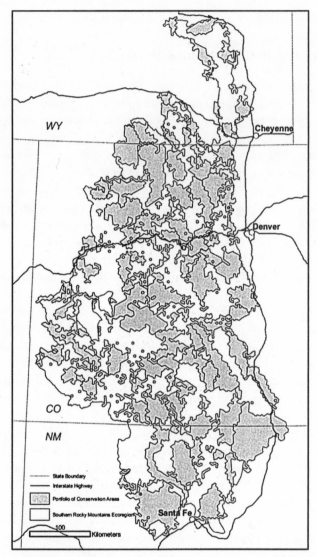

Figure 8.7 A portfolio of conservation areas for the Southern Rocky Mountains ecoregion. The conservation planning effort identified 188 different conservation areas covering approximately 50% of the ecoregion. Conservation areas ranged in size from approximately 1214 hectares (3000 acres) to over 404,000 hectares (1 million acres). (From Southern Rocky Mountains Ecoregional Assessment Team 2001.)

design paid little attention to the requirements of conservation targets for long-term persistence. In many, if not most, cases, if conservation targets are to persist over the long term, conservation areas will need to be considerably larger and more actively managed to sustain ecological processes and disturbance regimes than they have been in the past. The real costs of this management, along with bringing much larger areas under conservation attention, will almost certainly lower the flexibility of conservation practitioners in designing networks of areas to conserve biodiversity. In addition, paying more attention to persistence, viability, and integrity will lead to less efficient designs of networks of conservation areas (Pressey and Logan 1998, Margules and Pressey 2000). Although it will clearly remain desirable for conservation organizations and government agencies to achieve the most conservation at the lowest cost in either acquisition or management activities (i.e., to be efficient), there appear to be no good reasons for trading off gains in the long-term persistence of biodiversity for short-term gains in efficiency.

Like the planning principles discussed above, algorithms are increasingly being used to help identify important areas for biodiversity conservation in regional planning exercises, especially in parts of the United States, Australia, and South Africa. Yet, except for Australia, where these algorithms were developed in many cases by scientists who work for natural resource agencies, the adoption of these tools is far from widespread. Prendergast et al. (1999) suggested that there are several reasons why these algorithms have not been more widely adopted as planning tools. In many cases, land managers are simply unaware of them; they remain known primarily to academic scientists through scientific journals. In other cases where managers are aware of them, their use is precluded by low levels of funding, a lack of understanding of the purpose of the tools, or general opposition to a systematic approach to conservation. Experiences in TNC's ecoregional planning projects suggest that there may be another reason as well (Groves et al. 2002). Managers and conservation practitioners who do not understand how the algorithms work may be distrustful of their results and less likely to support the implementation of a regional conservation plan based on such results.

In spite of these shortcomings, area selection algorithms have proven to be useful tools. But conservation planners need to keep in mind that they are tools, the results of which must be interpreted and applied with ample caution. As Prendergast et al. (1999) observed, they are not a panacea; however, area selection algorithms and other similar models are becoming increasingly sophisticated and more representative of real–world situations.

For example, Margules and Pressey (2000) identified four spatial constraints on the selection of conservation areas that have now been incorporated into area selection algorithms: costs, commitments, masks, and preferences.

Costs refer to the fact that it is now possible to measure the opportunity costs in terms of foregone production of some lands in favor of biodiversity conservation. Commitments are areas in a planning region, such as existing conservation areas, that have already been or will be committed to conservation purposes. Masks are areas that are excluded from consideration as conservation areas, typically agricultural areas, that have been committed to some sort of development (Howard et al. 2000) or areas that are not feasibly restorable for the purposes of biodiversity conservation (Heijnis et al. 1999). Preferences are areas that a priori are preferred over others for conservation, such as areas of low human populations or areas previously identified as containing important conservation targets.

All four of these spatial constraints were incorporated into an area selection algorithm for a biodiversity conservation plan in Papua New Guinea (Nix et al. 2000). More recently, researchers from South Africa (Reyers et al. 2002) have modified area selection algorithms to incorporate the additional features of environmental gradients and turnover in species composition (beta diversity; see Chapter 1), while Briers (2002) has incorporated connectivity among conservation areas into these algorithms.

Perhaps the most important ingredient to putting the principles of conservation planning into practice is ensuring that those individuals who have a major stake in the outcome understand, are engaged in, and are supportive of both the process and the results. Lacking such engagement and support, even the best-intended applications of the principles outlined here may have limited success. In NGOs, scientific staff members are usually the ones who have primarily participated in the planning process to this juncture. However, it will usually be other staff, such as program directors, government relation specialists, and communication and development personnel, who will be heavily involved in the implementation of the plan's results. Therefore, it is critical that they have a general understanding of the planning process and are supportive of the conservation areas selected. In situations where area selection algorithms have been used to help the planning team, extra efforts may be needed to explain these tools to overcome the "black box" effect they sometimes have on team members. Similarly, in governmental organizations, professional planners and various types of scientists will be heavily involved in the preparation of natural resource management plans. Yet it is line officers and managers, such as district rangers in the U.S. Forest Service, who will make the most significant decisions about plan implementation. As with the NGO community, these officers and managers must clearly understand the planning process in order to be supportive of taking conservation actions that are identified.

Landscape Emphasis

One of the great debates in conservation biology was known as SLOSS—single large or several small reserves (Noss and Cooperrider 1994). The issue was whether it was better to conserve biodiversity in a single large reserve or in several small reserves. Although many scientific papers were published on the topic, in the end the debate was recognized by many as unproductive and intractable on several counts. First, the size and number of conservation areas will depend on the needs of the conservation targets. Second, given the rapid and substantial losses of natural habitats, it remains impractical to think that conservationists and natural resource managers would ever have the luxury of making such choices anyway. Third, biologists from both sides of the debate concluded that both large and many conservation areas will be needed to conserve biological diversity (Soulé and Simberloff 1986).

Recall from Chapter 4 that biodiversity occurs at multiple spatial scales and multiple levels of biological organization (see Figure 4.1). On the basis of these observations, Poiani et al. (2000) identified three types of functional conservation areas: functional sites, functional landscapes, and functional networks. Functional sites are designed to conserve a small number of ecosystem-, community-, or species-level targets at one or two scales below the regional scale. In the United States, early TNC preserves that were established to protect imperiled local-scale species or communities are examples of functional conservation sites. Functional landscapes, by contrast, seek to conserve a large number of ecosystem, community, or species targets at all scales below regional. The conservation targets are intended to represent or serve as surrogates for all the elements of biodiversity at the landscape scale. Functional networks are an integrated system of functional sites and functional landscapes designed to conserve regional-scale species such as wide-ranging carnivores.

Ecological theory, corroborated by some empirical findings, suggests that larger spatial scales, often referred to as the landscape scale, should be the preferred scale for selecting conservation areas (Schwartz 1999, Poiani et al. 2000, Noss 2002). These landscape areas are likely to harbor greater numbers of species, contain larger populations of species that have a higher likelihood of maintaining population viability, minimize the deleterious edge effects of smaller conservation areas, conserve species with larger area requirements, and are generally more capable of sustaining ecological processes and natural disturbance regimes (Schwartz 1999, Soulé and Terborgh 1999). In addition, in areas that are high in beta diversity, meaning that there is high turnover in species composition along relatively short distances, a landscape approach will help capture this beta diversity. This preponderance of evidence in favor of identifying and selecting conservation areas at the landscape scale is undeniable. Conservation planners should heed this advice and emphasize the selection and establishment of Poiani et al.'s (2000) functional landscapes whenever possible.

Yet in many parts of the world, the high degree of habitat loss has left mostly small patches of natural habitat and few landscape-scale conservation areas. In these situations, conservationists should not abandon small conservation areas, because they can be effective for conserving many conservation targets (Shafer 1995, Schwartz 1999). Many species are, in fact, naturally restricted to small patches (e.g., cliff habitats, Carolina bays of the southeastern United States, serpentine soils). Turner and Corlett (1996) argued that small patches of tropical forest are likely to conserve a great deal of diversity for decades, even though they may no longer function to conserve animal species with large home ranges. Similarly, remnant patches of prairie habitat in the central United States continue to conserve a high diversity of plant and animal species, and few truly endangered species are known in the region (Schwartz 1999). In other cases, it may be possible to combine small conservation areas into larger assemblages with corridors and buffer zones (e.g., Gurd et al. 2001 for mammals in eastern United States). Perhaps the best advice with regard to size in the selection of conservation areas comes from ecologist Mark Schwartz (1999): "More is better" but "small is not bad." In all likelihood, the most effective conservation programs will continue to identify, establish, and maintain conservation areas at a variety of spatial scales that can collectively support a full array of species, community, and ecosystem targets.

At the outset of this chapter, I noted two tasks: selecting a portfolio of conservation areas and designing these into a network of conservation areas. No matter how big or small conservation areas are, biologists and planners must address the interrelationships among them. Intuitively, most conservationists understand that a system of conservation areas, each appearing as an island in a sea of non-natural cover, is not likely to be very successful in the long run. Consequently, planners must address such issues as how close or far apart conservation areas should be located from one another, to what extent they should be connected, and what the land surrounding these areas should look like. This next section will discuss issues and make recommendations for transforming a portfolio of conservation areas into a true conservation network.

Networks of Conservation Areas

As nearly everyone recognizes, natural landscapes around the world are increasingly fragmented by human-dominated landscapes. In some parts of the world, such as the Great Plains in the central United States, habitat loss to agriculture has been so severe that even the fragments of natural habitat are often difficult to find. In other areas, such as the Cascade or Coastal Range Mountains in the Pacific Northwest of the United States, the checkerboard pattern of clearcut timber harvest stands out as a more obvious example of habitat fragmentation. Biologists have long recognized the potential impact of habitat loss and fragmentation on the selection and design of conservation areas. Jared Diamond (1975) first proposed a set of principles for faunal conservation based primarily on the theory of island biogeography (Figure 8.8). Those principles stated that, all else being equal:

A: Larger reserves (conservation areas) are better than small ones.
B: A single large reserve is better than a group of small ones of similar area.
C: Reserves closer together are better than reserves far apart.
D: Reserves clustered compactly are better than reserves in a line.
E: Reserves connected by corridors are better than unconnected ones.
F: Round reserves are better than long, thin ones.

These principles generated a great deal of debate in the scientific community, especially principle B, which initiated the previously mentioned SLOSS debate. Nevertheless, a number of these principles have stood the test of time. In the conservation strategy that was adopted for the northern spotted owl (*Strix occidentalis caurina*), Jack Ward Thomas and colleagues (1990) noted five concepts of reserve design, which were later adopted and expanded upon by Noss and Cooperider (1994) and Noss et al. (1997). Several of these guidelines have their origins in Diamond's original thinking about reserve design. Although the first two guidelines have already been considered in this chapter and others, taken together, they have much to offer conservation planners in envisioning a network of

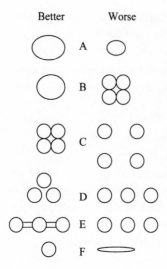

Figure 8.8   Principles for the design of faunal reserves. These principles were first advanced by Diamond (1975) on the basis of the theory of island biogeography. Although the second principle (B) generated a great deal of debate in ecology (known as SLOSS—single large or several small reserves), most of these principles have stood the test of time and have been incorporated into today's thinking about the design of networks of conservation areas. (See the text for the wording of the six principles.) (Modified from *Biological Conservation*, Volume 7, J. Diamond, The island dilemma: lessons of modern biogeographic studies for the design of natural preserves, pages 129–146, copyright © 1975, with permission from Elsevier Science.)

conservation areas in which the whole is greater than the sum of the individual parts.

1. *Species well-distributed across their native range are less susceptible to extinction than species confined to small portions of their range.* Species that are distributed in multiple populations across the variety of environmental regimes and habitats they naturally occupy will be less susceptible to the stochastic processes that can lead to extinction (see Chapter 7). In any given year, some populations may be subject to natural catastrophes such as floods and fire, abnormally high levels of predation, or human-related threats such as habitat loss and degradation. At the same time, however, if there are a sufficient number of populations appropriately distributed, then the species will be in the best possible position to hedge its bets against extinction. In essence, this is the rationale for the setting of conservation goals in Chapter 6.

2. *Large blocks of habitat containing large populations of target species are superior to small blocks of habitat containing small populations.* A previous section of this chapter discussed the advantages of placing emphasis on landscape-scale conservation areas. Not only will larger populations probably be superior to smaller ones as this guideline indicates, but larger blocks of habitat are more likely to contain area-sensitive species and intact disturbance regimes and ecological processes, which, in turn, are more likely to support conservation targets with higher levels of viability and integrity.

3. *Blocks of habitat close together are better than blocks far apart.* The fundamental premise behind this guideline is that the closer patches or blocks of

suitable habitat are located relative to one another, the more likely it is that individuals will be able to move among these blocks and patches. This increased movement allows for the persistence of metapopulations or effectively increases the size of a single, interbreeding population. The reasoning behind this guideline is that individual animals dispersing or traversing unsuitable habitat often experience high levels of mortality. Recent simulation studies by Fahrig (2001) support this notion, suggesting that second to reproductive rates, the most important factor affecting extinction is the rate of emigration. Successful emigration, in turn, can be increased through improvements to matrix habitat and/or by reducing the amount of matrix habitat that species must move through.

Prevailing wisdom among conservationists over at least the last 20 years has been to follow this guideline and locate new conservation areas as close as possible to existing areas. Some of the area selection algorithms discussed previously (e.g., Sites) have features that promote this idea of close proximity. But as biologist Shafer (2001) recently pointed out, there is an alternative viewpoint. Putting some distance between conservation areas, so that populations or occurrences of conservation targets are effectively independent of one another, has the advantage of safeguarding against natural catastrophes. This may be especially important for threatened or endangered species where a single event such as a storm or disease could catastrophically drive a species to extinction.

4. *Habitat in contiguous blocks is better than fragmented habitat.* Ecologist Larry Harris from the University of Florida was one of the first scientists to address the deleterious effects of habitat fragmentation on biota in his 1984 book *The Fragmented Forest.* Since that time, the effects of habitat fragmentation have been widely studied in many species and ecosystems around the world. In an overview of habitat fragmentation, Noss and Csuti (1997) considered fragmentation to have two components: a loss or reduction in a particular type of habitat and a division of the remaining habitat into smaller, more isolated patches.

Among the negative effects of fragmentation are elimination of some species by chance, reduction in movement and dispersal for others leading to isolation, reduction in population size leading to local extinctions in some cases, introduction of competitor and predators from adjacent disturbed landscapes, and disruption of ecological processes. The patchwork landscape resulting from clearcut timber harvest that is evident from an airplane in the mountainous regions of the Pacific Northwest is often pointed to as the "poster child" for the negative effects of habitat fragmentation. All else being equal, fragmentation studies in these old-growth forest ecosystems and elsewhere suggest that it is far better to conserve these forests in one or a few large blocks of habitat than small, isolated patches. Simulation studies by Fahrig (1997, 2002) temper the implication, however, that the arrangements of habitat patches can mitigate for overall habitat loss. Her studies suggest, in fact, that the negative effects of habitat loss far outweigh those of fragmentation and that the deleterious effects of fragmentation that lead to local extinction of species only do so when habitat has been reduced below a 20% threshold.

5. *Blocks of habitat that are roadless or otherwise inaccessible to humans are better than roaded and accessible blocks.* Roads and the maintenance of roads are known to have a number of negative effects on species, communities, and ecosystems ranging from increased air and water pollution to the spread of exotic species and direct and indirect mortality (see reviews by Jones et al. 2000 and Trombulak and Frissell 2000). Two recent studies in the United States examined the values of U.S. Forest Service roadless lands for conserving biodiversity. DeVelice and Martin (2001) assessed roadless, national forest lands across 83 ecoregions in Alaska and the eastern and western continental United States. They concluded that including these lands in a network of conservation areas would increase representation of biophysical or environmental features not currently represented sufficiently in a reserve system, increase the coverage of lower-elevation conservation areas, and increase the number of spatially large conservation areas necessary for wide-ranging and area-sensitive species to survive.

Strittholt and Dellasala (2001) assessed the importance of National Forest roadless lands for their contribution to biodiversity conservation in a single ecoregion, the Klamath–Siskiyou in northern California and southern Oregon. Their findings were similar to those of DeVelice and Martin (2001). Roadless areas complemented the biodiversity found in existing wilderness areas, adding a substantial number of biophysical habitat types, nearly a 33% increase in special elements (e.g., occurrences of rare species), and over a 40% increase in watersheds important to conserving aquatic biodiversity. These studies and others clearly show the conservation value of areas that are relatively free of roads and, conversely, the negative impact roads can have on biodiversity.

In Chapter 7 on viability and integrity, networks of roads were a key variable used in suitability indices to indirectly assess the viability and integrity of potential conservation areas. When planners and conservationists have such choices or flexibility, roadless areas are likely to maintain species populations with higher viability and contain examples of communities and ecosystems with higher ecological integrity than roaded areas. However, it is important for conservation biologists and planners to be able to identify how much and which portions of these roadless areas are most important in their contributions to biodiversity conservation.

6. *Interconnected blocks of habitat are better than isolated blocks, and dispersing individuals travel more easily through habitat resembling that preferred by the species in question.* Concerns over habitat fragmentation led to the development of ideas on how to best link or connect such fragments of habitats. Whether or not linkages or corridors, as they often are called, are beneficial in mitigating the effects of habitat fragmentation remains a controversial issue in ecology and conservation biology.

Corridors and Linkages

Natural landscapes are generally more connected than landscapes altered by humans, and corridors are essentially a strategy to retain or enhance this natural connectivity. Therefore, those who would destroy the last remnants of natural connectivity should bear the burden of proof of proving that corridor destruction will not harm target populations. (Beier and Noss 1998)

Defenders of landscape connectivity in general and wildlife corridors in particular sometimes fall back on the naturalistic premise that the pre-agricultural world was naturally connected and of a piece. . . . Such an argument is too facile, however, for there are times when the restoration or imposition of connectivity may be more harmful than fragmentation. (Dobson et al. 1999)

As these quotes suggest, the promotion of corridors and connectivity to mitigate or abate the effects of habitat fragmentation is far from a universally accepted conservation strategy (see Simberloff et al. 1992 for an early review of the potential detriments of corridors). Before addressing the controversies about corridors, I will explain exactly what is meant by corridors and connectivity, how they can enhance biodiversity conservation, and the scales at which they can be considered. The discussion that follows is almost exclusively terrestrial in focus. However, connectivity is clearly a critical issue in the conservation of freshwater ecosystems (see Chapter 10), especially for migratory fishes and other migratory aquatic fauna. Similarly, Chapter 11 highlights the growing interest in connectivity among marine conservation areas as well as the need to think critically about the connections among terrestrial, freshwater, and coastal marine ecosystems.

Corridors have been defined in many different, and often confusing, ways (Dobson et al. 1999). For example, artificial habitats, such as power-lines, greenbelts or greenways (Flink and Searns 1993), and railroad rights-of-way, have been referred to as corridors. Natural habitats that do not necessarily connect habitat patches, such as some riparian areas, have also been subsumed under the rubric of corridors. For the purposes of regional conservation planning, two definitions have been proposed that appear most useful. First, a *corridor* can be defined as a linear landscape element that provides for movement among habitat patches, but not necessarily reproduction (Rosenberg et al. 1997). Second, in a review article on corridors, Beier and Noss (1998) defined a corridor as "a linear habitat in a dissimilar matrix, that connects two or more larger blocks of habitat and that is proposed for conservation on the grounds that it will enhance or maintain the viability of specific wildlife populations in the habitat blocks [patches]." The key point of these definitions is that corridors are relatively linear features that connect habitat patches and enhance the movement of specific species through them.

Some authors prefer to use the words "linkages" and "landscape linkages" in place of corridors (e.g., Noss and Cooperrider 1994, Bennett 1999). These same authors assert correctly that the conservation goal should be to improve connectivity among isolated habitat patches and that corridors are only one mechanism for doing so. For example, improving the quality of matrix habitat surrounding habitat patches could improve connectivity for many species (Noss and Cooperrider 1994, Fahrig 2001) without specifically establishing a corridor. The remainder of the discussion in this section will focus on the conservation value of corridors (used synonymously with linkages) for linking patches of habitat. I discussed the issue of improving the quality of matrix habitat in Chapter 7 under the topic of landscape context, one of the three criteria used to assess viability and integrity, and will address it further below (see "The Semi-Natural Matrix").

Noss (1991) described three types of corridors that occur at different spatial scales: fencerow, landscape mosaic, and regional. Fencerow corridors connect small patches of habitat that are in close proximity and often consist of narrow rows of trees or shrubs. Britain's hedgerows that connect small, isolated woodlots are a good example of this type of corridor, which generally benefits animals with small home ranges or relatively short daily movements and animals that prefer habitat edges (e.g., squirrels and other small mammals). Landscape mosaic corridors connect major landscape features and are large enough to conserve both edge species and species whose habitat requirements are more restricted to interior habitats, such as some neotropical migratory birds (e.g., Faaborg et al. 1993). Regional-scale corridors are spatially the largest type of corridor, connecting conservation areas within a regional portfolio or network of such areas. For the purposes of this book, this is the corridor of interest for regional conservation planning.

Even among the detractors of corridors, nearly all scientists agree that functional corridors must be designed with specific species, landscapes, and ecological processes in mind, and that the goals of any corridor must be explicitly stated and analyzed (Noss and Cooperrider 1994, Dobson et al. 1999). Corridors should serve two purposes: (1) ensure the daily or seasonal movements of particular species so that all of a species' life history requirements are met and (2) facilitate dispersal from a natal site to the home range of a breeding adult (Dobson et al. 1999). It is this latter function that is of most interest from a regional planning perspective.

The most extensive and useful reference to date on corridors and linkages is a publication by IUCN's Forest Conservation Program, *Linkages in the Landscape*. In this book, author Andrew Bennett (1999) summarizes the conditions in which corridors are most likely to be effective in promoting connectivity:

- Significant parts of the landscape have been highly modified.
- Species are habitat specialists or depend upon undisturbed habitats.
- Maintaining ecological processes requires some continuity of habitat.
- Species have limited movement capability in relation to the distances among suitable habitat patches.
- A goal is to maintain exchanges of individuals among adjacent populations or metapopulations.

Bennett (1999) also provided a checklist of issues that biologists should be concerned with in the design and management of corridors from a wildlife conservation perspective. Here is an abridged version of that checklist: purpose of the proposed linkage, current status and tenure of the linkage, species the linkage purports to benefit, requirements of the species in relation to the proposed linkage, conditions necessary for the linkage to be functional, management and monitoring of the linkage, and lessons learned that will benefit others.

Several examples of proposed and actual corridors illustrate how they provide connectivity at a regional scale among conservation areas. Noss and colleagues have been some of the strongest proponents for establishing such linkages. One of the first proposed linkages was Pinhook Swamp along the Georgia–Florida border, which was proposed to provide connectivity between the Okefenokee National Wildlife Refuge and the Osceola National Forest (Figure 8.9). This linkage was to benefit wide-ranging species such as the black bear (*Ursus americana*) and red-cockaded woodpecker (*Picoides borealis*). More recently, Noss and colleagues (1999b) prepared a biodiversity conservation plan for the Klamath–Siskiyou ecoregion. This plan identified specific aquatic and terrestrial linkage zones for wide-ranging species such as the fisher (*Martes pennanti*) and salmonid fishes (Figure 8.10). In Costa Rica, government and conservation organizations

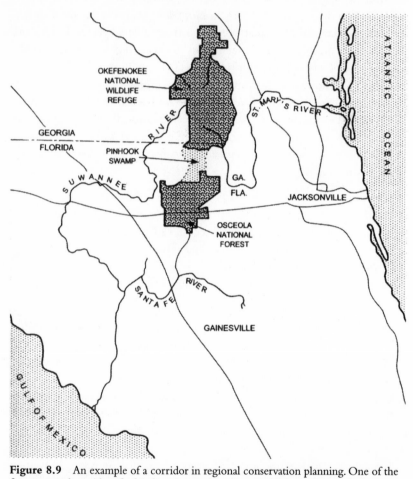

Figure 8.9 An example of a corridor in regional conservation planning. One of the first proposed corridors for biodiversity conservation was Pinhook Swamp between Okefenokee National Wildlife Refuge and the Osceola National Forest in the southeastern U.S. (From Nodes, networks, and MUMs: preserving diversity at all scales, R. Noss and L. Harris. *Environmental Management* 10:299–309, 1986, copyright © Springer-Verlag.)

worked together to establish a corridor between La Selva Biological Station in the Atlantic Lowlands and Braulio-Carillo National Park along the continental divide (Bennett 1999). This corridor provides migratory habitat for over 75 bird species that exhibit altitudinal migrations and conserves nearly the entire watershed for two major river systems.

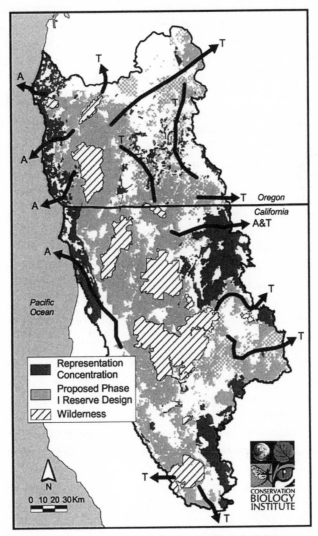

Figure 8.10   An ecoregional plan for the Klamath–Siskiyou ecoregion. This plan identifies important linkages for terrestrial (T) and aquatic (A) conservation targets. (From Noss et al. 1999b. Reprinted by permission of the Natural Areas Association.)

Hoctor et al. (2001) proposed a linked reserve system for the state of Florida in which the linkages help ensure, in principle, the long-term viability of wide-ranging species such as the Florida panther (*Felis concolor*) and the black bear. The state of Maryland is designing a similar system of green infrastructure hubs (conservation areas) and corridors as part of a regional biodiversity conservation strategy (Weber and Wolf 2000). Maddock and Benn (2000) proposed important linkages among existing conservation areas in Northern Zululand, South Africa, to enhance the functionality of ecological processes in the different vegetation types found within these conservation areas.

Perhaps the boldest proposed corridor project to date is the Mesoamerican Biological Corridor (Kaiser 2001), which proposes to link parks and other managed tropical forestlands from Chiapas, Mexico, to Darian, Panama. Despite nearly national US$1 billion in governmental aid to fund the project, progress in establishing the corridors themselves remains slow and modest. Additional examples of corridor projects are provided in Bennett (1999).

As mentioned earlier, whether or not corridors provide conservation benefit is a debatable topic. Although the empirical evidence is thin, Beier and Noss (1998), in an extensive review of corridors, documented 10 studies that provided persuasive evidence of cases in which corridors improved connectivity and presumably the viability of populations connected by corridors. In addition, they pointed out that no studies have demonstrated negative effects from corridors. Nevertheless, a number of tradeoffs should be considered with regard to proposed corridors (Dobson et al. 1999):

- When a fixed amount of land or resources is available for conservation, establishing a corridor must often be weighed against the cost of enlarging an existing conservation area (Rosenberg et al. 1997). National Forest management plans in the United States often present such tradeoffs in the various land management options that are proposed in these plans. Recent work by Fahrig (2001) suggested that establishing a corridor instead of enlarging a conservation area is usually a poor tradeoff from the perspective of biodiversity conservation.
- There are real costs associated with the management of corridors that must be evaluated relative to the costs of maintaining core conservation areas, including restoration costs.
- Because corridors often contain a great deal of edge habitat, there may be negative effects on species using these corridors, such as invasions by exotic species, and increased predation and parasitism on native species.

• Corridors may allow the undesired movement of pests, fire, and disease across otherwise isolated habitat patches.

The jury is still out on the conservation benefits of corridors. The evidence to date suggests that conservation planners must weigh several issues. First, corridors should be designed to benefit specific conservation targets (usually species) and the ecological processes that sustain these targets. Second, the proposed biological benefits of a corridor need to be carefully analyzed against what could be detrimental effects, including movements of disease and unwanted species. Third, there may be other strategies that should be considered for achieving the desired benefits of a corridor, such as artificial movement of species or improvements in the overall quality of matrix habitats. Finally, and most importantly, if choices exist between enlarging existing conservation areas or establishing a corridor, recent studies suggest that enlarging an existing conservation area will most often be the wiser conservation choice.

In the face of limited biological information for assisting in the design of a network of conservation areas, landscape ecologists Richard Forman and Sharon Collinge (1996) have advanced what they termed a "spatial solution" to conserving biodiversity in landscapes and regions. They have hypothesized that their alternative approach makes good ecological sense and will conserve the "bulk of nature and natural processes." Their solution has four essential features:

1. Maintain a few large patches of natural vegetation. These patches are needed for species that do not prosper along habitat edges and have relatively large home ranges, and are similar in concept to the functional landscapes of Poiani et al. (2000).
2. Maintain vegetated corridors along streams and rivers. The benefits of maintaining these riparian corridors are well known and have been extensively published.
3. Establish connectivity between some large patches. The pros and cons for so doing have already been outlined.
4. Maintain bits of nature across the matrix. As previously discussed, small patches of habitat (similar to Poiani et al.'s [2000] functional sites) will maintain some species but have a number of disadvantages compared to large patches. Yet some patches of natural habitat, even if small, are almost certainly better than none.

Although overly simplified from a biodiversity standpoint, these guidelines may prove useful in portions of the world in which very little is known about the biodiversity other than the different types of major ecosystems that dominate the large remaining patches in a matrix of otherwise semi-natural or anthropogenically converted habitats. In several places in this book, I have discussed the matrix—the area outside conservation areas. In the following section, I consider the value of the matrix to conservation in more detail.

The Semi-Natural Matrix

Although much can be done to conserve biodiversity within a network of conservation lands and waters, few biologists or planners would agree that these areas alone will be sufficient to conserve a region's biodiversity. Some conservation biologists would argue that the majority of the world's biodiversity occurs in and needs to be conserved in what ecologist Jim Brown has referred to as the "semi-natural matrix" (Franklin 1993). Although the matrix of lands and waters outside conservation areas varies tremendously from areas of intensive human use (e.g., urban) to areas of more minimal use (e.g., lands with natural cover that are grazed or waters that are heavily fished), it is the latter type that forms the semi-natural matrix.

These lands and waters of the semi-natural matrix, according to Franklin (1993), have been overlooked from a conservation perspective. They provide habitat for many species, increase the effectiveness of conservation areas by acting as buffers, and contribute toward connectivity among conservation areas. Chapter 7 highlighted the importance of landscape context—essentially, the semi-natural matrix—for the long-term integrity and viability of target species and communities that occur within conservation areas. Much has been written about conserving biodiversity in "managed landscapes" or "off-reserve" lands (Szaro and Johnston 1996, Hale and Lamb 1997). Although it is beyond the scope of this book to review that literature, there is little doubt that a semi-natural matrix of natural land cover where ecological processes and natural disturbance regimes still function, even if degraded to some extent by human uses, will contribute to the health and welfare of conservation areas established primarily for the purposes of biodiversity conservation.

The implications for conservation biologists of the importance of the semi-natural matrix to conservation are considerable. Planners must not only work hard to identify, design, and establish a network of conservation areas, they must also be concerned about the fate of lands and waters outside conservation areas. A reasonable question from any conservation practitioner at this point might be: How should the conservation community identify and manage new and existing conservation areas and simultaneously be concerned about the conservation of the matrix? Although not entirely satisfying, the most practical answer to this question is that planners and biologists should focus first and foremost on those lands and waters that surround conservation areas. Within the landscape context of conservation areas, they should be specifically concerned with the management of the semi-natural matrix in relation to the targets (species, communities, and ecosystems) of attention. For example, if the targets of concern are freshwater mussels and native fish species, planners need to ensure that the upstream watershed continues to be managed to maintain high water quality and a natural flow regime. If the target is a forest ecosystem that requires a natural fire regime to maintain it, the concern should be a fire management policy in the matrix that is consistent with the needs of the conservation area.

Once one or more alternative designs of a network are in place, the planning team is in a position to judge how well this potential network will achieve the target-based goals that were established for the project (see Chapter 6). The following section addresses these issues and provides examples of how planning teams have performed.

Measuring Success: To What Degree Does the Network Achieve the Target-Based Goals?

Chapter 6 outlined the steps for setting quantitative target-based goals to address this question: How much is enough? Recall that conservation goals have two components, representation and quality. The representation component consists of a numeric goal (e.g., how many occurrences or what percentage of a target should be conserved) and a distributional factor that relates to how the target should be distributed across the planning region. Once a network of conservation areas has been assembled, planners can then evaluate how well that design achieves the goals.

Table 8.3 summarizes target-based information on conservation goals for the Southern Rocky Mountains Ecoregional Plan. For all of the targets combined, including freshwater and terrestrial ecosystems and species, the plan met the conservation goals for 37% of the targets. In general, the proposed network of conservation areas did a good job of "capturing" ecosystem-level targets, but not nearly so well with a number of species groups (e.g., plants = 13%).

It is important to consider why a goal may not have been obtained for a particular target. There are four possible explanations. First, there may no

Table 8.3 *Success Rates of Targets Meeting Conservation Goals for the Southern Rocky Mountains Ecoregion*

Target	Total Number of Targets	Number of Targets Meeting Goals	Percent of Targets That Met Goals
Terrestrial Ecosystems	39	35	90%
Aquatic Ecosystems	107	98	92%
Rare Plant Communities	79	26	33%
Species			
Amphibians and Reptiles	8	4	50%
Birds	23	6	26%
Fish	9	3	33%
Invertebrates	132	23	17%
Mammals	28	6	21%
Wide-Ranging Mammals	6	3	50%
Plants	177	23	13%
Species Total	383	68	18%
Total	608	227	37%

From the Southern Rocky Mountains Ecoregional Assessment Team 2001.

longer be sufficient conservation opportunities in the planning region because of habitat conversion, degradation, or other causes. Second, the region may be poorly inventoried for a particular taxonomic group; there may actually be more populations or occurrences that could be conserved, but they are yet to be discovered. Such is often the case with aquatic ecosystems, which have generally not been surveyed to the extent of terrestrial ecosystems. Third, there may be occurrences of targets that are not of sufficient viability or integrity to include in the portfolio of conservation areas. Fourth, target occurrences may not meet the distributional requirements of the goal. It is often the case, for example, that species ranges have been reduced and they are no longer distributed over regions that they once inhabited.

A review of how well conservation goals were met should suggest several courses of action for a team. In some cases, a planning team may want to undertake inventory efforts for groups of targets for which there is little information in the planning region (see Chapter 5). In other cases, there may be populations or occurrences of targets that are degraded but for which it is feasible to undertake restoration efforts to improve the viability or the integrity of one or more occurrences (see Chapter 7). In regions where substantial opportunities remain for conservation, it may be worthwhile for a planning team to examine several alternative portfolios to determine which areas do the most satisfactory job of achieving goals and which are most feasible to implement. Some area selection algorithms (outlined earlier) make examining several alternative network designs relatively easy. Finally, it may be necessary for a team to conclude that a target has been sufficiently reduced or eliminated from a region and that it is no longer a prudent investment to try to conserve that target.

Full Circle: The Four-R Framework Revisited

Chapter 2 outlined a framework for the planning process that I have developed throughout this book. Referred to as the Four-R Framework (Shaffer and Stein 2000), it emphasizes the selection and design of a network of conservation areas that is representative, resilient, redundant, and restorative. Having worked through the major steps of the planning process and reviewed principles and criteria for selecting conservation areas that have been developed by other scientists and conservationists, we can now return to this framework and determine whether it remains a useful construct for conservation planning.

Recall that *representative* refers to representing the variety of ecosystems and the environmental gradients across which they occur in conservation areas. Earlier in this chapter, after reviewing principles and criteria advanced by many researcher and planners, I concluded that representativeness is an overarching principle critical to the biodiversity planning process. *Resilient* (also termed *persistence* or *functionality*) refers to targets having sufficient integrity and viability to remain extant through both natural and human-cased disturbances; this is also an overarching principle. *Redundant* refers to the setting of conservation goals so that several occurrences of a target are represented in conservation areas, thereby helping ensure that the target will remain viable in the planning region over the long term. Setting such goals influences the size and design of the portfolio, and enables conservation planners to evaluate how effective their final design of a network of conservation areas will be in achieving the target-based goals. *Restorative* refers to identifying areas where it may be feasible to restore or otherwise improve the viability and integrity of degraded occurrences of targets. In a world where the burgeoning human population continues to alter, degrade, and destroy natural habitats, conservation planners and practitioners will increasingly face situations where restoration may be the only way to save some species and ecosystems from irreparable losses.

Although there are other important criteria and principles of importance (e.g., flexibility, complementarity), the concepts, tools, and steps
developed in this biodiversity planning process thus far suggest that the
principles embedded in the Four-R Framework have much to recommend to conservation planners. Keeping this framework in mind should
help planners design a network of conservation areas that has the best possible opportunity for conserving the full array of a region's biodiversity for
generations to come.

Key Points in Selecting and Designing a Network of Conservation Areas

1. Using a GIS, map data on conservation target occurrences. Include
   only those target occurrences judged to be of satisfactory viability or
   integrity (or feasibly restorable). Alternatively or in addition, assemble
   GIS data layers (e.g., suitability indices) to help evaluate the potential
   viability or integrity of target occurrences. Other data layers such as
   those on land ownership and management, stream networks, road
   networks, and other administrative data should also be available for
   the area selection process. You should also choose a type of selection
   unit (e.g., watershed, hexagon, etc.) to which to attribute these data in
   a GIS environment at this point.

2. Decide whether or not the circumstances warrant the use of a computerized area selection algorithm to help identify conservation areas.
   In most cases, such algorithms will make the selection process more
   efficient, transparent, and repeatable. There are several publicly available algorithms to choose from, depending on the characteristics of
   the data.

3. Unless detailed information is readily available with which to delineate detailed boundaries of conservation areas, the selection of conservation areas should focus on generalized areas identified through
   selection units, such as watersheds, or other digitally available grid systems to which data on conservation targets have been attributed. It
   may be helpful to refer to these places as areas of biodiversity significance or some similar name, as opposed to a specific name of a conservation area. The details of how these conservation areas will be
   individually designed and managed are more appropriately considered in a separate process, referred to as site conservation planning or
   conservation area planning (see Chapter 13).

4. Ensure that targets are represented in conservation areas across the environmental gradients in which they occur by using a stratification of ecoregions (see Chapter 2) and/or by using digital environmental data in combination with data on target occurrences.

5. Where conservation options remain in the planning region, explore a range of alternative portfolios of areas that meet the goals of the project. These alternatives will provide flexibility for meeting other goals and objectives for the use of land and water, many of which are often mandated by law and policy.

6. Landscape-scale conservation areas should be emphasized in the selection process. These areas are likely to contain more conservation targets, more viable species' populations, and intact ecological processes and disturbance regimes compared to spatially smaller areas.

7. Pay close attention to who is involved in this critical part of the planning process. Decision makers, such as program directors, and other program staff, such as government relations and communication personnel should be involved in this stage of the planning process, in addition to scientific and planning staff members. Such involvement will help gain support for and build credibility in the final product.

8. The portfolio of conservation areas should be envisioned as a network whose net effect is greater than the sum of the individual parts (conservation areas). In most cases, it will be desirable to locate new conservation areas as close as possible to existing ones. Potential conservation areas with significant amounts of fragmented habitat should be avoided if possible. Corridors for linking key conservation areas should be considered, but only on a target-specific basis and after various tradeoffs have been evaluated.

9. The products of this step in the planning process should be a map that identifies the final design of the network of conservation areas and a table that summarizes the conservation areas that have been selected as well as the conservation targets that occur in these areas.

CHAPTER NINE

Safeguarding Nature's Investments: Setting Priorities for Action among Conservation Areas

There are many views about how to best identify priority conservation areas. To some extent, this diversity is welcome as it arises from attempts by people with varying backgrounds to solve different problems in different parts of the world.

—CHRIS MARGULES
AND ROBERT PRESSEY (2000)

Most regional or ecoregional assessments of biodiversity are likely to identify scores of conservation areas. In some cases, there may be as many as several hundred areas pinpointed, depending on such factors as the size of the planning region, the percentage of the region remaining in natural habitat, the types of conservation targets, the overall goals, and the methods for delineating conservation areas. Although by virtue of its selection in the planning process, each conservation area is important and worthy of management attention, not all areas are of equal conservation value or in need of attention with the same degree of urgency. At first glance, the challenge of conserving all of the identified areas may seem overwhelming, especially for any single organization or agency. But through a practical approach to priority setting, this challenge can be winnowed down to an ambitious set of objectives, which, if undertaken by the conservation community as a whole, is within its collective reach.

In *The Sinking Ark*, ecologist Norman Myers (1979) first suggested the need for conservationists to consider a "triage strategy." The origin of this strategy was the World War I battlefield where doctors were overwhelmed by more wounded than they could care for. They placed each wounded soldier in one of three categories: those who would clearly be helped by medical attention, those who could probably survive without it, and those who were likely going to die with or without such attention. In working with endangered species, Myers (1979) recommended that conservationists employ a similar strategy. We can expand this line of thinking to conservation areas. At this point in the planning process, conservation areas in Myers's third category—areas not likely to persist even with a considerable infusion of human resources—have either been eliminated or assigned to a "feasibly restorable" option. The challenge, then, appears to be how to best distribute conservation efforts between those conservation areas that would clearly benefit and those that may persist if nothing is done. As Myers observed, these are difficult decisions that involve many factors, from biological to economic, cultural, and political. The discussion and criteria outlined in this chapter should help conservationists and planners wrestle with these decisions and allocate resources most effectively.

Recall from Chapter 1 that setting priorities for biodiversity conservation is occurring at many scales, from the global and continental exercises of the World Wildlife Fund, Conservation International, and others, to the myriad efforts occurring at the scale of conservation area by numerous organizations. Even at the scale of the individual conservation area, planners must set priorities for which targets are going to be the focus of attention. The purpose of this chapter is to assess how planners can set priorities among conservation areas in an ecoregional or regional conservation plan. In practice, there are not enough resources to work on conserving all the important places identified in the portfolio at the same time, even with the involvement of many organizations and agencies. This situation forces setting priorities for action among the numerous places that need attention.

Several criteria can help planners determine which conservation areas are the highest priorities for action:

- Some measure of the biodiversity or conservation value of an area.
- Threats at conservation areas.
- The degree to which the conservation targets in an area have been conserved elsewhere (see the principle of complementarity, Chapter 8).
- The number of rare or endangered species (sometimes included in the biodiversity value criterion).
- The quality or ecological condition of a conservation area (sometimes included in the conservation value criterion).

- The ability to restore a degraded area (e.g., increase its ecological integrity, return a target population to viable levels).
- The feasibility or possibility of achieving conservation at a particular place and the potential leverage that taking action at one area may provide for achieving conservation at other areas (a multiplier effect).

The two most commonly used criteria in setting priorities are biodiversity or conservation value and threats. These two criteria have been defined and, in some cases, named differently by various conservation planners and biologists. For example, The Nature Conservancy (TNC) defines *conservation value* in its ecoregional planning work as "a criterion based upon the number, diversity, and health of conservation targets" (Groves et al. 2000a, Valutis and Mullen 2000). The World Wildlife Fund (WWF) uses a criterion referred to as biological distinctiveness to rank candidate priority conservation areas in its ecoregional assessment work (Dinerstein et al. 2000). The *biological distinctiveness* of an area is defined by its representation of endemic taxa or rare communities, its level of species richness, and the presence of unique ecological or evolutionary phenomena.

In a similar vein, threats as a criterion for setting priorities have been assessed in many different ways. Again, the work of TNC, WWF, and Pressey and colleagues is illustrative. The Nature Conservancy has emphasized two aspects of threats in its ecoregional planning efforts (Groves et al. 2000a, Valutis and Mullen 2000). First, it focused on *critical threats* as those that are likely to destroy or seriously degrade conservation targets at many or most places within the conservation area where they occur. Threats, then, can be assessed by their degree of severity and, in theory, are directly tied to specific conservation targets. In practice, most regional planning efforts will not have detailed, site-based knowledge of threats that are impinging on individual conservation targets. Consequently, most threat assessments at this scale are focused more generically on the conservation area itself instead of any individual conservation targets. For example, planners and biologists may know that a particular area is threatened by invasion by an exotic species, but they probably will not know which particular species or communities might be most susceptible to negative impacts by this exotic species.

The second important aspect of threats considered by TNC is urgency: how likely it is that that the threat will have a negative impact on a conservation area in the near future (2–4 years). In a similar approach, the WWF has categorized threats into three broad groups: conversion, degradation, and wildlife exploitation (Dinerstein et al. 2000). Within each of those categories, it assesses the degree of urgency and the severity (how widespread the threat is within a conservation area) of threats to help in setting priorities among candidate conservation areas ("priority areas," in WWF parlance). Pressey and colleagues have articulated the threat criterion as one of "vulnerability" (Pressey et al. 1996a, 2000, Margules and Pressey 2000). *Vulnerability* can be defined generally as the risk of a conservation area or planning unit being "transformed by extractive uses" (Margules and Pressey 2000). In practice, Pressey and colleagues have defined vulnerability in quantitative terms, such as the percentage of private or lease-held portions of a planning unit that has been cleared of native vegetation (e.g., Pressey et al. 2000) and, by corollary, is vulnerable to similar activities in the future.

What should be obvious by now is that the criteria of biodiversity values and threats have been interpreted and applied in different ways to set area-based or project-specific priorities for conservation. In the following section, I review in more detail how some of the more common priority-setting criteria have been applied. I conclude with some recommendations for their use in regional conservation planning that deal, in part, with the variability in how these criteria have been applied.

Priority-Setting Criteria

Biodiversity Value

A conservation area may be biologically valuable for many reasons. It could harbor the populations of one or more endangered species, contain the highest-quality example of a community or ecosystem target, or represent home to numerous conservation targets, both terrestrial and aquatic. Both TNC and WWF have used qualitative ranking schemes to evaluate biodiversity value.

The Nature Conservancy rates biodiversity value on the basis of two factors: the diversity and number of conservation targets at a conservation area, and the "biodiversity health" of those target occurrences. Within each of these factors, a value of high, medium, or low are assigned to each conservation area. For the number and diversity of targets, these qualitative ratings are based on the following definitions (Valutis and Mullen 2000):

High = Relatively large number of targets in comparison to other areas in the ecoregion and the area contains both aquatic/terrestrial targets and targets at different spatial scales.

Medium = Moderate number of targets or includes both terrestrial/aquatic targets and targets at different spatial scales.

Low = Low number of targets relative to other areas in the ecoregion or includes both terrestrial/aquatic targets or targets at different spatial scales.

The biodiversity health factor is similarly defined with a qualitative ranking system based on the criteria of size, condition, and landscape context (see Chapter 7). TNC's approach to evaluating biodiversity value places a premium on the number of targets at a conservation area, the quality and condition of those target occurrences, at what spatial scales those targets occur, and whether there is a mix of aquatic and terrestrial targets. Although the criteria for evaluating biodiversity value are clear, they are not necessarily explicit or rationally explained. There is considerable room for ambiguity in how any individual planner interprets such relative terms as "moderate number of targets." In addition, it is not clear which factor (number of targets versus condition of targets) is more important in determining the overall biodiversity value or if they are weighted equally. For example, does an area with a high-quality example of an aquatic ecosystem have a higher biodiversity value than one that harbors more targets but of lesser quality? Finally, it is not intuitively obvious why simply having both aquatic and terrestrial targets in a conservation area makes it more valuable for biodiversity conservation than an area that contains only terrestrial or only aquatic targets.

The World Wildlife Fund (Dinerstein et al. 2000) defines biodiversity value in its ecoregional visioning efforts on the basis of two factors: biological distinctiveness and landscape integrity. These two factors are similar to, yet somewhat different from, those used in TNC's approach. As noted above, biological distinctiveness is defined on the basis of such criteria as endemism, species richness, representation of habitat types, and unique ecological phenomena. Landscape integrity is defined by the presence of large blocks of nonfragmented habitat. These two factors are combined in an integration matrix (Table 9.1). The matrix qualitatively ranks each of the two factors to identify priority areas in a five-level rating system: I, highest-priority areas that form the core of an ecoregional strategy; II, high-priority areas; III, areas of regional (but not global) priority; IV, areas of importance to state or local governments; V, lower-priority areas that contain widely distributed communities or common species assemblages.

As with TNC's approach, the criteria used to arrive at these various ranks are clear, but their definitions are not explicit. The use of qualitative

Table 9.1 *A Matrix for Ranking Candidate Priority Conservation Areas*

Ranks are I to V, with I being the highest priority and V lowest, for illustrative purposes only. Local teams can best determine these ranks.

Landscape Integrity and Biological Distinctiveness	High in Endemism or Rare Communities	Moderate Endemism, High Richness	Low Endemism, High Richness	Low Endemism, Low Richness	Protects a Single Threatened Species	Contains Representative Habitats or Communities
Intact habitat block	I	I	I	II	II	III
Relatively intact multiple large blocks	I	I	II	II	III	III
Relatively intact multiple medium-sized blocks	I	II	II	II	III	V
Incomplete large blocks	I	II	III	II	IV	IV
Incomplete medium-sized blocks	II	III	III	III	V	V
Incomplete small blocks	II	III	IV	IV	V	V
Restorable large blocks	III	IV	IV	IV	V	V
Restorable medium-sized blocks	III	IV	V	V	V	V
Restorable small blocks	IV	IV	V	V	V	V

Modified from Dinerstein et al. 2000.

terms such as "low" or "moderate" leaves considerable room for interpretation, although the relative importance of such factors as endemism and richness in determining the final ranks is clear (Dinerstein et al. 2000). Unlike TNC's system, the WWF ranking approach was designed solely for use in an expert workshop setting.

Pressey and colleagues (1994) have developed a more quantitative approach to estimating biodiversity value with the concept of irreplaceability (see Chapter 8). This concept is predicated on the notion that for many regions of the world, several alternative designs for combining potential conservation areas into a system will achieve the target-based goals of a regional conservation plan. In essence, determining an area's irreplaceability involves exploring all the possible combinations of areas needed to design a portfolio of conservation areas, then calculating the percentage of times an individual area appears in the different possible combinations. For example, if there are 100 different possible combinations of areas in which to establish a portfolio and an individual area appears in 40 of these combinations, its irreplaceability is 40% (see Figure 8.2). If a conservation area contains the only known population of a rare species, then it might appear in every possible combination, and its irreplaceability would be 100%. For relatively small numbers of conservation areas and very small numbers of targets, these sorts of calculations are mathematically manageable. However, for a portfolio of conservation areas that might contain several hundred areas, numerous conservation targets, and millions of possible combinations of sites, a statistical approach can be used to predict irreplaceability (Ferrier et al. 2000).

Using the concept of irreplaceability, as advanced by Pressey and colleagues, to define biodiversity value has a number of advantages. First, it can be quantitatively defined and allows for direct comparison among areas because each potential conservation area (*selection unit* in Chapter 8) receives an irreplaceability score. Second, and most important, it is specifically tied to targets contained within areas and the goals that have been established for those targets. Third, it can be used to evaluate alternative portfolios of conservation areas and is especially valuable as an interactive tool in negotiations over which areas in a region will be developed and which will be maintained as conservation areas (see Pressey et al. 1995). Fourth, it can be combined with a measure of threat or vulnerability to graphically establish conservation priorities (see below).

At the same time, using the irreplaceability approach has at least one serious weakness: It assumes there are several combinations of conservation areas that will achieve the desired target-based goals. In many parts of

Figure 2.2. A map of the world's ecoregions, as described by the World Wildlife Fund. This map categorizes 867 ecoregions within 14 biomes and 8 biogeographic realms. A key to the terrestrial ecoregions delineated in this map, sources of information on these ecoregions, technical descriptions, and digital data are available at www.worldwildlife.org/science. (From Olson et al. 2001. Terrestrial ecoregions of the world: a new map of life on Earth. *BioScience* 51:933–938. Copyright, American Institute of Biological Sciences.)

1

Figure 2.1. A map of Nature Conservancy ecoregions of the United States and adjacent regions. This map recognizes 69 ecoregions in the conterminous U.S., 1 in Hawaii, and 11 in Alaska. This map has been modified from the ecoregional maps of Bailey (1995) and the ECOMAP (1993) working group of the U.S. Forest Service. The Nature Conservancy is preparing biodiversity conservation plans for each of these ecoregions.

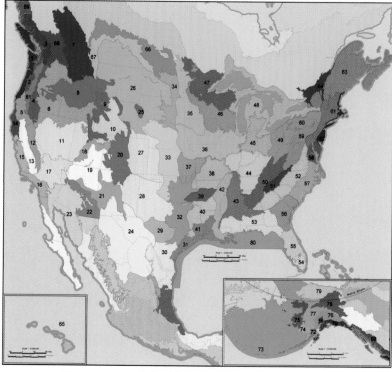

Ecoregions of the United States of America

1 Pacific Northwest Coast
2 Puget Trough - Willamette Valley - Georgia Basin
3 North Cascades
4 Modoc Plateau and East Cascades
5 Klamath Mountains
6 Columbia Plateau
7 Canadian Rocky Mountains
8 Middle Rockies - Blue Mountains
9 Utah-Wyoming Rocky Mountains
10 Wyoming Basins
11 Great Basin
12 Sierra Nevada
13 Great Central Valley
14 California North Coast
15 California Central Coast
16 California South Coast
17 Mojave Desert
18 Utah High Plateaus
19 Colorado Plateau
20 Southern Rocky Mountains
21 Arizona–New Mexico Mountains
22 Apache Highlands

23 Sonoran Desert
24 Chihuahuan Desert
25 Black Hills
26 Northern Great Plains Steppe
27 Central Shortgrass Prairie
28 Southern Shortgrass Prairie
29 Edwards Plateau
30 Tamaulipan Thorn Scrub
31 Gulf Coast Prairies and Marshes
32 Crosstimbers and Southern Tallgrass Prairie
33 Central Mixed-Grass Prairie
34 Dakota Mixed-Grass Prairie
35 Northern Tallgrass Prairie
36 Central Tallgrass Prairie
37 Osage Plains/Flint Hills Prairie
38 Ozarks
39 Ouachita Mountains
40 Upper West Gulf Coastal Plain

41 West Gulf Coastal Plain
42 Mississippi River Alluvial Plain
43 Upper East Gulf Coastal Plain
44 Interior Low Plateau
45 North Central Tillplain
46 Prairie-Forest Border
47 Superior Mixed Forest
48 Great Lakes
49 Western Allegheny Plateau
50 Cumberlands and Southern Ridge and Valley
51 Southern Blue Ridge
52 Piedmont
53 East Gulf Coastal Plain
54 Tropical Florida
55 Florida Peninsula
56 South Atlantic Coastal Plain
57 Mid-Atlantic Coastal Plain
58 Chesapeake Bay Lowlands
59 Central Appalachian Forest
60 High Allegheny Plateau
61 Lower New England/Northern Piedmont
62 North Atlantic Coast

63 Northern Appalachian-Boreal Forest
64 St. Lawrence–Champlain Valley
65 Hawaiian High Islands
66 Aspen Parkland
67 Fescue–Mixed Grass Prairie
68 Okanagan
69 Alaska Coastal Forest and Mountains
70 Gulf of Alaska Mountains and Fjordlands
71 Cook Inlet Basin
72 Alaska Peninsula
73 Bering Sea and Aleutian Islands
74 Bristol Bay Basin
75 Beringian Tundra
76 Alaska Range
77 Interior Alaska Taiga
78 Yukon Plateau and Flats
79 Brooks Range Tundra Coastal Plain
80 Northern Gulf of Mexico
81 West Cascades

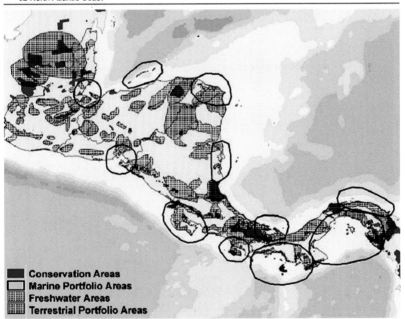

■ **Conservation Areas**
□ **Marine Portfolio Areas**
▦ **Freshwater Areas**
▦ **Terrestrial Portfolio Areas**

Figure 2.5. An example of a completed portfolio of conservation areas, the principal product of the seven-step planning process, for the Central America region. The portfolio includes terrestrial, freshwater, and coastal marine conservation areas. This map represents a combination of multiple ecoregions.

3

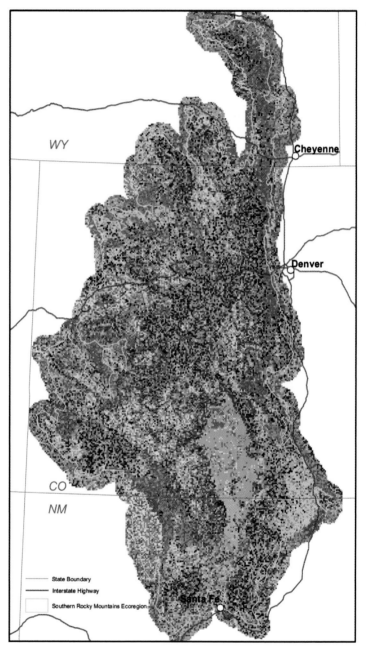

Figure 4.5. Ecological Land Units (ELUs). Environmental variables such as geology, elevation, and landform were used in the Southern Rocky Mountains Ecoregional Plan to produce a composite layer of ELUs. These ELUs were used to help "capture" the environmental variation in the ecosystem-level conservation targets. The different colors represent unique combinations of landform (e.g., canyon, cliff, ridge top), surficial geology (e.g., granite, shale), and elevation classes. The combinations of these three classes of variables resulted in 830 ELUs for the ecoregion. The figure is illustrative of the variability in ELU type but does not display all of the variability that resulted from these analyses. (From Southern Rocky Mountains Ecoregional Assessment Team 2001.)

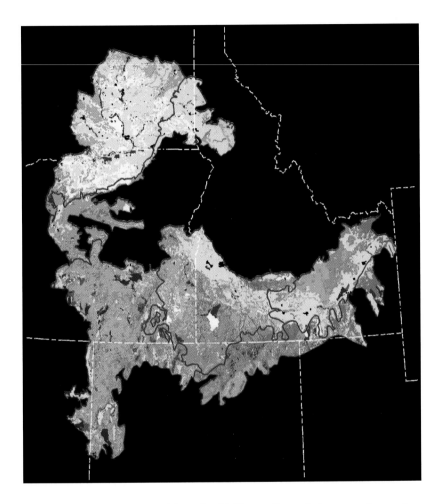

Figure 5.2. Vegetation types of the Columbia Plateau. By overlaying maps of vegetation or land cover with maps of land management status (see Figure 5.1), planners and biologists can assess the degree to which ecological communities or ecosystems are already represented in existing conservation areas. Such an analysis also identifies "gaps"—target communities or ecosystems that still need representation in conservation areas. (Map courtesy of D. Stoms, Department of Geography, University of California, Santa Barbara.)

Figure 8.6. Terrestrial ecosystems of the Southern Rocky Mountains ecoregion. By combining information on the distribution of conservation targets such as this map of the terrestrial ecosystems of the Southern Rocky Mountains Ecoregion with information on environmental gradients, planners can select conservation areas that conserve these targets across a range of environmental conditions. (From the Southern Rocky Mountains Ecoregional Assessment Team 2001.)

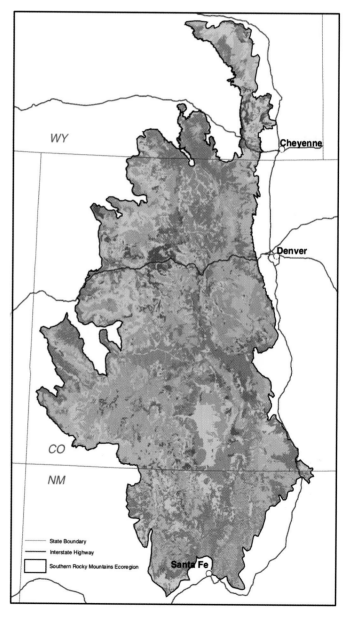

Key for figure 8.6

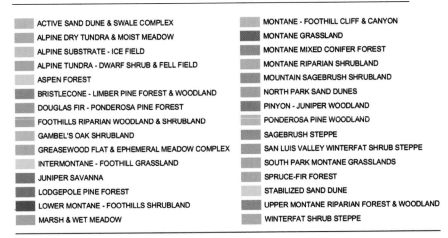

	ACTIVE SAND DUNE & SWALE COMPLEX		MONTANE - FOOTHILL CLIFF & CANYON
	ALPINE DRY TUNDRA & MOIST MEADOW		MONTANE GRASSLAND
	ALPINE SUBSTRATE - ICE FIELD		MONTANE MIXED CONIFER FOREST
	ALPINE TUNDRA - DWARF SHRUB & FELL FIELD		MONTANE RIPARIAN SHRUBLAND
	ASPEN FOREST		MOUNTAIN SAGEBRUSH SHRUBLAND
	BRISTLECONE - LIMBER PINE FOREST & WOODLAND		NORTH PARK SAND DUNES
	DOUGLAS FIR - PONDEROSA PINE FOREST		PINYON - JUNIPER WOODLAND
	FOOTHILLS RIPARIAN WOODLAND & SHRUBLAND		PONDEROSA PINE WOODLAND
	GAMBEL'S OAK SHRUBLAND		SAGEBRUSH STEPPE
	GREASEWOOD FLAT & EPHEMERAL MEADOW COMPLEX		SAN LUIS VALLEY WINTERFAT SHRUB STEPPE
	INTERMONTANE - FOOTHILL GRASSLAND		SOUTH PARK MONTANE GRASSLANDS
	JUNIPER SAVANNA		SPRUCE-FIR FOREST
	LODGEPOLE PINE FOREST		STABILIZED SAND DUNE
	LOWER MONTANE - FOOTHILLS SHRUBLAND		UPPER MONTANE RIPARIAN FOREST & WOODLAND
	MARSH & WET MEADOW		WINTERFAT SHRUB STEPPE

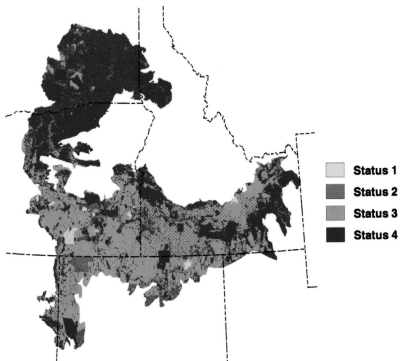

Status 1
Status 2
Status 3
Status 4

Figure 5.1. The status of land management in the Columbia Plateau ecoregion. Only 4% of the Columbia Plateau ecoregion falls within status 1 and 2 lands that are considered adequate for the conservation of biological diversity. The different categories of land management codes were developed by the U.S. National Gap Analysis Program. The World Conservation Union (IUCN) has developed a similar set of codes for protected areas, such as national parks, that have some degree of formal legal protection. (From *Precious Heritage* by The Nature Conservancy, copyright © 2000 by The Nature Conservancy. Used by permission of Oxford University Press.)

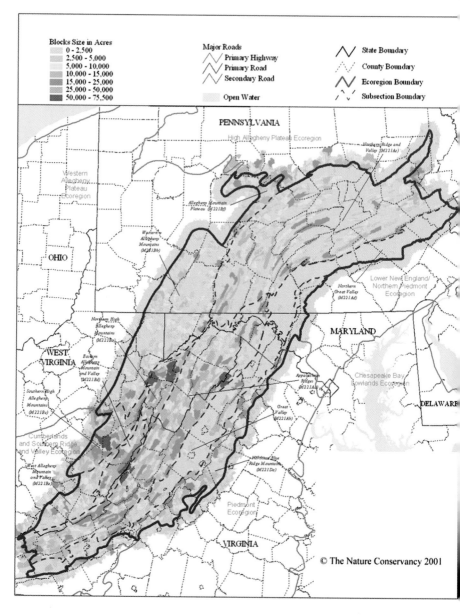

Figure 8.3. An example of identifying roadless blocks of forested habitat in ecoregional planning. Blocks of natural forest cover that were largely unroaded formed the basis for selection of many conservation areas in the Central Appalachian Forest Ecoregional Plan in the eastern U.S. (From The Nature Conservancy 2001.)

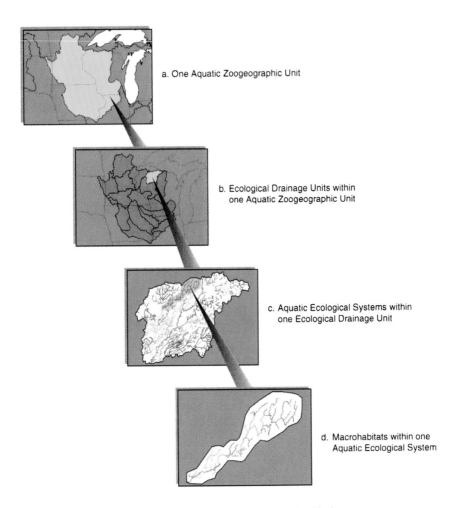

Figure 10.2. A four-tiered hierarchy classification framework of freshwater ecosystems. This hierarchy influences characteristics of the finer spatial scale units.

9

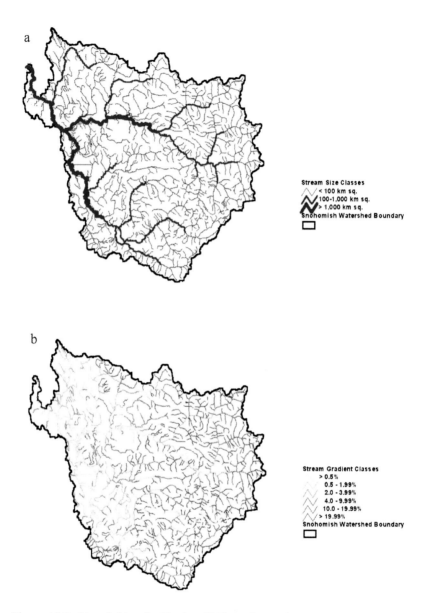

a

Stream Size Classes
/\/ < 100 km sq.
/\/ 100-1,000 km sq.
/\/ > 1,000 km sq.
Snohomish Watershed Boundary
☐

b

Stream Gradient Classes
> 0.5%
0.5 - 1.99%
2.0 - 3.99%
4.0 - 9.99%
10.0 - 19.99%
> 19.99%
Snohomish Watershed Boundary
☐

Figure 10.3. Macrohabitat classification. (a) Three classes of stream size based on drainage areas. (b) Six classes of stream gradient. (c) Eight classes of geology that influence the source and chemical characteristics of stream water. (d) Stream macrohabitats based on the patterns of variables shown in parts (a)–(c) in conjunction with stream elevation and patterns of stream connectivity (connected to lakes, large rivers, etc). (e, on page 12 of insert) Aquatic ecological systems classified for watersheds ≤100 km² based on cluster analysis of macrohabitat membership.

c

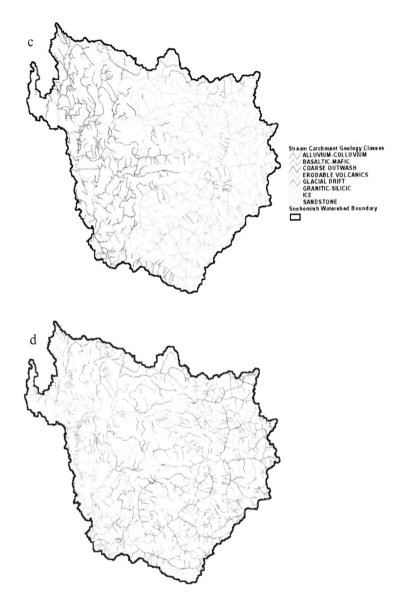

Stream Catchment Geology Classes
ALLUVIUM-COLLUVIUM
BASALTIC-MAFIC
COARSE OUTWASH
ERODABLE VOLCANICS
GLACIAL DRIFT
GRANITIC-SILICIC
ICE
SANDSTONE
Snohomish Watershed Boundary

d

Figure 10.3 (continues)

11

(Figure 10.3 continued)

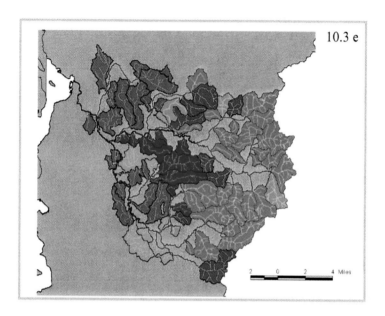

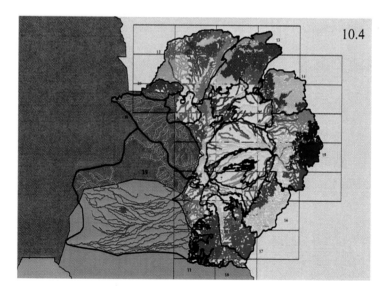

Figure 10.4. Ecological Drainage Units (EDUs) and Aquatic Ecological Systems in the Brazil portion of the Pantanal floodplain wetland. The black outlined polygons are EDUs. The different-colored highlighted streams are different aquatic ecological system types within the EDUs. Aquatic ecological systems are not shown for the Paraguay and Boliva portions of the Pantanal.

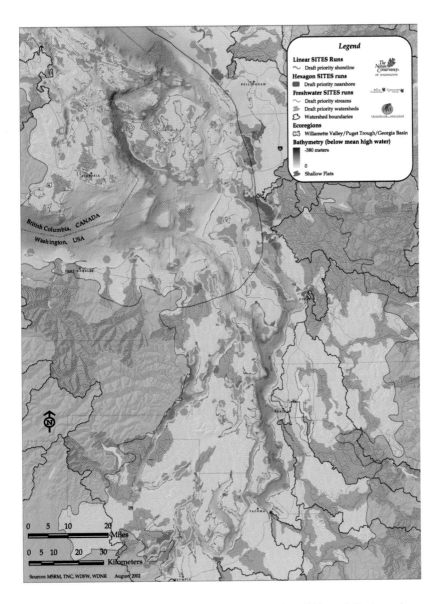

Figure 11.7. Integrating priority areas across the Puget Sound/Georgia Straits region. To aid in the development of an integrated set of priority areas, the draft marine priority areas (represented as yellow lines and blue hexagons, see Figure 11.4) were compared to draft freshwater priority areas (pink watersheds and blue stream reaches) on a watershed-by-watershed basis. Areas where there is greatest overlap among freshwater and marine areas (and terrestrial areas, not shown here) are given extra weight as priority areas. In addition, when the same conservation goals could be met by multiple priority area configurations, preference was given to configurations that best aligned freshwater and marine areas.

Figure 11.3. Priority marine conservation areas from TNC's northern Gulf of Mexico regional plan.

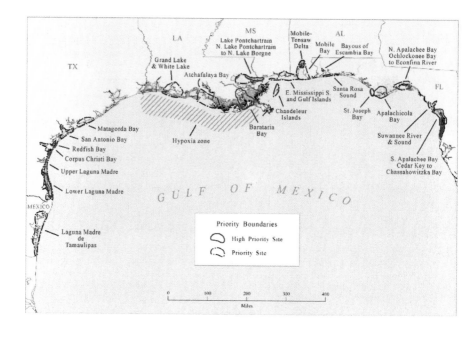

Figure 11.4 (opposite page).
Development of a Puget Sound/Georgia Straits regional plan. (a) A stepwise analysis was done in Sites (an area selection algorithm) to identify potential conservation areas. Data on the forage fish targets (e.g., herring and sand lance) were run through Sites first, and areas that met the goals for these species were then locked into the model (red hexagons) for subsequent runs. Data for lingcod, rockfish, and the rocky reef ecosystems were placed into the model next, and further areas were locked in (green hexagons). This procedure was repeated for seabirds and marine mammals (purple hexagons), and then for the rest of the fine-filter targets (blue hexagons). These results were then combined with separate Sites analyses on the linear shoreline ecosystem data (yellow segments). (b) These Sites results were evaluated and developed into large ecologically meaningful sites or seascapes.

14

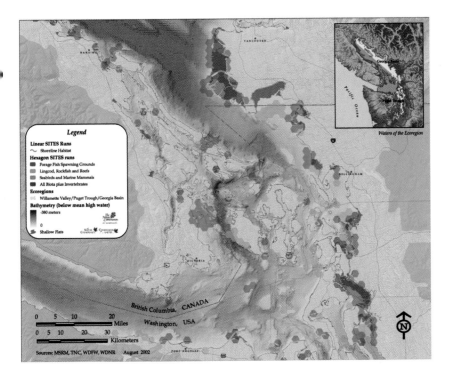

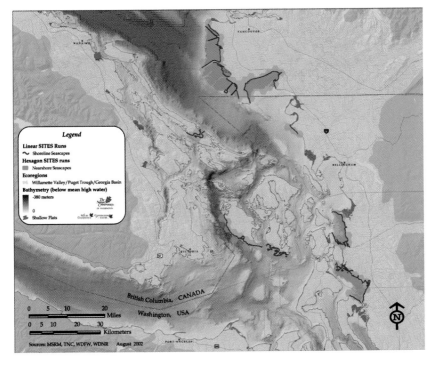

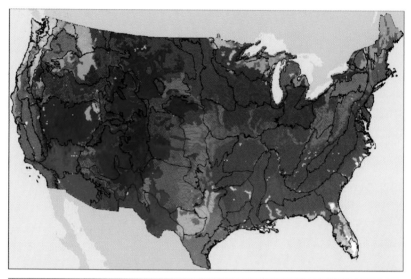

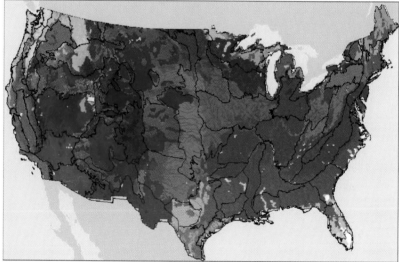

Figure 12.1. Ecoregions for the conterminous United States objectively derived from physical and climate attributes. Figure 12.1a uses current climate attributes. Figure 12.1b uses Hadley Model HAD2CM climate attributes for 2100. Colors are assigned based on a principal component analysis so that ecoregions with similar abiotic attributes have similar colors regardless of where or when they occur (Hargrove and Hoffman 1999, 2000). Boundaries of the modified Bailey Sections, used by The Nature Conservancy as ecoregions for conservation planning in the conterminous United States, are delineated in black. Differences between Figure 12.1a and Figure 12.1b are highlighted in Figure 12.2.

the world, this is clearly not the case. Natural habitats have been degraded or destroyed to the point that there are few, if any, options remaining for conserving biodiversity. As Chapter 8 demonstrated, no TNC ecoregional plans in North America achieved their target-based goals, and the gap between the goals and what was realized through each plan was substantially greater in those ecoregions with less remaining natural cover. Strictly interpreted by the original definition (Pressey et al. 1994), nearly all conservation areas in some of these planning regions would be 100% irreplaceable and the metric of irreplaceability would therefore be of little value for setting priorities among conservation areas. In addition, estimating irreplaceability requires the use of specific area selection algorithms that are designed to calculate irreplaceability on a dynamic basis as some areas are developed and others are reserved (e.g., C-Plan algorithm, Pressey 1999). For reasons discussed in Chapter 8, it may not always be feasible or practical to use such algorithms.

A more practical approach to using irreplaceability as a measure of biodiversity value may be to apply it in a post hoc fashion once a set of conservation areas has been identified. Noss and colleagues (2002) defined the irreplaceability of 43 conservation "megasites" (aggregations of several conservation areas) in the Greater Yellowstone area of North America on the basis of an area's contribution to nine target-based criteria. These criteria included viability benchmarks for focal species and target-based goals for imperiled local-scale species, vulnerable and declining bird species, coarse-scale aquatic species, plant communities, wetland habitats, geoclimatic classes, aquatic habitats, and habitats for focal species. For example, planners established a goal for the planning region to conserve 25% of the area of each wetland vegetation type. Then they scored each conservation area on the basis of how well it contributed to the planning region's goal for wetlands. They compiled similar scores for each criterion for each conservation area, then added them up to get a cumulative irreplaceability score for each megasite. In this manner, each megasite was ranked in terms of its irreplaceability on the basis of its contribution to goals for eight different types of targets, including but not limited to a single type of conservation target such as vulnerable and declining species.

There are, of course, many other ways to assign biodiversity value to a conservation area. The examples from TNC, WWF, and Pressey's work illustrate approaches that are commonly being used around the world, but they are not intended to be an exhaustive list. Nevertheless, they are sufficient to establish the importance of biodiversity value as being one of the primary criteria used to help set conservation priorities.

Threats

The assessment of threats in conservation planning is a critical step in developing appropriate conservation strategies. A recent analysis of recovery plans under the Endangered Species Act revealed that a basic lack of understanding of the nature of threats facing threatened and endangered species may be negatively influencing the recovery process (Lawlor et al. 2002). The assessment of threats in regional planning may take on a variety of forms, depending on the level of available information.

In some parts of the developed world, there is a great deal of knowledge about the status and distribution of flora and fauna and on threats to these biota. In these situations, it may be appropriate to design a detailed threats assessment as part of the regional or ecoregional planning effort. For example, in Great Lakes ecoregion planning effort (The Nature Conservancy 2000a), the team developed a detailed database on the threats to targets at each conservation area in the portfolio, the management of the area, and the potential broad strategies for abating threats and improving management (Table 9.2). In most parts of the world, however, our knowledge of area-specific threats to biodiversity will be general and limited.

From a regional planning perspective, an assessment of threats to individual conservation areas should serve two specific purposes: (1) identifying conservation areas that are in most urgent need of attention to abate a current or imminent threat and (2) identifying threats that recur across multiple conservation areas and may best be addressed at a scale greater than the individual conservation area (see Chapter 13).

In practice, there are several points to be made about conducting a threats assessment of conservation areas. First, most information on threats to conservation areas is likely to be derived from expert opinion and not from any sort of quantitative analysis (see Chapter 5). Second, although it is always more useful to obtain information on threats to actual conservation targets than to the conservation areas themselves (i.e., the threat may be impinging on a nontargeted element of biodiversity), a simple identification of the major threats to the conservation area should suffice for the purposes of regional planning. Third, some threats will probably go unnoticed in a regional assessment, but they will be discovered in a more detailed assessment at the project or area level. For example, a regional assessment may easily detect that an important aquatic ecosystem is threatened by an altered hydrological regime from a dam but may fail to observe a more difficult-to-detect threat, such as drawdowns of groundwater that are beginning to negatively affect stream flows. Finally, threats can be said to have both *stresses* and *sources* (Poiani et al. 1998, The Nature Conservancy

Table 9.2 *Threats to Ecoregional Conservation Area: Great Lakes Coast A*

	Threat					
Severity	Urban and Residential Development without Population Growth	Industry	Recreation (includes hiking, biking, skiing, camping, etc.)	Aquatic/Hydrologic Alterations	Exotic Species	Fire Suppression
Very High (The projected impact would cause destruction or elimination of a conservation target and/or supporting processes.)						
High (The projected impact would cause serious degradation of a conservation target and/or supporting processes.)	HIGH	HIGH				
Medium (The projected impact would cause some or uncertain degradation of a conservation target and/or supporting processes.)			MEDIUM	HIGH	HIGH	HIGH
Low (The projected impact would cause slight deterioration of a conservation target and/or supporting processes.)						

(table continues)

Table 9.2 *(Continued)*

Scope	Threat					
	Urban and Residential Development without Population Growth	Industry	Recreation (includes hiking, biking, skiing, camping, etc.)	Aquatic/Hydrologic Alterations	Exotic Species	Fire Suppression
Very High (All conservation target occurrences would be impacted, or 90–100% of the area of the site would be impacted.)						
High (Many of the conservation target occurrences would be impacted, or 50–90% of the site would be impacted.)	HIGH	MEDIUM	MEDIUM			
Medium (A few of the conservation target occurrences would be impacted, or 10–50% of the site would be impacted.)				MEDIUM	HIGH	HIGH
Low (One conservation target occurrence would be impacted, or less than 10% of the site would be impacted.)						

Immediacy

Very High (Impact has already been realized.)

High (Projected impact will be realized within 5 years.) MEDIUM MEDIUM MEDIUM HIGH HIGH HIGH

Medium (Projected impact will be realized within 10 years.)

Low (Projected impact will be fully realized after 10 years.)

Irreversibility

Very High (Impact is permanent and not reversible.)

High (Impact is reversible, but not practically feasible.) MEDIUM LOW MEDIUM MEDIUM MEDIUM HIGH

Medium (Impact is reversible with a commitment of additional resources.)

Low (Impact is easily reversed at a low cost.)

271

From The Nature Conservancy 2000a.

2000b). For example, our priority stream ecosystem may be threatened by increased sedimentation (the stress), but the source of that sedimentation could be road building, new developments, improper forestry operations, or inappropriate agricultural practices. It is unlikely that a regional assessment will ascertain all or even the most important sources of some stresses. Again, they would emerge during more detailed planning at the scale of the conservation area (see Chapter 13).

One approach to characterizing threats in regional planning is through the assessment of the vulnerability of a conservation area to extractive resource uses. Pressey and colleagues have been responsible for advancing the concept of vulnerability (Margules and Pressey 2000, Pressey et al. 2000, Pressey and Taffs 2001a, 2001b). Vulnerability can be defined quantitatively, as Pressey and colleagues have done, or more subjectively based on expert opinion (Noss et al. 2002). For example, in an analysis of nearly 1500 landscapes across all of New South Wales, Australia, Pressey et al. (2000) estimated vulnerability to clearing as "the percentage of private and leasehold portions of each landscape without vegetation, reflecting the desirability of the landscape for intensive uses on tenures historically subject to clearing." In a later analysis restricted to the western portion of New South Wales, Pressey and Taffs (2001b) examined vulnerability to clearing and cropping separately and defined four classes of vulnerability (zero, low, moderate, high) based on an assessment of the suitability of land for clearing or cropping in relation to soil impacts.

Once an assessment of vulnerability has been conducted, this information can be combined graphically with data on irreplaceability to help establish priorities for conservation action (Figure 9.1). The graph can be subdivided into four quadrants to help establish priorities (in descending order):

Quadrant (priority) I: Conservation areas with the highest values for both irreplaceability and vulnerability.

Quadrant (priority) II: Areas with high vulnerability but lower values of irreplaceability, because either their targets are already conserved elsewhere or there is the potential to conserve them in other new conservation areas.

Quadrant (priority) III: Areas with high values of irreplaceability but a lower risk of loss (e.g., areas not likely to be cleared for agriculture or forestry).

Quadrant (priority) IV: Areas that are not presently thought to be vulnerable and are generally replaceable (targets are already conserved elsewhere or can be).

Noss and colleagues (2002) also used the irreplaceability–vulnerability approach to setting priorities for the Greater Yellowstone region (Figure

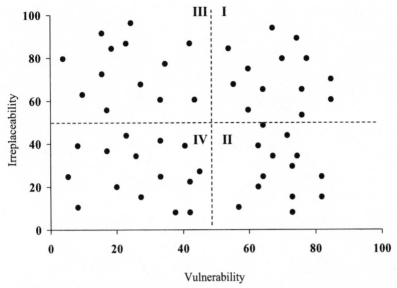

Figure 9.1   The irreplaceability–vulnerability framework for setting conservation priorities in ecoregional planning. The graph shows hypothetical data points for 50 conservation areas. Some of the most vulnerable areas are likely to be degraded or destroyed over time. As this happens, the irreplaceability of other areas will increase. Conversely, as some areas are conserved, the irreplaceability of others will decrease. Vulnerability will change over time as new threats appear on the landscape. Quadrants I–IV represent different priority levels for conservation action with I being the highest priority and IV the lowest. (Reprinted by permission of *Nature*, Margules and Pressey, Volume 405:243–253, copyright © 2000, MacMillan Publishers Ltd.)

9.2). As indicated earlier, they defined irreplaceability somewhat differently than Pressey and colleagues, and their assessment of vulnerability was based on expert opinion. Their four-quadrant evaluation of conservation priorities differed significantly in one area. Compared to Margules and Pressey (2000), they placed higher priority on areas of high biodiversity value (highly irreplaceable) even if those areas were thought to be less vulnerable (quadrant III of Margules and Pressey became quadrant II of Noss et al.). Their rationale was that conserving areas of high biodiversity significance while they are still ecologically intact makes good sense, as it is generally more cost-effective to do so when these areas are not highly threatened. Few areas, particularly large functional landscapes, will remain extant without some attention. How planners and conservation biologists balance the expenditure of effort between threatened (and often somewhat degraded areas) areas and areas that are less threatened (but often in better ecological condition) is difficult to answer and may hinge, in part, on feasibility,

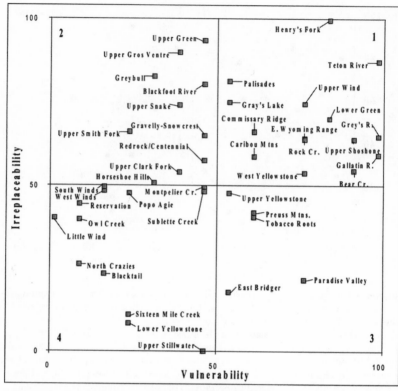

Figure 9.2 A graph of irreplaceability versus vulnerability for conservation areas in the Greater Yellowstone ecosystem. Although similar to the graph of Margules and Pressey (2000) in Figure 9.1, note that there is a significant difference in how the priority quadrants were assigned, with priority quadrant III in Figure 9.1 being assigned a higher priority. (From Noss et al. 2002. A multicriteria assessment of the irreplaceability and vulnerability of sites in the Greater Yellowstone Ecosystem. *Conservation Biology* 16:895–908. Copyright, Blackwell Publishing, Inc.)

leverage, and local opportunity (see sections that follow). However, as Chapter 12 on marine planning suggests, conservation biologists and planners would be well-advised in some cases to focus more attention on the "low-hanging fruit" (i.e., biologically significant, less vulnerable areas).

The Nature Conservancy's ecoregional planning projects have conducted threats assessments in several different ways. Typical of these projects, however, is the approach of the Southern Rocky Mountain Ecoregional Assessment Team (2001). Team members evaluated threats to conservation targets at conservation areas based on two factors: severity and urgency. Severity represented the degree of damage that a target could be expected to sustain over a 10-year period. Planners assigned a rating of high, medium, or low depending on the anticipated degree of degradation. For urgency, they also assigned a rating of high, medium, or low depending on the time frame in which the threats would probably have their greatest impact. They summarized the threats for all conservation areas in the portfolio, according to these ratings for severity and urgency (Table 9.3). They wrote brief descriptions to characterize the most pervasive threats across the planning region. These analyses were useful in suggesting which threats were recurring and, by extension, which threats might be the best candidates to address with multiarea conservation strategies (see Chapter 13).

In a somewhat analogous manner to the Pressey combination of irreplaceability and vulnerability, the Southern Rockies team combined information on threats with information on biodiversity value (i.e., conservation value) to establish three generalized categories of conservation priorities referred to as Activity Levels 1, 2, and 3 (Table 9.4):

Table 9.3 *Summary of Threats to Conservation Targets at Conservation Areas (Ranked High for Severity and Urgency)*

Threat	No. of Areas with High Severity and Urgency	% of Areas with High Severity and Urgency	No. of Areas Impacted by Threat	% of Areas Impacted by Threat
Parasites/Pathogens	54	29%	56	29%
Development–Residential	39	21%	81	43%
Fire Management Practices	33	18%	148	79%
Mining Practices	19	10%	95	51%
Road/Utility Corridors	12	6%	67	36%
Invasive/Alien Species–Plants	11	6%	71	38%
Management of/for Certain Species	10	5%	40	21%
Recreational Use	8	4%	104	55%
Incompatible Forestry Practices	8	4%	43	23%
Ditches, Dikes, Drainages, Diversions	7	4%	46	24%
Invasive/Alien Species–Animals	7	4%	24	13%
Development–Commercial	6	3%	11	6%
Incompatible Grazing Practices	3	2%	78	41%
Dam/Reservoir Operation	3	2%	67	36%
Recreational Vehicles	3	2%	38	20%
Development–Recreational	3	2%	13	7%
Oil or Gas Exploration	3	2%	8	4%

From the Southern Rocky Mountain Ecoregional Assessment Team 2001.

Table 9.4 *Information Used by the Southern Rockies Ecoregional Assessment Team to Set Priorities for Action among Conservation Sites*

For illustrative purposes, only a sampling of the sites is shown here. The final scores for each category were subsequently used to place sites in one of three activity levels: high, medium, or low.

Site Name	State	Conservation Value: Uniqueness			Conservation Value: Landscape Integrity			Threat: Urgency			Threat: Severity			Total Score
		Raw Score	Weight	Final Score	Raw Score	Weight	Final Score	Raw Score	Weight	Final Score	Raw Score	Weight	Final Score	
Agua Caliente	NM	2	2	4	3	1	3	2.0	2	4.0	2.0	1	2.0	13.0
Animas River	CO	2	2	4	2	1	2	1.8	2	3.6	2.1	1	2.1	11.7
Archuleta Creek	CO	3	2	6	1	1	1	2.0	2	4.0	2.0	1	2.0	13.0
Baldy Chato	CO	2	2	4	2	1	2	2.5	2	5.0	3.0	1	3.0	14.0
Baldy Cinco	CO	2	2	4	1	1	1	2.5	2	5.0	3.0	1	3.0	13.0
Beaton Creek East	CO	3	2	6	3	1	3	2.0	2	4.0	1.5	1	1.5	14.5
Beaver Creek–Lone Cone	CO	3	2	6	3	1	3	2.0	2	4.0	2.0	1	2.0	15.0
Bennett Creek–South	CO	3	2	6	2	1	2	2.8	2	5.6	2.3	1	2.3	15.9
Berthoud Pass	CO	2	2	4	2	1	2	1.6	2	3.2	2.0	1	2.0	11.2
Big Dominguez River	CO	3	2	6	1	1	1	2.1	2	4.2	1.7	1	1.7	12.9

Billy Creek Uplands	CO	3	2	6	2	1	2	1.6	2	3.2	1.6	1	1.6	12.8
Black Mountain	CO	3	2	6	2	1	2	1.5	2	3.0	2.5	1	2.5	13.5
Box Elder Creek	WY	3	2	6	2	1	2	1.7	2	3.4	1.7	1	1.7	13.1
Brush Creek at Cannibal Point	CO	3	2	6	1	1	1	1.7	2	3.4	2.3	1	2.3	12.7
Burning Mountain	CO	3	2	6	1	1	1	2.0	2	4.0	2.0	1	2.0	13.0
Butler Creek	CO	3	2	6	2	1	2	1.3	2	2.6	2.0	1	2.0	12.6
Butterfly Haven	CO	3	2	6	3	1	3	1.2	2	2.4	1.4	1	1.4	12.8
Canyon Largo	NM	3	2	6	1	1	1	2.0	2	4.0	2.0	1	2.0	13.0
Carnero Creek	CO	2	2	4	2	1	2	1.9	2	3.8	2.0	1	2.0	11.8
Castle Peak	CO	2	2	4	2	1	2	2.3	2	4.6	2.1	1	2.1	12.7
Cattle Creek	CO	3	2	6	1	1	1	2.0	2	4.0	1.3	1	1.3	12.3
Chacon Canyon	NM	3	2	6	1	1	1	3.0	2	6.0	3.0	1	3.0	16.0

Southern Rockies Ecoregional Assessment Team 2001.

Activity Level 1: Areas ranking high for conservation value and threat; need high level of conservation action within the next 10 years.

Activity Level 2: Areas ranked moderately for conservation value and threat; need monitoring of threats and status of targets with more moderate level of conservation action in the next 10 years.

Activity Level 3: Areas ranked low for conservation value and threat; need only a low level of conservation action over the next 10 years.

These Activity Levels were determined by summing values for conservation value, urgency, and severity of threat.

In their research on measuring conservation success, Salafsky and Margoluis outlined an approach to assessing threats based on their extensive work with integrated conservation and development projects in the Asia and Pacific realms (Salafsky and Margoluis 1999, Margoluis and Salafsky 2001). Although their methodology, referred to as the Threat Reduction Assessment (TRA) approach, is most appropriately applied at the scale of the individual conservation area, the criteria they have used to assess and rank threats are applicable at a broader regional scale. The TRA approach has seven basic steps, which culminate in an index that assesses how well the key threats to an area are being abated. For the purposes of a threats assessment in regional planning, the first four steps are most pertinent:

1. Define the project area spatially and temporally.
2. Develop a list of all direct threats to biodiversity, but guard against including every possible threat.
3. Rank the threats for each project based on three criteria: area, intensity, and urgency. Area refers to the percentage of a site that will be affected by the threat, intensity refers to the degree of impact on the site (e.g., modest degradation of targets or total elimination), and urgency refers to how soon the threat will have its negative impacts.
4. Add the scores for each criterion for each threat to get a total ranking of threats for the project or conservation area.

The TRA approach and threat assessment methods of The Nature Conservancy use similar criteria to rank threats in a pseudo-quantitative manner that enables planners to compare threats across different conservation areas. Both approaches rely heavily on expert opinion.

The irreplaceability–vulnerability approach outlined by Pressey and colleagues and the ecoregional approach to threats assessment illustrated by TNC's Southern Rockies team are conceptually similar. They both combine criteria of biodiversity value (e. g., irreplaceability) and threats (e.g., vulnerability) to arrive at conservation priorities. The approach of

Pressey and colleagues is more quantitative, less subjective, and less dependent upon expert opinion. However, it has some disadvantages as well. Some threats are not easily mapped, and there are many situations in which such mapped information is not available. In these cases, it will be difficult to develop any rigorous indices of vulnerability. In addition depending on the selection units used in conservation planning or the choice of conservation targets, estimates of cleared land may be poor predictors of future vulnerability of either a conservation target or a specific conservation area. The suitability indices referred to in Chapter 7 on integrity may be just as good, if not better, predictors of vulnerability because they look at several factors in addition to the amount of natural land cover (e.g., road density, human population density). As previously mentioned, defining irreplaceability in such quantitative terms may be precluded by the lack of appropriate data or algorithms. As with much of the work of conservation planning, the choice of methods will depend on available data, planning team capacity, budgets, and time.

Feasibility

In addition to biodiversity value and threats, two other criteria are helpful in setting priorities among conservation areas in the portfolio: feasibility and leverage. *Feasibility* refers to the probability that a conservation organization or natural resource agency will achieve some degree of conservation success at an area in the portfolio (Groves et al. 2000a, Valutis and Mullen 2000). It is related to several factors, including the capacity and effectiveness of conservation institutions and agencies in a region, whether it is technically possible to abate specific threats, whether the political will exists to take action in the region, and the financial costs of so doing.

For many reasons, it may not be feasible to achieve even a modicum of conservation success at certain areas. For example, some populations of endangered species may be so reduced in size that the probability of their recovery, no matter what conservation actions are taken, is very low (see Myers's conservation triage discussion at the beginning of the chapter). In other situations, abating a threat to an area may not be possible. Such is the case in some places where exotic species are already well established and nearly impossible to eradicate. In yet other cases, it may be technically possible to abate a threat, but the political or socioeconomic circumstances in the area make successful abatement highly unlikely. The rise of the so-called Anti-Environmental or Wise Use movement in the western United States provides several good examples. Because of strong opposition from these groups, federal natural resource agencies have been unable to enforce some laws or have done so only with great difficulty. A case in point was a closure of a tertiary road in northern Nevada by federal officials in the late 1990s to protect the habitat of an endangered trout species from sedimentation, and a local community's efforts (the infamous "shovel brigade") to reopen and rebuild that road. In some situations, local conservation organizations or agencies are either nonexistent or so limited in capacity that there is effectively no organization to implement conservation strategies and actions. The best laid plans will certainly go astray when there is no one or no organization to implement them.

It is clear that there are several situations where it may not be feasible to actually implement conservation action and abate threats at a conservation area. Rather than be considered a criterion with equal importance to biodiversity value and threats, feasibility may best be viewed as a filter with which to further winnow the list of most important conservation areas. Areas that are deemed not feasible for conservation action remain on the list of important places, but there is no sense of urgency until institutional, socioeconomic, or political circumstances change such that conservation action becomes plausible. In many cases, regional planning teams will have sufficient information about a proposed conservation area to determine whether it may be feasible to undertake actions there. But for many locations, a feasibility assessment will not be possible until more detailed information is gathered during the site or conservation area planning process (see Chapter 13). In general, conservation practitioners are better off determining the feasibility of taking action at a place sooner instead of later. In practice, once an organization or institution makes an initial investment in a project, it becomes increasingly difficult to withdraw from that project, even if the feasibility of abating significant threats and achieving real conservation success appears unlikely. Investment in the project by local donors and buy-in by stakeholders and partners are strong inertial forces that can sustain even projects that are likely to fail.

Leverage

In the context of priority-setting criteria, *leverage* refers to the ability to affect conservation at several areas by taking action at an individual area (Groves et al. 2000a, Valutis and Mullen 2000). Despite its intuitive appeal as a criterion, leverage can easily be misapplied, or even abused, by program managers who would exaggerate the leverage value of a project that otherwise may rate poorly with respect to biodiversity value and threats.

In reality, there will be only a few demonstrable situations in which taking action in one place will truly catalyze partnerships, funding, or other mechanisms to accomplish conservation at other areas.

One such situation arose on the Green River in Kentucky (U.S.), a stream ecosystem harboring several species of endangered fish and mussels that were continually threatened by altered flow regimes from an upstream impoundment. There, TNC worked with the dam operator, the U.S. Army Corp of Engineers, to establish a more natural hydrologic flow regime downstream of a dam that would benefit endangered fish and mussels. The working relationship on the Green River eventually led to a broader agreement between TNC and the Army's Corps to work at several dams operated by the Army Corps around the United States to improve flows and aquatic integrity in priority stream ecosystems.

Other Criteria

Two more criteria are often at play in priority setting: (1) the degree to which a particular conservation target is already receiving some degree of protection or conservation effort in the planning region and (2) the opportunity (politically, financially, or otherwise) to accomplish conservation. If a particular type of conservation target, such as a representative ecosystem, is already conserved to some extent within the planning region, it may be more appropriate (depending on other factors, such as threats and biodiversity value) to focus conservation action on other types of conservation targets.

Opportunity is a more nettlesome criterion, but nevertheless an important one. For example, an opportunity to purchase land within a particular conservation area may arise only occasionally, because of turnover in land ownership and other factors. As a result, conservation practitioners will have to carefully weigh that opportunity, even if the conservation area in question may not rank highly by other priority-setting criteria. Changes in governments similarly create opportunities to undertake conservation efforts areas on public lands.

★ ★ ★

All of the criteria discussed in this section have been applied in numerous ways to set priorities for action in conservation areas. Unfortunately, analyses with which to judge the efficacy of one application or interpretation over another have been few and far between. Some protocols for establishing priorities place greater weight on one criterion than another, although there have been no published sensitivity analyses to demonstrate the effect of that weighting on the selection of priority conservation areas. Given this conundrum, what can a conservation planner do? The best advice may be fourfold:

1. Be explicit and transparent about which criteria are used so it will be clear and repeatable to any user of a plan how the priorities were established.
2. Avoid the use of vague terminology or ratings whenever possible. For example, if a qualitative rating scheme is used to rank a particular criterion (high, medium, low), ensure that such ratings are defined clearly and unambiguously.
3. Conduct sensitivity analyses to examine the effects of including or excluding particular criteria or to assess the impact of placing greater emphasis on some criteria over others (i.e., weighting), and provide a rationale for why particular criteria or weighting schemes were used.
4. Understand that any set of criteria for setting priorities is to some degree subjective, biased, and value-laden. Decision support tools and software may help planning teams decide which criteria are most important to use and in what fashion (see below). The most effective set of criteria will be those that reduce subjectivity and clarify known biases and values.

The criteria I have discussed to this point all relate to current conditions at a conservation area. The next big step for conservation planners and biologist is to use tools that are becoming available for modeling and predicting future threats and conditions.

Extra Credit: Assessing Future Threats

Although indices to vulnerability and suitability are tools that can shed some light on future threats to biodiversity, other types of analysis may be more informative. Usually these indices are based on current or even dated information. What may be more insightful are actual predictions of future land development. White and colleagues (1997) examined the impact on biodiversity of six future landscape scenarios for Monroe County, Pennsylvania (U.S.). These included a "plan trend" alternative based on implementing a 1981 county comprehensive plan, a "buildout" alternative that assumed full development in various land-use zones, and four other alternatives that attempted to balance conservation and development. They used remote sensing to derive habitats for vertebrate species and combined the habitat data with information on area requirements to estimate species richness and habitat abundance for all terrestrial vertebrates. Not surprisingly, their models predicted that present trend and buildout alternatives comprised greater risks to habitat abundance for most species than alternative scenarios that incorporated conservation. They concluded that their methods were useful in predicting future threats to biodiversity in the region.

In a somewhat similar effort, Steinitz and colleagues (1997) examined alternative future conservation strategies for the region of Camp Pendleton, California. Criteria in their analyses included vertebrate species richness, potential habitat for nine representative vertebrate species, natural areas needed for connectivity, existing ecological reserves, important riparian zones, floodplains, prime agricultural areas, scenic and historic routes, and fire risk intensity. Analyses such as those of White, Steinitz, and their colleagues are most appropriately applied at the scale of conservation areas because they are usually too detailed and too expensive for most regional or ecoregional projects. However, they may be useful exercises for selected parts of a planning region (i.e., counties) that are likely to experience the greatest levels of human development.

A more practical example of an analysis directed at future threats in a regional planning context involves projections of future housing density. In the Southern Rocky Mountains ecoregion, residential development is occurring at a faster rate than population growth, due primarily to rural "ranchette" development and low-density suburban growth. Theobald (2000) predicted future housing densities through 2020 and 2050 (see Southern Rocky Mountains Ecoregional Assessment Team 2001, Appendices 3 and 4). Theobald's projections indicated that by 2020, about 25% of the land area in this ecoregion will consist of suburban and urban landscapes, including ranchette development. Habitat fragmentation, increased recreation, and increased roads from this type of development are all serious threats to biodiversity. The ability to predict where this sort of development will most likely occur is potentially a tremendously valuable tool for helping determine priorities for action among conservation areas in this ecoregion and for portfolios in other ecoregions experiencing a similar threat. Such information is an important complement to an assessment of current threats that may or may not be a good indicator of future conditions.

As noted at the outset of this chapter, there are many different criteria with which to establish conservation priorities, and these criteria may be defined and weighted in different ways that will affect the priority-setting outcomes. At times, conservation planning teams may need help reaching agreement on which criteria and in what manner they will be used to set priorities. For these situations, decision support tools are available to help teams decide which criteria are most important for their needs and understand the underlying reasons for their choices.

Decision Support Tools

The world is an increasingly complicated place. Decisions about the future disposition of land and water use are also increasingly complex. Most areas of remaining natural habitat that are not formally or legally protected are under some degree of threat from various types of developments. Indeed, in many parts of the world, even legally protected conservation areas are threatened. Although most conservation biologists and planners appreciate the need to establish conservation priorities, they disagree on how to proceed. In particular, many planners and biologists do not agree on what criteria should be used to set biodiversity priorities and how they should be applied in a priority-setting process (Figure 9.3). Fortunately, formal tools can be useful.

Many of these tools fall under the umbrella of multicriteria decision analysis (MCDA), which generally involves choosing among a number of alternatives (e.g., alternative priority areas for conservation) based on how those alternatives rate against a set of criteria. For example, the Analytic Hierarchy Process (AHP) is a form of MCDA that enables users to assess the relative weight of using different criteria to arrive at a decision (Saaty 1994). It has been widely applied in academia, government, and private corporations in areas of strategic planning, policy making, resource allocation, and risk assessment. Expert Choice is one of several software packages (www.expertchoice.com) that uses the AHP for structuring decisions with a model, setting priorities among criteria through pairwise comparisons of criteria, and then understanding and reporting on decisions with graphical results and sensitivity analyses.

This decision support tool can help conservation planning teams determine which criteria they will use to select conservation areas. It can also be used to develop criteria for selecting priority conservation areas for action among the many areas that will emerge from a regional or ecoregional planning process. The California Legacy Project mentioned in Chapter 8 is using the AHP to help identify potential conservation areas to meet the terrestrial biodiversity goals of the project (H. Regan, personal communication). The team identified three primary criteria (similar to the biodiversity value and threats criteria mentioned above) and then a num-

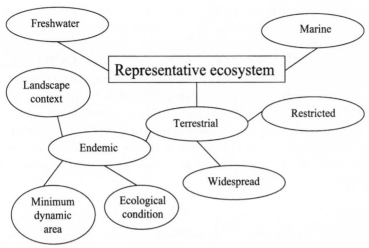

Figure 9.3   Criteria for assessing biodiversity value in regional planning. Determining how to best define a criteria of biodiversity value for setting priorities in regional planning can quickly become complicated. For example, assume that one part of biodiversity value is whether a class of targets—representative ecosystems—occur at a particular conservation area. Representative ecosystems can be broken down into at least three major types—freshwater, marine, and terrestrial—and some team members may place greater value on one type than another. Within each type, these ecosystems can be further evaluated by their range of distribution (widespread, endemic, restricted). In turn, each of those types can be assessed as to their ecological integrity by evaluating area, condition, and landscape context. Decision support tools such as the Analytic Hierarchy Process (AHP) can help sort out how much emphasis to place on each criterion and how changing that emphasis affects the overall criterion of biodiversity value and the setting of priorities for biodiversity conservation.

ber of subcriteria related to each of the primary criteria. Through a series of pairwise comparisons between all criteria, they then used AHP to determine their preferences for using certain criteria over others and the degree of consistency in how the team thought individually about each of the criteria and subcriteria.

For example, if a certain group of plant communities was a conservation target, team members may wish to consider sites that are rich in several communities, sites that contain rare communities, or sites that contain the best representative examples of all communities. With the AHP, team members could then determine whether they wanted to emphasize rarity, richness, or representativeness in pursuing the plant community goals. The AHP was used to determine which criteria the team regarded as more important in meeting the conservation goal, and it also provided a measure of the consistency of the resultant ranking of preferred criteria. At a later point in the planning process, AHP will be used to evaluate different alternative lands for protection based on different criteria and subcriteria for setting priorities.

The Analytic Hierarchy Process is only one of many MCDA methods (e.g., Hwang and Yoon 1981). Other MCDA methods that have utility for conservation priority setting are fuzzy MCDA (e.g., Slowinski 1998), attribute weighting (Pekelman and Sen 1974), and approximate ratio comparisons, to name only a few. For a comprehensive list of MCDA techniques, refer to the publications at the International Society for Multicriteria Decision Analysis Web site: http://www.mit.jyu.fi/MCDM/publ.html.

★ ★ ★

Throughout this chapter, I have reviewed a number of different criteria for setting conservation priorities in a regional planning process. I have provided several examples for how these criteria are being applied by various organizations. As the epigraph at the beginning suggested, there is no one set of right or wrong criteria for setting these priorities or one right or wrong way for applying them. Nevertheless, there are several key steps and recommendations based on the experience to date that planners should consider in setting priorities.

Key Points for Setting Priorities among Conservation Areas

1. Some measure of biodiversity or conservation value and an assessment of threats are the two most important and commonly used criteria in setting conservation priorities. Several different approaches to using these criteria are presented in the chapter, ranging from the more quantitative to more qualitative approaches based largely on expert opinion. Although it is highly recommended that planning teams use some form of these two criteria, specifically how they are applied will depend to a large degree on availability of data and on the expertise and capability of the planning team.

2. Although previously published papers on priority setting have recommended that the highest priorities for conservation action are those areas that are of the greatest biodiversity value and the most threatened, the recommendation here is that areas of high biodiversity value, whether currently threatened or not, should be the highest priorities for conservation action. Focusing only on the most threatened

places is often a focus on the most expensive conservation work. Many of the highest-quality remaining areas for conservation attention may not be under imminent threat today, but they almost certainly will face threats in the future. A dual focus on both areas that are highly threatened today as well as conserving areas that are not currently threatened, especially those places where the patterns and processes of biodiversity remain largely intact, is likely to be the most profitable long-term strategy for setting conservation priorities.

3. Although not biological in content, the criteria of feasibility and leverage are also important to consider in any conservation priority-setting process. These two criteria can help planners winnow down the set of highest-priority strategies and areas for action, though neither should carry the weight in the priority-setting process for individual conservation areas that biodiversity value and threats do.

4. No matter what criteria are used in the priority-setting process, the planning team should strive to be clear, transparent, and explicit about the selection and application of these criteria. Conduct sensitivity analyses to determine the effects of including or excluding certain criteria or the weighting of one or more criteria over others. MCDA may help determine which criteria to use in the priority-setting process and in what ways they should be applied.

CONSERVATION PLANNING FOR THE BIOSPHERE

The three chapters of
Part III focus on conservation planning
in freshwater ecosystems, marine ecosystems, and
all terrestrial ecosystems in the face of climate change. The
steps in each chapter closely parallel those developed in detail in
Part II of the book and provide additional specific information particular
to the subject of the chapters. Because of the specialized nature of these
chapters, each is written by a contributing author with expertise in
these fields: Dr. Jonathan Higgins for planning in fresh-water
ecosystems, Dr. Michael Beck for planning in marine
ecosystems, and Dr. Earl Saxon for conservation
planning in the face of climate change.

CHAPTER TEN

Maintaining the Ebbs and Flows of the Landscape: Conservation Planning for Freshwater Ecosystems

JONATHAN V. HIGGINS

> To reduce this problem [of the crisis-by-crisis
> protection of aquatic species] and to truly pro-
> tect biodiversity, ecosystems and habitats must
> be protected on a systematic basis, before they
> are so degraded that their constituent species
> become endangered.
>
> —PETER B. MOYLE
> AND RONALD M. YOSHIYAMA (1994)

Freshwater biodiversity is a significant component of life on Earth, yet it remains poorly understood, highly threatened, and inadequately represented in most biodiversity assessments. This chapter provides an overview of the status and trends of freshwater biodiversity, the types and levels of information available, and how those data have been applied to plans for conserving diversity. It focuses on how this information can best be used to ensure that freshwater conservation targets are represented in regional conservation plans. The paucity of information on freshwater biota, the limited work on classifying freshwater systems, and the need to rely on ecosystem-level targets are all reasons for providing a separate chapter on freshwater conservation planning. In addition, the upstream/downstream nature of aquatic systems, watershed (catchment) boundaries, and the types of threats affecting aquatic systems add dimensions to freshwater conservation planning that are missing from terrestrial systems. However, as we shall see in the sections ahead, the general guidance on targets, goals, integrity, and selection of conservation areas provided in previous chapters is equally applicable to the aquatic and terrestrial realms.

Freshwater ecological systems cover 0.8% of Earth's surface and harbor approximately 12% of the animals and 2.4% of all known species in the world (McAllister et al. 1997, Revenga et al. 2000). This significant component of biodiversity is in a crisis. Comprehensive global assessments of the status of freshwater biodiversity are difficult because of the lack of data for many countries and most taxa (Revenga et al. 2000). However, the few quantified estimates that exist are staggering (Allan and Flecker 1993). More than 20% of the freshwater fish species in the world have become extinct, or are threatened or endangered, including 63% of the species in South Africa and 42% in Europe (Moyle and Leidy 1992). Fifty-five percent of the extinctions of fish, birds, and mammals worldwide in the past 400 years have been freshwater-dependent species (McAllister 1997).

The status of freshwater fauna in the United States is indicative of the circumstances worldwide (Figure 10.1). Of all terrestrial and freshwater plant and animal groups, mussels, crayfish, stoneflies, fishes, and amphibians are the most threatened taxa. Over two-thirds of the species of freshwater mussels, half the crayfish species, almost half of the stoneflies, and more

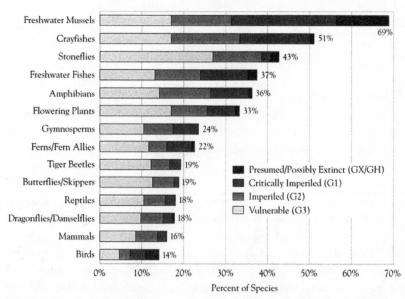

Figure 10.1 Percentages of U.S. plant and animal species at risk. Obligate inhabitants of freshwater habitats, such as freshwater mussels, crayfishes, stoneflies, freshwater fishes, and amphibians, are the most imperiled faunal groups in the United States. (From *Precious Heritage* by The Nature Conservancy, copyright © 2000 by The Nature Conservancy. Used by permission of Oxford University Press.)

than one-third of the freshwater fish and amphibian species are considered extinct, imperiled, or vulnerable (Stein et al. 2000).

The rate at which freshwater species are being lost is unknown in most places because of insufficient historic documentation. Where there are data, the picture is bleak. Miller et al. (1989) estimated that 10% of freshwater fish species became extinct in North America during the past century. North American freshwater fauna are projected to disappear at a rate five times faster than terrestrial fauna, and at a rate equal to that of tropical forests (Ricciardi and Rasmussen 1999).

Climate change, exotic species introductions, overharvesting, habitat alteration and loss, pollution, and alterations to natural flow regimes from channel and bank alteration, irresponsible land use, and dams have had significant negative impacts on aquatic systems and their resident biota (Miller et al. 1989, Ward and Stanford 1989a, Allan and Flecker 1993, Doppelt et al. 1993, Mills 1993, Richter et al. 1997a.) While there is a suite of common global anthropogenic threats, the degree of impact of each type of threat may differ regionally (e.g., McAllister et al. 1997, Revenga et al. 1998, 2000). For instance, in Africa and South America, overfishing is noted as a major source of impact, while it is now a minimal threat in the United States. On the other hand, alterations of natural flow regimes and non-point-source pollution remain as dominant sources of impact worldwide.

Assessing the status and trends of human impacts on freshwater ecological systems provides an indirect measure of the status of freshwater biodiversity. Revenga et al. (2000) summarized the condition of freshwater systems and patterns of human modifications to them at a global scale. They found that worldwide, the number of dams has increased sevenfold in the past 50 years, water withdrawals increased six times during 1990–1995, water quality has degenerated in most regions, and historic trends in commercial fisheries for well-studied rivers showed dramatic declines over the course of the twentieth century. Of 106 major watersheds worldwide, 46% are modified by at least one dam over 150 m (492 ft) in height with a reservoir capacity that exceeds 25 km^3 (15.53 mi^3) (Revenga et al. 1998). There are 306 such dams in the world, with 56 additional ones planned for construction. These statistics belie the fact that there are millions of smaller dams. Forty-two of the 106 watersheds evaluated have lost more than 75% of their original forest cover. Eighty-two of the watersheds have less than 5% of their land area protected, and the most species-rich watersheds are the least protected.

The Status of Information on Freshwater Habitats and Biodiversity

An evaluation of the types of available information for freshwater conservation planning can be divided into two types: physical or environmental attributes and biotic attributes.

Physical Attributes

The types and scales of existing information on aquatic habitats are diverse and inconsistent. The initial framework for investigating patterns of freshwater habitat and biota requires maps of watersheds, streams, and lakes. Watersheds are the basic unit that define the boundaries of the landscape that determine zoogeographic patterns, the natural characteristics of aquatic ecological systems, and many of the types and degrees of impact to them. Worldwide, very large watersheds have been mapped for major river basins of the world by Revenga et al. (1998), the U.S. Geological Survey (2001), and the World Conservation Monitoring Centre (2001). However, there is no comprehensive global map of watersheds for rivers or lakes at a spatial scale useful for regional conservation planning. High-resolution maps are available for many countries and regions within them. Watersheds can be delineated for river systems and lakes at a variety of scales using digital elevation models (DEMs). Many countries have fine-scale, high-quality digital hydrography data that allow Geographic Information System (GIS) analysis. However, most countries represent streams and lakes simply as line work at a coarse scale that does not allow for extensive GIS analysis at fine scales of resolution. The U.S. Geological Survey (USGS) makes available data on stream hydrography and ancillary data on elevation, slope, aspect, flow direction, and flow accumulation that are useful for gaining an understanding of general patterns of stream systems and the landscapes they flow through (U.S. Geological Survey 2001).

In addition to knowing the boundaries of freshwater ecological systems, planners need a way to structure the basic information on the natural and anthropogenic patterns and processes within and between ecological systems. Conservation planning requires an understanding of ecosystem types and the landscape context in which they occur. A great deal of energy in conservation science has focused on developing classification systems to define and organize information about the diversity and distribution of ecosystem types. Such classification systems are used to develop surrogates to "capture" biota not tracked as specific conservation targets (see the discussion of fine and course filters in Chapter 4). They also provide descriptions of the landscape useful for representing species and community targets in conservation areas across environmental gradients. Terrestrial ecology has a long history of vegetation and ecosystem classification, and many of these systems are being used as basic information for conservation planning. This long and productive history does not exist for aquatic ecology, and the need for freshwater ecosystem classification has become apparent with the recent focus on conservation planning.

Although a commonly accepted classification of global freshwater ecosystems has not been available, several regional classification systems provide insight into the characteristics of these ecosystems (see Hudson et al. 1992, Leach and Herron 1992, Naiman et al. 1992, and Maxwell et al. 1995 for reviews). These classification systems enable planners to move forward with conservation planning despite the lack of data on aquatic species and communities. They are based on the best available knowledge and on pragmatism. While many have skeptically asked whether these classification approaches describe aquatic ecosystems appropriately and therefore provide conservation planners with useful information, the testing of the classification frameworks is a bit of a catch-22. If data on species and communities were sufficient to directly conduct fine-scale conservation planning, classification systems could be tested for their efficacy, and they might not have to be relied on as surrogates. Classification frameworks are based on current knowledge and data, and will be improved as more information becomes available.

Classification systems are hypotheses regarding pattern. Systems useful for conservation planning depict potential biodiversity patterns at multiple scales. Using classification systems should result in more comprehensive and precise plans. Tests of the contribution and value of classification systems should evaluate whether the planning outcomes that resulted from using them are more comprehensive than those driven solely by currently available data and expert knowledge.

Planners at continental scales use several ecoregion and ecosystem-level classification frameworks to evaluate freshwater ecosystems and their biodiversity. Some ecoregions represent regional patterns of the environment, such as climate and physiography, that influence the characteristics of aquatic ecosystems (e.g., temperature and hydrologic regimes and habitat structure). Bailey (1995, 1996) described a hierarchy of ecoregions for the world and finer-scale ecoregions for North America (see Chapter 2). Bailey's ecoregions are based on climate and physiography as expressed through major vegetation cover types. Even though they are commonly thought of as "terrestrial" ecoregions, they are defined by attributes that characterize freshwater ecosystems as well. Omernik (1987) also classified and mapped an ecoregion hierarchy for North America based on physiography, land use, soils, and natural vegetation. Fine-scale ecoregions (level IV) are being generated as part of this hierarchy (e.g., Chapman et al. 2001). These ecoregions represent areas defined by a combination of natural and anthropogenic patterns in surface waters and are used to establish a framework for water quality monitoring and standards (Omernik 1995). The level VI ecoregion characterizations provide helpful information on physical and biological stream and lake characteristics and sources of anthropogenic impacts.

To develop a framework for a global assessment of ecoregional priorities for freshwater conservation, the World Wildlife Fund–United States identified freshwater ecoregions of North America (Abell et al. 2000), based in part on hierarchical patterns of freshwater biota from Maxwell et al. (1995). Similarly, Olson et al. (1997) delineated ecoregions of Central and South America and the Caribbean based on patterns of drainage, climate, physiography, and zoogeography. Freshwater ecoregions of other continents are also being developed by the World Wildlife Fund.

To adequately capture the diversity of habitats and ecosystems within planning regions such as these and ecoregions described in Chapter 2, we need a systematic way to portray them. Abell et al. (2000) identified eight major habitat types in North America, and Olson et al. (1997) described nine types for Central and South America and the Caribbean. There are many regional and global freshwater habitat classification frameworks that offer more detailed distinctions. For example, Maxwell et al. (1995) developed a general framework for classifying riverine and lake ecosystems for North America. The Nature Conservancy (TNC) has also developed methods for classifying and mapping the patterns and processes of aquatic ecosystems at multiple scales. These methods, which have been applied in North, Central, and South America, will be described in more detail later in this chapter.

Biotic Attributes

Despite their critical status, freshwater organisms have not gained the same attention as other elements of biodiversity. Freshwater biodiversity is grossly underrepresented in the literature on conservation assessment and management (Boon 1992, McAllister et al. 1997, Abell 2002). The dearth of data, characterization, and mapping and a lack of understanding freshwater biota are the result of several factors inherent to aquatic ecosystems, and these present significant challenges to conservation planning (Fausch et al. 2002).

First, freshwater species are not readily seen and cannot be evaluated from remote sensing or aerial photography. The composition, distribution, and abundance of aquatic biota are estimated through subsampling, which is organized around a few simple environmental patterns and generally occurs at easy access points such as stream-road intersections. As a result, most aquatic ecosystems are not well sampled across their environmental gradients, and the data represent disjunct sections of lake and river systems. Second, aquatic biota are mobile, are elusive, and have distinct diurnal and seasonal distributions. Third, many species have distinct life stages that are not well studied. Fourth, freshwater systems generally change volume and structure within and between years, causing changes in distribution and abundance of biota. These factors make it difficult to define and map the attributes that are vital to conservation planning.

Freshwater species are not well documented in many countries. Where freshwater census data exist, they are generally focused on threatened and endangered species or those of commercial and recreational importance. Fish species richness and endemism have been compiled for major river basins by Revenga et al. (1998) and for North American freshwater ecoregions by Abell et al. (2000). At the regional scale, information on species is varied. Continental and country atlases provide regional patterns and ranges of freshwater fishes (e.g., Scott and Crossman 1973, Hocutt and Wiley 1986, Page and Burr 1991). In North America, the Natural Heritage Program (NHP) network, Conservation Data Centers, state agencies, and NatureServe track detailed information and exact location data for globally rare and endangered species covering a range of taxa (Stein et al. 2000). These data often contain historic and current information and vary in their taxonomic comprehensiveness. Abell et al. (2000) summarized species-level patterns and levels of endangerment for North America within the freshwater ecoregional framework, and Master et al. (1998) evaluated the patterns of rare and endangered fishes and mussels for the United States.

Information on aquatic communities or species not listed as threatened or endangered is generally not included in biodiversity conservation plans and assessments. However, in the United States, several state-level "fishes of" books (e.g., Trautman 1981, Etnier and Starnes 1993) and natural resource agencies (e.g., Illinois Natural History Survey, Wisconsin Department of Natural Resources, Ohio Environmental Protection Agency) provide more comprehensive locational data for a broader range of freshwater taxa. These data include common species that can often be represented in conservation plans by targeting biodiversity at higher levels of biological organization, such as communities or ecosystems (see Chapter 4). Many agencies monitoring the status and trends of surface fresh waters in the United States have been developing programs that use biotic communities of fish and/or invertebrates to assess the status and trends of surface fresh waters. These data are useful in developing an understanding of the types and distributions of aquatic assemblages. The Nature Conservancy has developed a classification framework for aquatic communities at a fine spatial scale (Lammert et al. 1997). Since its original development, this framework has been modified for application to ecoregional planning and will be discussed later in this chapter.

The Complexity of Threats to Freshwater Biodiversity

Aquatic ecosystems remain vital for human water consumption, irrigation, transportation, fishing, and sewage transport. These, among other characteristics, make aquatic systems prone to a different set of impacts than terrestrial systems. The dynamics and connectivity of freshwater systems, and the fact that they are embedded in and derive many of their characteristics from their landscapes, result in a complex relationship between anthropogenic threats and their impact on freshwater biodiversity (Ward and Stanford 1989b, Pringle 2000). Anthropogenic impacts to aquatic systems occur at many spatial scales, and their effects on ecosystem processes and biodiversity depend on the landscape setting, type of ecosystem, and kind of impact. Changes to hydrologic, nutrient, temperature, and sediment regimes have resulted from land- and water-use changes and dams. These coarse-scale changes have affected fine-scale riverine and lacustrine habitats, as have finer spatial-scale impacts, such as shoreline revetment, woody debris removal, and a variety of human-made structures in streams and lakes.

Aquatic ecosystems commonly experience instantaneous changes to habitat volume, quality, temperature, and nutrient levels as a result of dam operations and water withdrawals. There are no globally comprehensive data on dams, water withdrawal, or water flow (Revenga et al. 1998, 2000), but databases are available for countries and states (e.g., National Inventory of Dams, http://crunch.tec.army.mil/nid/webpages/nid.cfm). Surprisingly, the status of some data gathering is worsening. For instance, stream flow data are becoming less available in some areas because fewer monitoring stations are being funded.

The connected nature of various habitats in aquatic ecosystems, and the complex and mobile life histories that have evolved in many aquatic biota, presents challenges to evaluating impacts. Many freshwater biota use a variety of habitats in lakes, streams, and even oceans to complete their life cycles. Most obvious are fishes that migrate from oceans to streams, streams to oceans, and within freshwater ecosystems. This type of environmental partitioning often makes it difficult to determine specific threats because the type and level of impact on biota are often discerned at different places or life stages. The issue of connectivity and disruption of habitat accessibility is complex. Beyond direct impacts to migration from dams and levees, alterations of natural flow regimes from dam operations, water withdrawals, and various land uses can also preclude access from rivers to natural, seasonally connected habitats such as floodplains, lakes, and wetlands. Pringle (2000) provided more detailed information on the downstream, upstream, lateral, and vertical impacts to aquatic ecosystems associated with connectivity disruption.

Point- and non-point-source pollution present significant problems worldwide. Water quality conditions can cause dramatic changes to biotic assemblages. Point-source discharges from sewage and industrial facilities often change freshwater systems to open sewers, with little natural fauna remaining. Non-point-source pollution, such as sediments and chemicals from agriculture, is widespread and has a tremendous long-term impact not only on local aquatic habitats and biota but on aquatic systems and organisms hundreds of miles downstream. National databases such as the Environmental Protection Agency's Index of Watershed Integrity provide watershed-scale data on point and non-point impacts. Regional assessments can be fairly comprehensive and are now being based on biotic attributes as well as physical and chemical attributes (Environmental Protection Agency 2000b). State-level water quality assessments are stream-reach-specific and provide finer spatial scale information (e.g., Illinois Environmental Protection Agency 1996).

Non-native species often have significant effects on freshwater biodiversity, can quickly spread to connected habitats, and are also very difficult to eradicate once they are established (Mills et al. 1993, Sala et al. 2000, Lodge 2001). Many intentional and unintentional introductions of freshwater biota have permanently changed biotic composition and ecosystem dynamics through habitat alteration and biotic interactions (e.g., Moyle et al. 1986, Miller et al. 1989). For example, the zebra mussel (*Dressena polymorpha*) was unintentionally introduced to the Great Lakes via ballast water from transoceanic shipping (Mills et al. 1993). First discovered in Lake St. Claire in 1988, the zebra mussel is now widely distributed throughout the Great Lakes basin, as well as rivers of the central and eastern Mississippi River basin, many eastern river systems, and numerous unconnected inland lakes (http://www.fcsc.usgs.gov/Nonindigenous_

Species/ZM_Progression/zm_progression.html). Zebra mussels have had rapid, extensive impacts on the biological, chemical, and physical dynamics of freshwater ecosystems (e.g., Nalepa and Fahnenstiel 1995). These impacts have resulted in changes in biomass, primary productivity, benthic and planktonic species composition, water quality, and energy flow. The impacts from non-native freshwater species can go far beyond aquatic habitats, having cascading effects on terrestrial organisms, such as birds and bears, that depend on food from freshwater systems (Spencer et al. 1991).

Anthropogenic impacts to freshwater ecosystems can make them more susceptible to invasions. Habitat alterations, such as changes in hydrologic regime and riparian habitat, have made riverine systems more prone to invasion by non-native species (Minckley and Deacon 1991). The USGS maintains a database of nonindigenous aquatic species (http://www.fcsc. usgs.gov/Nonindigenous_Species/nonindigenous_species.html).

Given the level of alteration to freshwater systems, most conservation plans will require significant effort to evaluate the components (mainstem rivers, tributaries, headwaters, lakes, etc.) of drainage basins and to identify specific places for conservation and restoration within them. These plans must then consider the impacts to the components and their biodiversity, and provide strategies to abate them. These efforts must occur not only for the biodiversity that resides within a given freshwater ecosystem, but for the nonaquatic biodiversity that depends on freshwater ecosystems for its own health and welfare. For instance, salmon are essential sources of energy and nutrients for terrestrial vertebrates and forest systems (Willson and Halupka 1995, Willson et al. 1998). Many coastal marine organisms never use freshwater systems, but they depend upon freshwater inputs to maintain appropriate salinity gradients and levels of nutrients and food sources (see the example of the Dead Zone in the Gulf of Mexico in Chapter 11).

Because restoration will be necessary in many freshwater ecosystems, analyses of the current conditions and integrity of freshwater biodiversity will also be an important component of any regional conservation plan. Identifying the places that will act as sources for biodiversity in restored areas is critical to reestablishing the natural characteristics of these systems (Frissell 1997; see also the discussion of ecological restoration in Chapter 7).

Global and Continental Analyses of Freshwater Biodiversity

Several global assessments of freshwater biodiversity have used existing data and expert opinion to guide priority setting at a variety of scales. The United Nations Environment Programme (UNEP) and the World Conservation Monitoring Centre (WCMC) used species richness and endemism of fishes and selected invertebrates to identify 136 rivers, lakes, watersheds, and countries as places of special importance for freshwater diversity (Groombridge and Jenkins 1998). Revenga et al. (1998) identified conservation priorities at the scale of major basins worldwide. The ecological criteria for evaluating basin "value" were fish species richness and endemism and endemic bird areas. Duker and Borre (2001) summarized recent global priority-setting exercises for lakes and developed a framework to identify priority lakes for freshwater biodiversity conservation action. The World Resource Institute (WRI) conducted a global evaluation of freshwater priorities using species data and levels of human impact to identify specific river and lake basins in need of conservation attention. Although these global assessments are not comprehensive, the WRI analysis many did identify specific places on the ground in need of conservation action.

As previously noted, the WWF identified global and continental priority ecoregions for conservation. These ecoregional assessments are generally conducted at a finer spatial scale than the large basin assessments described above. Similarly, Master et al. (1998) evaluated 2011 watersheds in the contiguous United States for the conservation of rare and endangered fishes and mussels at a finer spatial analysis than the larger-scale assessments mentioned above. Results from these global and continental-level priority assessments have been useful in drawing attention to some of the most threatened large basins worldwide, and watersheds on a continental scale, that support high numbers of endemic and threatened species. The next step is to identify the suite of specific places within these large basins and watersheds that will conserve the full array of freshwater biodiversity and not just the endemic or vulnerable species components.

To conserve freshwater biodiversity as a whole, comprehensive assessments of rivers and lakes *within* ecoregions and basins need to be conducted. Such assessments of freshwater biodiversity are being conducted currently at the state, drainage basin, and ecoregion levels. The Gap Analysis Program in the United States (see Chapter 5) was developed to assess the patterns and levels of protection of biodiversity on a state-by-state basis, culminating in a national roll-up to indicate the gaps in protection of biodiversity (Jennings 2000). Although Gap Analysis programs historically focused their assessments on terrestrial fauna and flora, a nascent Aquatic Gap Program has recently emerged and is being piloted in Missouri and exported to several other states. In addition, both TNC and the WWF are identifying aquatic conservation priorities as part of their ecoregional planning processes (Abell et al. 2002, Groves et al. 2002). The remainder of this chapter will outline an approach to setting freshwater conservation priorities within ecoregions, largely based on the methods and experiences of TNC.

The Regional Planning Process

A six-step process for conservation planning for freshwater ecosystems is described next. It essentially parallels the planning process outlined in detail in Part II of this book.

Establishing a Classification Framework (Step 1)

What should planners use as a framework to start a regional planning process? What are the most meaningful assessment units, and what types of environmental patterns and processes must they pay attention to? Watersheds are natural units to evaluate freshwater biodiversity because they create zoogeographic range boundaries. However, ecoregions also correspond to patterns of distributions of freshwater biodiversity (Maxwell et al. 1995). They provide a spatial framework for delineating ecosystem patterns, and watersheds may be used to study aspects of freshwater ecosystems within this ecoregional framework (Omernik and Bailey 1997). Consequently, it is important for planners to use both ecoregional and watershed boundaries at different scales in planning. Regardless of the type of planning or assessment units, basins and ecoregions need to be subdivided in order to capture environmental variation (see Chapter 2). Several researchers have illustrated that drainage basin and physiography are both important determinants of freshwater biodiversity patterns within study regions (e.g., Pflieger 1989, Angermeier and Winston 1999).

The Nature Conservancy has developed a hierarchical classification framework that uses both ecoregional and watershed boundaries to define and map patterns of freshwater ecosystems at multiple scales (see Figure 10.2, in color insert). This hierarchy is based on the frameworks of Frissell et al. (1986), Tonn (1990), Moyle and Ellison (1991), Maxwell et al. (1995), and a body of research that show hierarchies of influence on the abiotic and biotic patterns of freshwater ecosystems. Understanding patterns and processes of freshwater ecosystems and landscapes at multiple scales is critical for conserving freshwater biota (Frissell et al. 1986, Fausch et al. 2002). In addition, conservation efforts must focus on multiple levels of biological organization if they are to successfully protect freshwater biodiversity (Moyle and Yoshiyama 1994, Angermeier and Schlosser 1995).

In the TNC framework, the upper-level classification needs (aquatic zoogeographic units and Ecological Drainage Units) are defined by continental patters of freshwater biodiversity. For North America, we have relied upon Zoogeographic Subregions defined by Maxwell et at. (1995) and freshwater ecoregions developed by the World Wildlife Fund (Abell et al. 2000) to obtain information on these upper-level units. Zoogeographic subregions are drainage basins that are distinguished by patterns of native fish distribution and account for climatic, geological, and biological history. In regions where maps on zoogeography do not exist, similar units can be developed with assistance from experts. The next level of the hierarchy is a regional stratification unit called the Ecological Drainage Unit. An EDU is a group of watersheds that share a common zoogeographic history as well as physiographic and climatic characteristics and represent an area of distinct ecosystem patterns and biotic characteristics. A variety of mapped information is used to define an EDU. In the United States, EDUs are defined by aggregating eight-digit hydrologic units (Seaber et al. 1987) that share common zoogeographic patterns and have similar climatic and physiographic patterns. Characteristics of regional climatic and physiographic patterns are defined by ecoregional Provinces, Sections, and Subsections (discussed in Chapter 2) and Omernik level IV ecoregions (e.g., Chapman et al. 2001).

Outside the United States, EDUs can be developed using comparable information. The finest scale of watersheds is used as the basic unit to agglomerate. Where fine-scale watersheds are not available, planners can use DEMs to create them. The finer scales of the classification framework, aquatic ecological systems and macrohabitats, are described next.

Identifying Conservation Targets (Step 2)

Freshwater conservation targets are selected at a variety of spatial scales (local to regional) and levels of organization, from species to communities and ecosystems. The majority of the targeted imperiled, threatened, and endangered species, or species of special concern, are mussels, crayfishes, and fishes. Some insects have been conservation targets, but this is not common. It is likely that other aquatic taxa are imperiled as well.

All life cycle stages and habitats must to be taken into account when planners define the spatial scale of a species, as adult distributions may be deceiving. For instance, adult mussels do not move great distances, but the dispersal range of the larval stage (glochidia) is determined by the range of the host fish. Fishes that have floating eggs can have their eggs dispersed substantial distances, and rearing habitats are often in different places from spawning habitats. Many wide-ranging species migrate and cover thousands of kilometers and perhaps dozens of ecoregions during their life cycle. For instance, American eels (*Anguilla rostrata*) migrate approximately 3000 km (1864 mi) from central Illinois to the Sargasso Sea. In such cases, wide-ranging species should be evaluated in the context of the multiple ecoregions or habitats (e.g. freshwater/ocean) in which they occur, independently of the spatial constraints of the study area (see Chapter 4).

Species distributional data are most commonly portrayed as polygons representing ranges or as sampling points on a hydrography map. Planners should be careful when working with polygon and point data and making inferences regarding the actual extent of lake and riverine habitat used by species. While range maps indicate the extent of a species distribution, they do not indicate that a species is distributed throughout the range. Sample point data do not indicate the range of a population or species within a river or lake. Discussions with experts can be useful in determining more precise extents of a population and/or species within a given ecosystem or region, and specific stream reaches can be highlighted in a GIS to more accurately represent spatial biotic patterns (Smith et al. 2002). Migratory species use distinct sections of riverine habitat for migration, spawning, and rearing. Data representing these distinctions are very useful for assessment and are most commonly available for fishes of commercial concern that migrate between rivers and oceans. Many freshwater species move within systems during different life stages, but these distinctions are not well documented or mapped.

Unlike their terrestrial counterparts, aquatic communities are generally not formally classified. However, a framework has been proposed (Lammert et al. 1997) and implemented in regional analyses in the United States (Higgins et al. 1998, Miller et al. 1998) and improved upon by Langdon et al. (1998). These reports provide sets of species that are diagnostic of repeating assemblages of aquatic species and the habitat types in which they are found. For instance, stonecat, northern hog sucker, creek chub, and horneyhead chub (*Noturus flavus, Hypentelium nigricans, Semotilus atromaculatus,* and *Nocomis biguttatus,* respectively) are diagnostic of an assemblage of 22 species of fishes generally found in low-gradient, medium-sized streams with low and moderate groundwater inputs in the lower peninsula of Michigan. Brook trout and slimy sculpin (*Salvelinus fontinalis, Cottus cognatus*) are diagnostic of an assemblage of five species of fishes that are found predominantly in moderate- and low-gradient small streams with high or moderate groundwater inputs in the same region (Higgins et al. 1998). Here are the steps for identifying communities:

1. Collect comparable data that represent comprehensive fish and/or macroinvertebrate assemblages.
2. Identify the samples that are from the least impacted sites and potentially contain the most natural assemblages. This step requires using available data on water quality, habitat and biotic assemblage data, condition, and land use and land cover.
3. Conduct a set of multivariate analyses to identify repeating and closely related groups of species assemblages. Angermeier and Winston (1999) provided a fairly comprehensive set of analytical methods that can be used.
4. Look for correspondence of assemblage types and environmental gradients using ordination techniques (McCune and Mefford 1995).

Results from these analyses should be interpreted with caution. Most riverine sampling efforts are limited to streams that can be waded. Larger rivers and smaller streams are often not included or are underrepresented in inventories, although some efforts are under way to assess large rivers as well (Environmental Protection Agency 2000a). Data from lakes generally focus on game fish, but even these data can show distinct assemblages (Schupp 1992). In addition, statistical analyses of presence/absence data will result in different patterns than analysis using relative abundance data. The relative proportions of species in unique community types may be drastically different, despite similar species composition.

In most situations, developing and implementing a freshwater community classification will require data that are unavailable, or it will demand resources beyond the scope of most regional conservation planning projects. An alternative to classification based solely on biological data is one in which freshwater ecosystems are classified and mapped using environmental patterns and processes that shape community structure. Similar approaches using abiotic-based classification models as targets have been used to capture terrestrial and marine species and communities across the environmental gradients that exist throughout a planning region (Ferrier and Watson 1997, Ward et al. 1999, Wessels et al. 1999, Groves et al. 2000a, Pressey et al. 2000). Freshwater ecological systems identified through this process can be used as conservation targets. As targets for conservation planning, these aquatic "ecological systems" represent the interrelation of stream, lake, and wetland networks, nutrient flow, energy exchange, and source/sink dynamics that are all key to the persistence of communities and individual species. Note that *ecological systems* is the name of a specific type of conservation target in freshwater conservation planning and should not be confused with the more generically and broadly applied term, ecosystems.

Patterns of environmental conditions that determine the characteristics of freshwater ecosystems and are known to influence biotic patterns are used to classify aquatic ecological systems. Typically, these classifications are limited to available mapped data, and they are developed using GIS data for stream hydrography, surficial geology, land surface elevation, and other ecologically relevant environmental factors. When fine spatial-scale data are available, a bottom–up approach to classifying and mapping freshwater ecological systems can be conducted. When only coarse spatial-scale data are available, a top–down approach is recommended.

A Bottom–Up Classification of Ecological Systems. Freshwater ecological systems can be defined and mapped by the types of fine-scale components that occur within them, referred to as macrohabitats (see Figure 10.2, in color insert). Macrohabitats are small to medium-sized lakes (or lake basins) and river valley segments that are relatively homogeneous with respect to local and regional environmental features. Freshwater ecological systems are catchments comprising repeating occurrences of a limited set of macrohabitat types. Thus, in a bottom–up approach, freshwater ecological systems are described by first defining and describing the macrohabitats that occur within a region and then identifying repeating groups of macrohabitat types that form freshwater ecological systems. This strategy provides detailed information on the components of ecological systems, which is useful for the planning and management of conservation areas.

This approach to mapping and classifying macrohabitats is based on the work of Seelbach et al. (1997), who described stream valley segments in Michigan, and Cupp's (1989) description of valley types. It is also similar in scale to channel units defined by Paustian et al. (1992) and comparable methods for mapping and classifying described in Stanfield and Kuyvenhoven (2002). As a first step, experts identify abiotic variables that shape regional freshwater system structure and function, and they seek digital data sources representing these variables. The variables that are typically used to define macrohabitats include hydrologic regime, waterbody size, connectivity, drainage network position, gradient, valley morphology, catchment geology, and elevation. While these variables do not fully explain the distributions of aquatic fauna, they have been shown to constrain or influence them by mediating a range of environmental factors (Poff 1997).

After experts have identified the suite of variables that are available in spatial data, ecologically significant classes of these variables are specified.

For example, stream segments may be sorted into a small number of gradient classes based on their range of slopes in the assessment region and those gradient classes that appear to shape biotic distributions. Both the variables selected to define macrohabitats and the classes into which these variables are broken are regionally specific. This is critical, because the environmental gradients and their influence on ecological pattern and process vary regionally.

Following the selection of environmental variables and class types, GIS tools (The Nature Conservancy 2000c) are used to attribute river segments and lakes in digital hydrography data with appropriate class values for each variable. Frequently, these tasks are accomplished using spatial data on hydrography, elevation, and geology (see Figure 10.3a–d, in color insert). When used with GIS tools, these data can be used to define stream gradients, upstream and downstream connectivity classes, stream order and link, watershed area, and geological composition of the contributing area, which can be subsequently used to model attributes such as stream and lake chemistry and hydrology. Unique co-occurring combinations of classes of all variables constitute each macrohabitat type (Figure 10.3d).

These automated GIS tools can be modified to work on any hydrography data that have flow directionality as an attribute. Where such hydrography data are not available, a bottom-up freshwater classification can be developed manually. In regions that have fairly simple levels of landscape complexity, manual attribution and delineation of the abiotic variables described above on maps or with a GIS can be carried out relatively easily. In complex landscapes, it is more time-consuming, unless the scale of valley segments is increased to be more pragmatic given the constraints and objectives of the assessment.

Manual attribution of valley segments and macrohabitat determination can be done following the methods in Higgins et al. (1998), where a relational database is used in conjunction with a GIS (e.g., ArcView). As segments are delineated based on a suite of environmental patterns, the segments are labeled with a unique identifying number, and the suite of habitat attributes (e.g., geology, gradient) are added to the database describing attributes of every segment. With this method, both GIS data and paper maps can be used to make decisions about macrohabitat boundaries and attributes. The resultant product is a GIS-coverage of macrohabitats and a relational database. Lake macrohabitats are classified using data sets similar to those described above and a separate set of GIS tools. These GIS tools derive attributes such as underlying geology, elevation, size, shoreline complexity, and number of surface connections.

In the final step of the bottom-up approach, aquatic ecological systems are defined by aggregating macrohabitat patterns (see Figure 10.3e, in color insert). There are two methods of aggregating macrohabitats into systems. The first is to visually assess the patterns of classification variables using the macrohabitat data. This process entails the observer deciding where significant breaks in environmental patterns exist and delineating boundaries of ecological systems manually. While subjective, it allows the evaluation of multiple factors and different weighting of factors in different circumstances. The second aggregation method is to use a statistical clustering algorithm to group watersheds based on their component macrohabitats. This approach is advantageous in that clusters are formed using an objective and repeatable method, data can easily be revised, and clusters can be designed to conform to a set of watershed size classes.

A Top-Down Classification of Ecological Systems. In many parts of the world, appropriate data are not available to permit a bottom-up freshwater classification approach. In such regions, top-down classifications may be substituted. In this approach, freshwater ecological systems are delineated directly from available digital data, paper maps, and other appropriate data sources, *without* an a priori classification of macrohabitat types. Using these resources, experts distinguish important environmental patterns and boundaries and begin to identify the patterns of diversity in aquatic ecosystems.

Ecological systems derived from top-down approaches are not intended to distinguish the fine-scale differences within riverine and lacustrine ecosystems, but to characterize the potential differences between them. A variety of strategies have been used for developing top-down ecological system classifications, depending on circumstances. Most frequently, ecological systems are mapped directly on the hydrography data, either manually or digitally, by evaluating patterns of hydrologic regime, physiography, geomorphology, vegetation, and drainage as observed on paper maps.

Figure 10.4 (in color insert) provides an example from the Pantanal, Brazil. Ecosystems were mapped in the Pantanal by first defining the variables most important to determine the characteristics and boundaries of the freshwater systems. In this case, they were temporal patterns of flooding, elevation, geology, landform, vegetation, gradient, and stream size. Initially, dramatic gradients, such as fall lines from large elevational changes and contrasts in geology, landform, and vegetation, provided places to draw boundaries of ecological systems. Further evaluation of geology as it relates to groundwater and water clarity, chemistry, and temperature were used to make finer demarcations among ecological system types. Tributaries were identified as distinct from mainstem large rivers because of the size of the habitat and the different seasonal hydrologic patterns. As each ecosystem type was delineated, descriptions of its characteristics were written. Repeating patterns of the characteristics were delineated and assigned the same ecological system type. After completing a draft map of ecological system types, experts reviewed and refined the types and boundaries in a workshop setting.

After identifying and mapping the spatial distribution and frequency of species, community, and ecological system targets, the next step in the planning process is to establish goals for representing these targets in a portfolio of conservation areas.

Setting Conservation Goals (Step 3)

Recall from Chapter 6 that goals have two components: representation and quality. The representation component describes how many or how much of the target needs to be conserved and how the target should be distributed across the planning region. For freshwater targets, EDUs are the stratification units to achieve the desired distribution of conservation targets across environmental gradients in the planning region (see Figure 10.2, in color insert).

Numeric goals for representation are established following the guidance in Chapter 6. For species and communities, relatively higher numeric goals should be established for endemic and endangered or vulnerable targets (Table 10.1). Goals for ecological systems are based on relative abundance. For instance, headwater systems are more common than large rivers, so the numeric goals will be different (Table 10.2). Goals for ecological systems have been developed by defining the minimum number of occurrences of these systems that need to be conserved per ecological drainage unit.

The goal-setting process should ensure that the necessary habitats, connectivity, and a sufficient number of populations to buffer against events causing local extinctions are incorporated into the goal statement. Goals for aquatic species targets can be derived in several ways. Experts on specific taxonomic groups can meet and develop the numeric and spatial distribution goals one species at a time. Information on minimum habitat size, connectivity, population size, and metapopulation dynamics would make goals more robust. However, this level of information is not available for most species targets. Consequently, sound professional judgment is most often used for groups of similar targets (see Chapter 5).

Table 10.1   *Examples of Goals per EDU for Representation of Species Targets in Conservation Areas*

Information on Global Rank categories and spatial scale categories is provided in Chapter 4. Chapter 6 provides definitions for the distribution category. Widespread = > 1 EDU historically occupied by species.

Common Name	Scientific Name	Target Category	Global Rank	Spatial Scale	Distribution	EDU Goal
Paddlefish	*Polyodon spathula*	Fish	G5	Regional	Widespread	1
Spectaclecase	*Cumberlandia monodonta*	Mussel	G2G3	Intermediate	Widespread	2
Tennessee dace	*Phoxinus tennesseensis*	Fish	G3	Local	Widespread	3
Mississippi flatwoods crayfish	*Procambarus cometes*	Crayfish	G1	Local	Limited	10
Flattened musk turtle	*Sternotherus depressus*	Reptile	G2	Intermediate	Limited	5

From Smith et al. 2002.

It is also important to consider the life history characteristics of species and the current landscape conditions. For example, species with local ranges may survive in a number of streams isolated by reservoirs, but because the streams may be functionally isolated, the number of streams containing such species (and, by inference, the conservation goal) may need to be relatively higher compared to situations where connectivity

Table 10.2   *Examples of Goals per EDU for Aquatic Ecological Systems from the Cumberland Southern Ridge and Valley Ecoregional Plan of The Nature Conservancy*

Minimum length refers to length of stream segments.

Aquatic Ecological System	EDU Goal	Minimum Length
Large Ridge and Valley rivers, origin in Blue Ridge and Ridge and Valley	1	40 km
Small Coastal Plain rivers, connecting downstream to saline embayments	2	15 km
High-gradient streams transitional from the Blue Ridge to Piedmont	3	5 km
Low-gradient, hydrologically unstable coastal plain streams in marls and calcareous clays	3	5 km
Oxbow lakes in Pleistocene alluvium	1	—

From Smith et al. 2002.

and metapopulation dynamics are still functioning. Schlosser (1991) and Fausch et al. (2002) discuss the landscape perspective of stream fish ecology, with valuable guidance to address spatial dynamics of stream systems and for setting goals. Similarly, Hilderbrand and Kershner (2000) address spatial requirements, metapopulation, and connectivity issues for inland cutthroat trout (*Oncorhynchus clarki*) conservation. When freshwater mussels are identified as targets, the goals for host fish species should also be addressed in a coordinated fashion, because the viability of the mussel species depends on the viability of the host fish, and the fish host can occur over a much larger spatial scale.

Evaluating Viability and Ecological Integrity (Step 4)

Once freshwater species and ecological system targets have been mapped, the next step is to evaluate the condition and viability of target populations and occurrences. This is done using a combination of expert knowledge, existing data, and GIS to address the size, condition, and landscape context criteria for target occurrences (see Chapter 7).

Size criteria for aquatic species targets are similar to those described in Chapter 7. Size criteria for ecological system targets can be defined as the minimum length/area needed to maintain the ecological processes and biota within the system type (see Table 10.2). Condition criteria are applicable to species, communities, and ecological systems. For species, they are the biological attributes of population density, age structure, sex ratios, and spatial distribution. For communities and ecological systems, species composition is one measure of condition and should be evaluated in terms of the proportions of native species as well as the proportions and types of exotic species. The Index of Biotic Integrity (IBI) and other measures of biological composition, such as the South African Scoring System, can be used to develop condition criteria for communities and ecological systems (e.g., Davis and Simon 1995, Simon 1999, Water Research Commission 2001). Landscape context refers to the environmental regimes (e.g., hydrologic, temperature, nutrients, sediment), physical habitat characteristics (e.g., woody debris, substrate types, habitats defined by geomorphology), and connectivity among habitats, as well as the status of sources of threats, such as dams, land use, and roads.

While local experts who are familiar with an area are the best source of information about a particular site, their knowledge is usually not comprehensive across the region of interest. Assessments of habitat quality, stream, lake, or watershed integrity, and priority conservation areas by government agencies, conservation organizations, or academic researchers also provide valuable biological and ecological data for evaluating viability and integrity of conservation targets. For instance, Moyle and Randall (1998) developed a watershed index of biotic integrity (W-IBI) using native ranid frogs, native fishes, native fish assemblages, anadramous fish, and fish abundance. Watershed-level assessments provide average values for all streams and lakes, and do not identify the specific, highest-quality examples within the watershed. However, these assessments do identify the most intact watersheds, which are generally among the highest conservation priorities. Reach-level data from natural resource agencies are generally not comprehensive, but they are more spatially specific and can highlight some of the best examples within watersheds. These data are very helpful when working in highly altered and fragmented landscapes, as conservation priorities are not often defined as whole watersheds.

Data on watershed characteristics, such as land use, land cover, roads, and dams, are useful descriptors of the landscape context as it relates to sources of threats and potential quality/integrity of freshwater ecological systems. For example, Jones et al. (2001) showed that spatial land-use data strongly correlated with sediment and nutrient loadings in streams. Recent research has identified relationships between an IBI and these characteristics. Moyle and Randall (1998) evaluated the association of human disturbance with W-IBI scores and found that low scores were correlated with the percentage of area containing dams, roads associated with streams, reservoir capacity, and percentage of each watershed that was historically fishless. Roth et al. (1996) showed that stream condition as estimated by an IBI and a Habitat Index (HI) was strongly correlated with land use at the watershed scale; riparian vegetation provided a weak indication of local condition. How broadly this finding can be applied across the variety of freshwater ecosystem types is not known.

The GIS tools used to generate watershed characteristics for classification purposes can also be used to generate a series of quantified quality indicators to describe the landscape context for the contributing area or local area of each ecological system occurrence, or for each of the smaller-scale macrohabitats. Evaluating the integrity of macrohabitats may be advantageous in regions with landscapes of heterogeneous patterns of impact. In these cases, there may not be many large-scale ecological systems in good condition, and the search may focus on smaller remnant pieces that can be used as centers for restoration (see Chapter 7). The highest-quality streams can be selected as high-quality examples of species targets, or if they meet the minimum length criteria, as examples of ecological systems. Typical indicators of quality or degradation are percent impervious area, percent agriculture, density of roads and road stream crossings, dams, and point sources of pollution.

While the state of current knowledge is insufficient to build a robust predictive model of the values of biotic integrity in all stream reaches using these indicators, these data are still useful as a filter or screen to evaluate the suitability of different areas for conservation. These sorts of filters can identify the most intact and most altered landscapes. However, these filters must be applied with some degree of caution. Streams with good surrounding land cover may have suffered from historic impacts, such as point-source pollution, past logging, or invasion by exotic species, that are not apparent from current land cover data. Harding et al. (1998) showed that the ghost of land use past (land-use history from 50 years ago) was a better predictor of stream fish and macroinvertebrate diversity than present conditions.

Indicators of landscape condition can be mapped individually (Figure 10.5a) or as an overall suitability index (see Chapter 7) based on combined values of indicators (Figure 10.5b). The robustness of using an overall indicator is debatable, as the impact of certain individual factors may be greater or lower than others, and an overall indicator represents an average value for all individual indicators. Assessing local patterns of impact and the cumulative impacts from the contributing area of a target occurrence is important, because different types of impacts have different levels of spatial influence. Cumulative impacts from land use affect water quality far from their sources (e.g., Dead Zone in the Gulf of Mexico; see Chapter 11), while the quality of riparian habitat may have a great impact on local conditions.

Identifying Conservation Areas (Step 5)

Identifying the portfolio of conservation areas can be accomplished in a series of manual steps or all at once in an automated manner. When done sequentially, the first step is to identify the areas that contain viable examples of species targets. While many areas contain a variety of combinations of species and community targets, those areas that have the highest levels of viability and ecological integrity are selected first. This process can be accomplished in an iterative way with review by a group of experts. It results in an array of areas that are considered "hotspots" (i.e., contain several species and community targets) in addition to those that may contain only one population of a target species. A review of the selected areas will

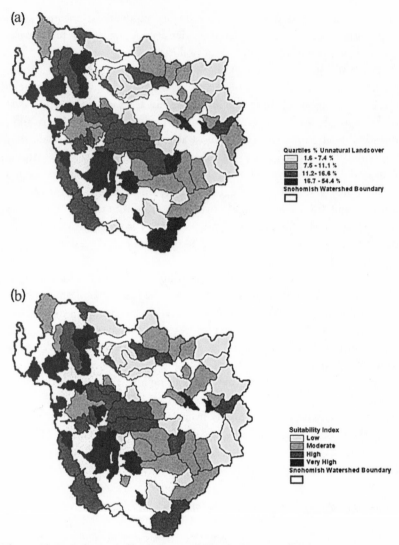

Figure 10.5   Indicators of landscape condition. (a) Patterns of landscape context/sources of threats shown in quartiles of percent unnatural land cover for aquatic ecological systems of < 100 km² in size. (b) Suitability index derived using a combination of land use/cover, dams, road crossings, and exotic species indicating degree of alteration in aquatic ecological systems of < 100 km².

ensure that endemic and threatened and endangered species are well represented and that goals are met to the greatest extent possible.

In addition, selected areas need to be embedded in a landscape that provides connectivity for separate life history needs, metapopulation dynamics (see Chapter 6), and resilience to random events like floods, droughts, and forest fires. Migratory fishes such as salmon need contiguous oceanic, migratory mainstem river, spawning, and rearing habitats accessible in order to complete their life cycle. Saunders and colleagues (2002) provided a framework for the selection of freshwater conservation areas that includes whole-catchment management, natural flow maintenance, and exclusion of non-native species.

Some examples of aquatic ecological systems and communities will be "captured" in conservation areas through a focus on species. However, the variety of ecological system targets is not likely to be represented through a species or community focus, and vice versa, since species targets are generally limited to rare and endangered taxa and do not include the wide array of common and representative taxa. The next step, then, is to identify those ecological system targets that still need representation in conservation areas after species and communities have been addressed. Subsequently, planners should identify the best remaining examples of ecological systems that have the highest degrees of integrity and best meet the numeric and distribution goals established for those targets. If there are multiple examples of some ecological system types potentially available for conservation purposes, those that provide needed connectivity for species or community targets, or are most complementary to types already represented in conservation areas, should be selected first.

An alternative is to use area selection algorithms (see Chapter 8). If an algorithm is used, watersheds are the most meaningful selection unit for freshwater targets, as opposed to hexagons or other types of polygons. Different scales of watersheds corresponding to the different-sized ecological systems are used. Species and community targets can be evaluated separately or they can be attributed to the watersheds, and the watersheds become the unit of analysis for the algorithm. This was done for fish and mussel species at a coarse scale by Master et al. (1998). These algorithms can be run on just freshwater targets, or together with terrestrial and marine targets being evaluated in the regional analysis. If terrestrial and aquatic target occurrences are analyzed together, the general preponderance of terrestrial data may result in a biased selection process that inadvertently excludes areas important to freshwater diversity. Comparing individual runs of these algorithms on freshwater targets alone with runs of freshwater, terrestrial, and marine targets together is advisable to evaluate how the outcomes may differ. The Middle Rocky Mountain–Blue Mountain Ecoregional Plan (Klahr et al. 2000) provides detailed information on the use of an area selection algorithm for selecting aquatic conservation areas.

Freshwater conservation areas can be displayed in several different ways, depending on the selection unit used and whether area selection algorithms are employed. Individual lakes and streams or river segments can be high-lighted. An alternative method is to highlight the watersheds for headwaters and creeks, and to highlight the mainstem sections of rivers and individual lakes. Groups of lakes and wetlands can be highlighted as polygons.

Designing a Network of Conservation Areas (Step 6)

Once conservation areas have been identified, connected corridors of streams and lakes in watersheds that have sufficient integrity to support conservation targets or are feasibly restorable are added to the existing portfolio. Designing a network of conservation areas in this manner for aquatic targets is especially critical, given the importance of connectivity to key ecological processes and species migration (see Pringle 2000, Saun-ders et al. 2002). Occurrences of certain ecological systems may provide needed connectivity between habitats for a species or community target but may not meet the minimum thresholds established for ecological integrity. Rivers with levees and streams with armored banks may not meet the integrity criteria for examples of their ecological system type, but they may suffice as connecting corridors. An example of a network of freshwater conservation areas designed with the methods outlined in this chapter is provided in Figure 10.6 (Smith et al. 2002).

Given the fact that the vast majority of river systems and lakes are degraded and fragmented to some degree, ecological restoration (see Chapter 7) must play a critical role in the conservation of freshwater diversity. Restoration may be as simple as developing an ecologically ben-eficial flow regime for regulated rivers or as difficult as implementing watershed-wide best management practices. In many cases, freshwater sys-tems have shown resilience and have responded rapidly to the reduction of impacts. To address altered flow conditions from dams, long-term historic gauge station data can be evaluated to characterized historic flow attrib-utes to set restoration goals (Richter et al. 1996, 1997b).

Aquatic ecosystems affected by hydrologic alteration or fragmentation often show rapid changes in habitat and channel morphology, vegetation, or movement and composition of fishes in response to changes in dam operations and flow regimes. Restoration efforts in the Kissimmee River (Toth 1995), the Upper Colorado River (Stanford 1994), and the Grand Canyon (Collier et al. 1997) are examples of positive responses from such "restoration" efforts. Systems affected by chronic pollution may take longer to respond, but they can still recover or change to a more desirable state (e.g., Lake Erie and Lake Ontario phosphorous levels) (Environment

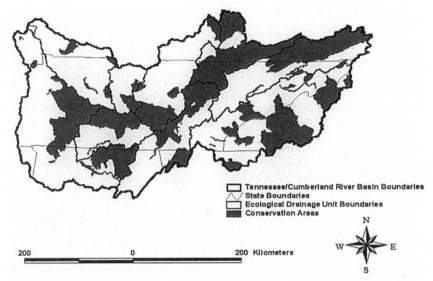

Figure 10.6   Conservation areas in the Tennessee and Cumberland River basins in the southeastern United States. This map represents 69 conservation areas stratified across nine EDUs identified using the series of steps discussed in this chapter. (From Smith et al. 2002.)

Canada and the U.S. Environmental Protection Agency 2001). Many "restoration" goals are more accurately defined as a set of ecologically sustainable goals or desired future conditions (see Chapter 7) that are generally a compromise of the management of aquatic systems for both native biodiversity and human use (Richter et al., in press).

Conservation and restoration of freshwater biodiversity cannot occur separately from terrestrial and marine efforts. While this chapter focuses on a framework for conducting regional freshwater conservation planning, it does so only for the intended purpose of bringing attention to attributes, concepts, and methods that are unique or important in conservation planning for aquatic biodiversity, and not to create more of a separation than already exists. The future of freshwater ecosystem conservation is integrally tied to the conservation of terrestrial and coastal marine systems. Both planning and conservation efforts for freshwater ecosystems need to be conducted collaboratively and concurrently with those for terrestrial and coastal marine within the same planning regions. This comprehensive approach is necessary to address the complex physical, chemical, and biological interrelationships of freshwater ecosystems with those defined as terrestrial and marine.

Key Points for Conservation Planning in Freshwater Ecosystems

1. Gather basic information, maps, and GIS data on patterns of biota, streams, lakes, watersheds, elevation, geology, vegetation, and land use in the study region. Refer to information on ecoregions being developed by the WWF and others to get a basic understanding of regional ecology and threats. Where maps of watersheds do not exist, they can be created using Digital Elevation Models.

2. Identify the list of species targets using existing classifications (e.g., IUCN Red List) and expert knowledge. Map the occurrence of these species targets using spatial data and expert knowledge at the scale that is known and available. If available, statistically evaluate biotic assemblage data to identify communities and map their distributions.

3. Classify and map Ecological Drainage Units (EDUs) and ecological systems using data on hydrography, elevation, geology, vegetation, and regional patterns of biota. Discuss models for classification with regional experts to ensure that the variables and classes being used are providing meaningful results.

4. Define conservation goals for species, communities, and aquatic ecological systems for each Ecological Drainage Unit.

5. Assess the types and degrees of impacts to conservation targets and watersheds using a variety of spatial data, data on conservation target condition, water quality, ecosystem integrity, and expert knowledge. Map the patterns of impacts separately and as a single suitability index for watersheds based on the aquatic ecological systems.

6. Identify areas that contain viable examples of the species and community targets. Conduct an analysis to see if those areas also contain high-integrity examples of aquatic ecological systems. Identify areas that contain examples of the ecological system types not captured in the areas selected with species and community targets. Add riverine and lake corridors that are necessary for connectivity between areas.

7. Represent the conservation areas on maps as small watersheds for stream and creek systems, buffered corridors along larger streams and rivers, individual lakes, and polygons of multiple lakes, streams, and wetland complexes.

8. Identify areas that are necessary for restoration to meet conservation goals or to maintain ecological processes.

CHAPTER ELEVEN

The Sea Around: Conservation Planning in Marine Regions

Michael W. Beck

The oceans, unlike forests, still look like the oceans after we've removed their contents, and even scientists are susceptible to being seduced to ignore phenomena that are out of sight.
—Carlton (1998)

The importance of marine diversity, the threats it faces, and the need for better conservation in the marine environment have become increasingly clear. Marine conservation may be as much as two decades behind terrestrial conservation. Fortunately, the identification of priority areas for marine conservation, through marine regional planning, is comparatively advanced. Indeed, many advances in regional planning in general have been made in marine environments. This chapter provides an overview of marine regional planning with an emphasis on points that would not be obvious in terrestrial planning efforts. However, between the two environments, there are mostly similarities in regional planning, and a separate chapter is therefore more a matter of accessibility, not necessity. The most important point is that marine planners should be aware of the methods discussed throughout the book. In turn, terrestrial planners need to be aware of planning advances in marine environments.

This chapter is primarily concerned with planning in coastal marine or nearshore environments. While there is no specific seaward boundary for these regions, the continental shelf is often a reasonable dividing line, as there are strong breaks between the shelf and the rest of the ocean when considering species' ranges, ecological processes, threats, and conservation strategies. In some places, strong current patterns, such as the California and Humboldt currents in northern California (U.S.), create the most obvious dividing lines between nearshore and offshore environments and have strong influence on coastal diversity patterns. Approaches for conservation in far offshore areas (e.g., the "high seas") are much less likely to be area-based than those nearshore. In nearshore areas, planners must pay greater attention to the importance of the integration of planning and action across terrestrial, freshwater, and marine environments.

A Brief Overview of Marine Diversity and Threats: Facts and Myths

A significant amount of the world's diversity is marine. At higher taxa levels (e.g., orders and phyla), most of the world's biological diversity is marine. This diversity is often overlooked in regional conservation and management plans, perhaps because the threats are not as obvious, nor are the losses in diversity as easily observed as they are in many terrestrial ecosystems. While not as apparently fragmented as many terrestrial environments in the world, the marine environment is highly threatened.

Underlying most of the threats to marine diversity are three main factors. First, burgeoning human populations along coasts, with their requirements for housing, food, and income, are causing harmful effects on nearshore estuarine and marine species and ecosystems. More than one-third of the world's human population lives in coastal areas, and that proportion is growing (United Nations Environment Programme 1999). In the United States, coastal counties make up only 11% of the land area in the lower 48 states, but population density in coastal counties is nearly five times that in the rest of the country. By 2010, 75% of the U.S. population is expected to live within 80 km (50 mi) of the coast. Coastal ecosystems have been and will be increasingly threatened by development and shoreline modification.

Second, even more distant human activities on land and in freshwaters have significant, although often overlooked, effects on coastal and marine ecosystems. Watersheds link the land to the sea, and such linkages can traverse very large distances (Goolsby et al. 2000, Mitsch et al. 2001). This link between land and sea is acknowledged but has not been well addressed by most federal agencies and nongovernmental organizations (NGOs). Estuaries may be some of the most anthropogenically degraded environments on Earth, in part because the harmful effects of misguided land and river management decisions accumulate downstream in estuaries (Edgar et al. 2000). For example, the excessive input of nutrients, particularly nitrogen, in the watersheds of the Mississippi River foster algal blooms that deplete oxygen and create a zone of hypoxia (often referred to as the "dead zone") in the summer off the coast of Louisiana. The primary source of this nitrogen is in the intensively farmed lands of Minnesota, Iowa, Illinois, Indiana, and Ohio (Figure 11.1) (Goolsby et al. 1999, 2000). To conserve diversity in the Gulf of Mexico, it is necessary to increase wetland restoration and develop more environmentally friendly farming practices on these lands in the American Midwest, some 1500 km (1000 mi) from the Gulf of Mexico (Mitsch et al. 2001). In Latin America and the Caribbean, incompatible development and farming practices have caused erosion and excessive sedimentation in coastal waters, threatening diverse mangroves and coral reefs.

Third, the exploitation and destruction of marine resources by humans is increasing. It is abundantly clear that the perceived limitlessness of the

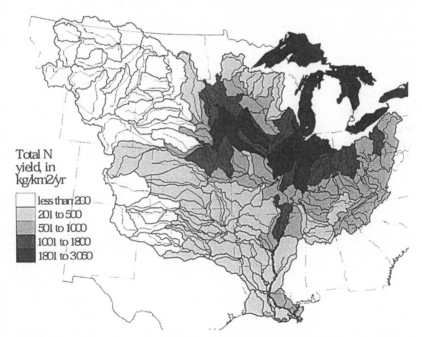

Figure 11.1   Input of nitrates in watersheds of the Mississippi Basin. These inputs help create the large zone of hypoxia or "dead zone" off the coast of Louisiana. (Modified from Goolsby 1999.)

resources of the seas is wrong. The most obvious direct exploitation is over-fishing. Fishing impacts include the direct take of targeted individuals and the often overwhelming indirect take of individuals as bycatch. Recent and historical overfishing have drastically altered ecosystems (Jackson 2001, Jackson et al. 2001). Overfishing is also drastically altering marine trophic structure. As fisheries deplete higher trophic levels (e.g., top predators) to economic and ecological extinction, their effort is increasingly directed at lower trophic levels. In other words, fisheries fish down food webs (Pauly et al. 1998). Some of the most devastating effects of fishing come from the destruction of ecosystems from such practices as trawling (i.e., scraping the bottom), blast fishing, and cyanide fishing, which kills coral reefs (e.g., Watling and Norse 1998). There are, however, many other serious threats to the marine environment, ranging from the extraction of mineral resources (e.g., oil) to the impact from shipping in ship waste, noise, and accidental spills and to the introduction of exotic species. Even excessive tourism can be detrimental, as from snorkelers and divers who damage reefs (e.g., Plathong et al. 2000) and cruise ships that dump waste.

The limitlessness of the seas is but one myth being overhauled that requires planning in the marine environment with more foresight; there are others. For example, we commonly assumed that there would be little genetic diversity in marine compared to terrestrial species, because the larvae of many marine species have the potential to disperse widely. As we look more closely and with better techniques, we find that there is significantly more genetic variation in marine species than previously presumed (e.g., Palumbi 1994, Shaklee and Bentzen 1998, Barber et al. 2000, 2002).

We also assumed that there would be few, if any, extinctions in the seas. Several recent studies dispel this myth. Global extinctions have occurred and continue to occur in the marine environment. We have seen extinctions of species such as the Stellar sea cow (*Hydrodamalis gigas*), West Indian monk seal (*Monachus tropicalis*), and Atlantic eelgrass limpet (*Lottia alveus alveus*). We may now be witnessing the extinctions of species such as the white abalone (*Haliotis sorensi*), barndoor skate (*Raja laevis*), and Texas pipefish (*Sygnathus affinis*) (Casey and Myers 1998, Carlton et al. 1999, Roberts and Hawkins 1999). Moreover, we are seeing greater incidences of local extinctions, such as those documented in the Wadden Sea, Netherlands (Wolff 2000). These local extinctions need to be taken much more seriously, because as we find increasingly greater local genetic variation, we may realize that these local extinctions were real extinctions of species and subspecies.

Finally, we had assumed that marine species and ecosystems are eminently restorable, which is almost certainly another myth. For example, it is commonly assumed that if we stop overfishing species, fish stocks will be able to rebound. In a few cases, such as Chesapeake striped bass (*Morone saxatilis*), there have been remarkable rebounds (Richards and Rago 1999), and these cases are widely promoted. However, a broader analysis of fished species suggests that rebounds are uncommon (Hutchings 2000). It is also clear that our success at restoring ecosystems such as salt marshes is limited (Minello and Webb 1997, Zedler 2000). Given our continuing inclination to drastically threaten and alter marine species and ecosystems and our limited ability to correct these mistakes, we must plan to conserve and manage the marine environment with significantly more forethought than in the past.

An Overview of Regional Planning in Marine Environments

A growing number of marine regional plans have been developed in recent years (Table 11.1). These include plans by the World Wildlife Fund (WWF) for the Sula-Sulawesi Seas, the Meso-American Reef, and the Nova Scotian shelf (Day and Roff 2000). The Nature Conservancy (TNC) has completed plans in the central Caribbean (Sullivan Sealey and Bustamante 1999), the northern Gulf of Mexico (Beck and Odaya 2001), the Cook Inlet, Puget Sound/Georgia Straits, and the Chesapeake Bay regions. The WWF and TNC jointly completed a plan in the Bering Sea (Banks et al. 2000). The Australian government is developing regional plans across the Great Barrier Reef (http://www.gbrmpa.gov.au/corp_site/key_issues/conservation/rep_areas/index.html) and elsewhere in Australia, such as Tasmania (Department of Primary Industries Water and Environment 2001). In California, a statewide marine planning exercise is currently being conducted that approximates the scale of regional planning even though the planning area is delimited by geopolitical boundaries.

Most of the formal planning in marine environments, however, has generally been done at scales smaller than regions within politically, not ecologically, defined units. For example, it is common and often mandated to have plans for states, countries, and federally designated areas (e.g., U.S. National Marine Sanctuaries), many of which are much smaller than ecological regions and do not have ecologically defined borders (e.g., Leslie et al. 2003). Many plans are done at the scale of individual bays or estuaries

Table 11.1 *A Review of Marine Regional Plans*

Region	Aims	Lead Group(s) and Partners	Targets	Examples	Goals	Tools/Methods	Data
Central Coast, British Columbia, Canada	MPA Network	The Living Oceans Society	Ecosystem, focal species	Kelp, geoduck beds, inlets, sea otter, marbled murrelet	A priori: No	Delphi, MARXAN analysis	Digital elevation model, fisheries, ground surveys
Bering Sea, Alaska, Russia	Biodiversity conservation, sustainable fisheries	WWF, TNC	Oceanographic ecosystem, species assemblages	Bathymetry, upwelling, polynyas, eelgrass, wetlands, fish, invertebrates	A priori: No	Delphi	Ground surveys
Northern Gulf of Mexico, U.S., Mexico	Biodiversity conservation	TNC	Ecosystem, imperiled species	Seagrass, oyster reefs, Gulf sturgeon, Florida manatee	A priori: Yes Numerical: In general, 20% of targets	Sites v1.0, Delphi	Trawl data, aerial photography, satellite imagery
Central Caribbean, multiple nations from Cuba to Venezuela	Biodiversity conservation	TNC	Ecosystem, oceanographic, species assemblages	Mangroves, coastal wetlands, Caribbean current, upwelling events, fish, invertebrates	A priori: No	Delphi	Digitized nautical charts
Eastern Africa	Biodiversity conservation	WWF	Ecosystem	Mangroves, sandy shores, coastal lakes	A priori: No		

Location	Goal	Organization	Approach	Features	Goals (a priori/numerical)	Tools	Data sources
Meso American Reef, Mexico, Belize, Guatemala, Honduras	Biodiversity conservation	WWF	Ecosystem, focal species	Coral reefs, mangroves, saltwater crocodile, manatees	A priori: No	Delphi, Sites VI.0, and data intensive	Aerial surveys, fisheries, ground surveys
Puget Sound, Georgia Straits, U.S.-WA, Canada-BC	Biodiversity conservation	TNC	Ecosystem, focal species	Salt marsh, kelp beds, eelgrass, forage fish spawning, grounds, rockfish	A priori: Yes Numerical: 30% of total shoreline, 30–60% of individual species		Aerial surveys, fisheries, ground surveys
Chesapeake Bay Lowlands, VA, MD, DE (U.S.)	Biodiversity Conservation	TNC	Ecosystem, focal species	Tidal flats, tidal marsh, blue crab, yellow perch	A priori: Yes Numerical: In general, 20% of existing targets	Delphi and data intensive	Aerial surveys, fisheries, ground surveys
Mid-Atlantic, U.S.	Biodiversity conservation, sustainable fisheries	NRDC	Ecosystem, focal species	Sargassum mats, boreal red coral, North Atlantic right whale, loggerhead turtles	A priori: No	Delphi, GIS software	N/A
Cook Inlet, Alaska	Biodiversity conservation	TNC	Ecosystem, indicator species	Eelgrass beds, kelp beds, harbor seal, dungeness crab	A priori: Yes Numerical: In general, 20% of existing targets	Delphi	Aerial surveys, ground surveys, fisheries

(e.g., U.S. National Estuary Program), which is ecologically sensible because bays and estuaries are reasonably independent ecological units. There are, however, few if any overarching plans to identify whether these are the most appropriate bays or estuaries to meet particular aims regionally. In the United States, the National Oceanographic and Atmospheric Administration (NOAA) identified marine ecoregional boundaries to inform the development of its National Estuarine Research Reserve (NERR) system (Clark 1982). The aim was to ensure that at least one NERR was placed in each region before adding multiple NERRs to regions. Formal regional plans were not done to site the NERRs.

The ultimate aim of any planning program needs to be clear from the start, and the lack of clarity and multiple aims confound many planning efforts (see Chapter 3). This is true for any type of planning, but many marine planning efforts appear to have been particularly troubled by conflicting and/or unclear objectives. Some planning, by charter, is only intended to identify marine reserves or Marine Protected Areas (MPAs). An MPA is generally defined as any area of the intertidal or subtidal terrain, together with its overlying water and associated flora, fauna, historical and cultural features, that has been reserved by law or other effective means to protect part or all of the enclosed environment. A marine reserve is usually identified as a more restrictive category of MPA and excludes some uses, often fishing. Most planning efforts, however, could have a wider mandate than the establishment of MPAs but are perceived either by the planners and/or stakeholders to be MPA or marine reserve plans. Marine plans by governments (e.g., British Columbia's Ministry of Sustainable Resource Management and NOAA Marine Sanctuaries) often have multiple and seemingly conflicting objectives, perhaps an unavoidable consequence of trying to satisfy multiple stakeholders.

Many planning efforts are flawed from the beginning by evaluating only one threat (usually overfishing) and developing only one strategy (MPAs). This myopic scope limits considerations of many potential targets (e.g., noncommercial species), threats (e.g., water pollution), and strategies (e.g., habitat restoration, pollution reduction). In addition, these MPA-focused plans tend to vilify powerful stakeholder groups (e.g., recreational and commercial fishing groups). A broader and more balanced view should be taken, and this is happening more often. Fishing is one important threat, and MPAs are one important strategy in the marine environment.

To the extent possible, regional planners should focus first on identifying the conservation areas that best and fully represent the biodiversity of that region. Only after these areas are identified should strategies for their conservation be addressed. The approaches for the conservation of these areas can vary. Effective coastal and marine conservation will often require approaches that address the many threats to the marine environment that arise in watersheds. For example, the conservation of some marine areas will require improvements in water clarity and quality through strategies from pollution abatement to best management practices for abating excess runoff of soils and nutrients from farms. Other conservation efforts will be directed in the coastal zone to address the increased hardening of shore-lines (e.g., jetties and seawalls) and loss of coastal habitats through strategies such as restoration, better management practices, and the acquisition of coastal and submerged lands. Still other area-based approaches will be nec-essary to abate threats from excessive recreational use, such as from boat anchors and cruise liners. Finally, marine reserves will also be of use in some areas.

The ultimate objective of the planning process outlined in this book is to identify a set of conservation areas that represent the full array of biodi-versity within a region and that are likely to persist over the long term (see Chapter 7). The remainder of this chapter makes recommendations for carrying out the major steps in the conservation planning process, as developed in previous chapters, for coastal marine environments.

Identifying Targets: Ecosystems, Species, and Aggregation Sites

The first step in regional planning is to identify conservation targets (Table 11.2). In marine environments, the most effective planning approach is to focus on marine ecosystems and the ecological processes that sustain them. This approach presumes that conserving a representa-tion of all the ecosystems will also conserve a representation of the diver-sity of species found in these ecosystems, an assumption that deserves more rigorous testing. Typical marine ecosystems include seagrasses, coral reefs, kelp beds, mangroves, salt marshes, tidal freshwater marshes, and sponge gardens.

Identifying ecosystems has always been a fuzzy concept, a point rein-forced by the lack of clear definitions for ecosystem, community, habitat, and similar terms (e.g., Whittaker 1975). Among marine ecologists, it has been common vernacular to use the terms *ecosystem* and *habitat* inter-changeably. The term *habitat*, in particular, has been used in multiple senses to describe the area used by an individual species (e.g., the habitat of the

Table 11.2   *Conservation Targets for The Nature Conservancy's Northern Gulf of Mexico Ecoregion Plan*

Ecosystems (and Subcategories)	Some Characteristic Species
Primary Ecosystem Targets	
Seagrass	*Thalassia testudinum, Syringodium filiforme, Halodule wrightii*
Tidal freshwater grasses	*Vallisneria americana, Potamogeton* sp., *Ruppia maritima*
Oyster reefs	*Crassostrea virginica*
Salt marsh	*Spartina* sp., *Juncus roemerianus, Distichlis spicata*
Polyhaline saltmarsh	*Spartina alterniflora, Juncus roemerianus, Distichlis spicata*
Mesohaline saltmarsh	*S. alterniflora, D. spicata, S. patens, Scirpus americanus*
Oligohaline saltmarsh	*Paspalum vaginatum, S. patens, Eleocharis* sp., *Sagittaria lancifolia*
Sponges and soft corals	Loggerhead sponges, vase sponges, sea fans, small hard corals
Tidal flats	Algae, polychaetes, bivalves
Tidal fresh marsh	*Scirpus* sp., *Typha* sp., *Cladium* sp.
Intertidal scrub/forest	*Avicennia germinans,* Iva sp., *Baccharis* sp.
Secondary Ecosystem Targets	
Muddy-bottom habitats	Polychaetes, amphipods, isopods
Coquina beach rock	*Donax* sp.
Beaches and bars	Shorebirds, mole crabs, amphipods and isopods
Serpulid worm reefs	Family Serpulidae
Imperiled Species	
Fringed pipefish	*Anarchopterus criniger*
Gulf sturgeon	*Acipenser oxyrinchus desotoi*
Diamondback terrapin	*Malaclemys terrapin* (ssp. *macrospilota, pileata, littoralis*)
Dwarf seahorse	*Hippocampus zosterae*
Opossum pipefish	*Microphis brachyurus lineatus*
Texas pipefish	*Syngnathus affinis*
Florida manatee	*Trichechus manatus latirostris*
Kemp's ridley turtle	*Lepidochelys kempii*

From Beck and Odaya 2001.

redfish) and for large areas of similar composition used by many species (e.g., seagrass habitat). Throughout this chapter, the term *ecosystem* is used to identify characteristic assemblages of plants and animals and their associated physical environment (e.g., marshes or oyster reefs). The term *habitat* is used in reference to the area(s) used by individual species. Modifiers are added to identify the particular habitats used by an animal. For example, the blue crab (*Callinectes sapidus*) has a seagrass habitat and a marsh habitat, and these refer to particular portions of seagrass and marsh ecosystems, respectively, used by the crab (e.g., Beck et al. 2001).

Once definitions are clear, it helps to have a consistent and reliable classification scheme to identify the different types of ecosystems. Although there is a growing number of classification schemes for marine ecosystems, they are less well developed than their counterparts for terrestrial environments. Most schemes are at coarse spatial scales (Cowardin et al. 1979, Davies and Moss 1999, Allee et al. 2000), but some schemes have been developed at finer resolutions for use within regions (e.g., Dethier 1992, Wieland 1993) or to focus within particular marine ecosystems, such as coral reefs or mangroves (e.g., Holthus and Maragos 1995, Twilley 1998, Mumby and Harbone 1999).

Ideally, classification schemes should be based on biological data; when this is not possible, surrogate data are used. In deeper water environments, typically beyond the depth range of 20–30 m for aerial imagery, classification schemes must usually use the more readily available abiotic data (see Chapter 4), such as sediment type, depth, slope, and temperature (e.g., Zacharias et al. 1998, Day and Roff 2000, Roff and Taylor 2000). These schemes rely on the assumption that certain physical factors control or are at least significantly correlated with repeating and characteristic assemblages of plants and animals. Such assumptions may not always be correct and are rarely tested. Consequently, in identifying ecosystem targets, planners should always place an emphasis on using biological data whenever they are available. As noted in previous chapters, a combination of abiotic and biotic-based targets will likely be most effective in conserving the full array of biodiversity in any given planning region.

Chapter 4 pointed out that not all biodiversity can be conserved through a focus at the ecosystem level. Those elements of biodiversity that are least likely to be represented by such a focus are endangered and imperiled species. Many of these species require individual attention because managing their habitats alone is necessary but insufficient for their conservation needs. In other words, they are declining faster than their habitats (e.g., Florida manatee, *Trichechus manatus latirostris,* and Kemp's ridley turtle, *Lepidochelys kempii*). It is also important to identify target species that are vital to the structure and function of ecosystems, because they are, for example, keystone species (Power et al. 1996) or ecosystem engineers that are crucial for creating or structuring ecosystems, such as oysters (Lawton 1994, Lenihan 1999). For all of these species, it is necessary to consider their role within ecosystems in each region; we cannot assume, for example, that a species is a keystone species in one region just because it is classified as such in another region. The role that species play in ecosystems varies geographically (Menge et al. 1994, Power et al. 1996).

Aggregation sites, which are usually associated with the physical convergence of water and land or of different water masses, are a third major type of target in marine systems. It is common in the marine environment to see large aggregations of species consistently occurring at particular places in space and time. These aggregations are often found in areas of high biological productivity or where different water masses meet (or both). Examples are the spawning aggregations of reef fish, and breeding congregations of seals and sea lions on haulout sites. Many spawning aggregations of reef fish occur at outer reef promontories where the inshore reefs meet oceanic currents (W. Heyman, TNC, personal communication). These currents presumably carry the spawned larvae to many new habitats for the settlement of juveniles. Upwellings, where cold, nutrient-rich water masses come up from depth to the surface, are another area known for abundant aggregations of species and are often areas with high productivity. Enhanced productivity and aggregations often occur in retention zones where water masses converge or are retained, such as near the mouth of bays (e.g., Chesapeake Bay) or in passages between islands. These areas of retention and convergence may be particularly important for larval transport, because larvae congregate actively and passively at these fronts (e.g., Pineda 1999).

In many marine plans, it is common to include fished species as targets, but this may not always be appropriate. Fished species are often included because there the substantial data on these species, although the quality, not just the quantity, of these data must be evaluated. Moreover, there is often concern about the potential threats from overfishing. For plans that are developed to identify representative areas of marine biodiversity, fished species should be included as targets only if they are truly imperiled (which is uncommon because most will go to economic extinction before they are truly imperiled) or have declined to such a degree that they affect overall ecosystem integrity. Some species can be truly imperiled by fishing, such as right whales (*Eubalaena glacialis*), white abalone (*Haliotis sorensi*), and stellar sea cows (*Hydrodamalis gigas*). Some fished species, such as oysters (e.g., *Crassostrea virginica*), green turtles (*Chelonia mydas mydas*), and sea otters (*Enhydra lutris*), have declined to such a degree that their low abundances may critically affect ecosystem integrity, (e.g., Jackson et al. 2001). Fished species should not be included as targets solely because of threats posed by fishing; this misses the point of identifying targets. Threats to biodiversity are addressed in the development of strategies for the conservation areas that are selected in the planning process (see Chapter 13).

Setting Conservation Goals

A conservation goal is characterized by the amount and quality of the target that should be represented in conservation areas across the planning region (see Chapter 6). There are many theoretical studies, and a growing number of empirical studies, examining how to set marine conservation goals, but as in terrestrial environments, much work remains to be done. Most of the recent work on species and ecosystem goals has been spurred by interest in identifying how large marine reserves need to be to conserve populations and ecosystems (for review, see Roberts and Hawkins 2000, National Research Council 2001). Much of this discussion revolves around two different objectives for setting goals in marine environments: fisheries management and biodiversity conservation. Goals associated with fisheries management aims are intended to identify the size of reserves that are necessary to conserve and possibly enhance the stocks of exploited species. Goals associated with biodiversity conservation objectives are intended to identify the minimum areas necessary to represent and conserve marine diversity in a region. The most important distinction between these objectives is that the areas (e.g., reserves) required to meet fisheries management goals will generally be much larger than the areas required to meet biodiversity goals. Stated another way, we can represent and protect biodiversity in smaller areas than will be required to reduce the risk of fisheries overexploitation and to increase fisheries yields (e.g., Hastings and Botsford 1999). The focus in this chapter is on the representation of biodiversity in conservation areas.

Numerous studies suggest that reserves may need to cover 10–40% of a region to be effective as a tool for biodiversity conservation (see reviews in National Research Council 1999, Roberts and Hawkins 2000). Turpie et al. (2000) suggested that a system of reserves that encompassed nearly 30% of the South African coast would be required to represent the known fish species on the coast. Ward et al. (1999) indicated that most of the marine taxa within Jervis Bay, Australia, would be accounted for only after 40% of the bay was contained within conservation areas. Most species–area curves suggest that the greatest losses of species richness will occur as remaining habitats decline below 20% (see Chapter 6, principle 3). Models show that the potential connectivity among marine reserves increases greatly as the amount of conserved areas approaches 30% (Roberts and Hawkins 2000). Similar to the terrestrial history of goal setting described in Chapter 6, some discussions about goals for marine conservation have been as politically motivated as they have been ecological. In 1998, for example, 1600 scientists and conservationists signed a statement indicating that we should aim to conserve 20% of the oceans by 2020.

Ecological knowledge and intuition is paramount in selecting goals, but a sensitivity analysis can also be informative. It may help to systematically vary the conservation goals for the targets and determine how this affects the size of the priority areas required to meet the goals (Figure 11.2). Curvilinearity in this relationship has important implications. Depending on the form of this relationship, the area required to meet goals can change dramatically. In some cases, small changes in goals (say from 20–30% in Figure 11.2) may require substantially larger areas (e.g., dashed line in Figure 11.2) or have little impact on the area required (e.g., dotted line). Sensitivity analyses should never be the final arbiter in identifying goals. Ecologically appropriate goals need to be set to determine the necessary areas/actions regardless of cost. However, careful consideration must be given to the higher costs that accompany certain goals, especially in the face of considerable uncertainty in identifying ecological goals.

Ideally, conservation goals should be based on historical or preexploitation estimates of the abundance and distribution of the targets. Unfortunately, goals often have to be based on current distributions (e.g., Beck and Odaya 2001). Many ecosystems and species have declined greatly in recent history. Setting goals on these diminished abundances ensures further degradation on a shifting baseline. The fact that every new generation of scientists and citizens sets ever lower goals for the "natural" state by referring only to recent personal memories of diversity and abundance is a real problem. Historical data can be difficult to find, but recent papers indicate that developing reasonable historical estimates of abundance in the marine environment is possible (Jackson 2001, Jackson et al. 2001, Wing and Wing 2001). (For a detailed discussion of the issue of historical context, see Chapter 6, principle 8.)

Most of the present exploration of goals focuses only on current conditions. Analysts have not given much consideration to what additional areas

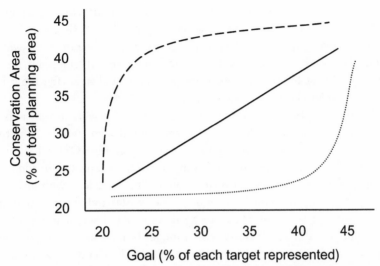

Figure 11.2 Hypothetical sensitivity analyses. The solid line indicates a 1:1 relationship in the goals set and the area required to meet those goals. The two curves represent different scenarios for how changes in goals can have vastly different effects on the area required to meet those goals. The dotted line suggests that small changes in goals may have little effect on the size of conservation areas required, while the dashed line suggests just the opposite.

may be necessary to buffer against natural and anthropogenic catastrophes. Alison and colleagues (2003) have examined the incidence of hurricanes in the Gulf of Mexico and oil spills on the west coast of North America to determine how much larger goals may need to be in order for diversity to be preserved in the face of future catastrophes. Their analysis suggests that considerations of potential catastrophes will likely increase goals by a factor of 1.1–1.25. In other words, if the intended goals are to include 20% of an ecosystem target without considering the impact of potential catastrophes, then actual goals should be adjusted to 22–25%.

Assessing Existing Conservation Areas

Defining protection and finding well-protected areas in the marine environment are difficult tasks (Jamieson and Levings 2001). While many large areas exist as marine sanctuaries, parks, and reserves, little direct management action is aimed at conservation of marine resources in these areas. In many of these seemingly protected areas, few activities are excluded (Jamieson and Levings 2001). Many NOAA Marine Sanctuary managers

have noted that the public assumes that these sanctuaries are highly protected, when in fact little harmful activity is prohibited. A similar situation
has been observed for Canadian marine sanctuaries (Jamieson and Levings
2001). History has shown that from a social and political standpoint, it can
be very difficult to change usage patterns in protected areas after the rules
or zones are in place.

In most regions, identifying existing marine conservation areas is a
complex endeavor. NOAA was recently mandated with identifying all the
marine protected areas in the United States (Executive Order 13158), a
job that turned out to be far more complicated than expected. In addition
to the problems identified above, there are multiple agencies with jurisdiction and protected area designations in the coastal and marine environment. Whereas this multiplicity of agencies is not different in terrestrial
environments, the agencies have had a longer time to develop more consistent approaches to their work and definitions. The fact that water masses
and species move more regularly between artificial jurisdictions in the
marine environment only adds to the complexity of the problem. In a few
regions, marine zonation, even though complex, has been clearly identified. For example, in many regions of the Great Barrier Reef, Australia,
several different use zones (e.g., areas for fishing, anchoring, diving, or
research) have been clearly identified on widely available navigation
charts.

In the developing world, paper parks are as serious an issue in the
marine as in the terrestrial environment, possibly even more so. The fact
that an area is declared a park or MPA "on paper" does not mean it is
being well managed from a conservation point of view. While an area may
be declared an MPA, it can be very difficult to limit access to marine areas
and enforce rules. On the whole, all existing marine conservation areas
need to be carefully scrutinized to ensure that the biodiversity within
them is being adequately safeguarded.

Assessing Population Viability and Ecological Integrity

As planners gather data on the distribution of the targets and note their
locations, they should try to include only populations of species and
examples of ecosystems that are likely to persist into the future. Doing
so will ensure a selection of conservation areas with targets that are
intact currently and will remain viable into the future. In the ideal
world, population viability analyses (PVAs) would be available for each

species. Formal analyses of viability are rare for marine species, and similar analyses of integrity are virtually nonexistent for marine ecosystems. One of the classic PVA models was for loggerhead turtles, *Caretta caretta*; this study indicated that one of the critical factors for this species was the number of males reaching adulthood (Crouse et al. 1987). Another recent analysis noted that the number of adult females was critical to the viability of the endangered North Atlantic right whale, *Eubalaena glacialis* (Fujiwara and Caswell 2001). Even the life of one adult female right whale matters.

Although PVAs for species and analyses of ecological integrity for ecosystems may not be available, there are often factors that can be used to "screen" or filter out areas that are not likely to have the best or most viable examples of species and ecosystems. Recall from Chapter 7 that these factors can be used individually or collectively in "suitability indices." Examples of such factors in marine systems are water quality; shoreline hardening (seawalls, jetties); indicators of high-use areas (docks, moorings, marinas); shipping lanes; and oil rigs. A typical use of these variables is effectively steering the selection of conservation areas away from places with low water quality, high use, or other possible types of degradation.

Selecting Conservation Areas

A number of different procedures have been used to identify conservation areas in marine regional plans. These range from stakeholder input to expert opinion to area selection algorithms. Most plans employ multiple methods.

The planning process in the Florida Keys National Marine Sanctuary for the Tortugas ecological reserves, while not a regional planning process per se, was an example of a plan that was largely driven by discussions among scientists and stakeholders in workshops. Stakeholders were eventually asked to draw and compare lines on maps (i.e., potential conservation areas) in the workshops (Haskell et al. 2000). There was remarkable consensus among the stakeholders for identifying and designating areas in the Tortugas, particularly given the previous high level of acrimony for identifying areas in the rest of the Florida Keys. Numerous senior NOAA managers have wanted to replicate this process.

The important lesson to be learned from the Tortugas designation was not the selection process itself, but the fact that the Tortugas was an ideal place for a marine reserve (i.e., a comparatively low-cost conservation area). The potential for consensus was greater because stakeholders were not as strongly invested in the area. The Tortugas are comparatively difficult and expensive to reach by fishermen and tourists, there was little coastal development, and portions of the lands and waters were already protected. While these ideal conditions may not exist in other regions, some of them can be found in areas in most regions. Much more consideration needs to be given to identifying the areas of low cost for conservation (i.e., "low-hanging fruit"). There seems to be an unfortunate tendency for managers and conservationists to focus first on the most difficult areas for conservation (i.e., "the battle zones").

Some plans use Delphi workshops, which involve gathering scientific experts who are asked to draw lines on maps to outline the most important areas for conservation of particular taxa. In the mid-Atlantic seaboard of the United States, the Natural Resources Defense Council led a Delphi workshop to identify potential priority areas (Natural Resources Defense Council 2001). The Bering Sea regional plan of TNC and the WWF (Banks et al. 2000) and the central Caribbean plan of TNC (Sullivan Sealey and Bustamante 1999) included systematic considerations of conservation targets and were data driven whenever possible. They were also strongly influenced by expert opinion in Delphi analyses. In California, the statewide process to identify marine reserves, directed by the Marine Life Protection Act, started first as a Delphi plan developed by scientists. This plan was discarded, and the state has started over with a purely stakeholder-driven planning process.

Area selection algorithms are being used increasingly to help planners identify conservation areas (see Chapter 8). Most marine planners are using algorithms developed by Australian ecologists Ian Ball and Hugh Possingham. The earlier and more terrestrially oriented version of their software is known as SPEXAN. (Working with scientists at the University of California, Santa Barbara, TNC changed the name to Sites.) This software has been further adapted for use in the marine environment as MARXAN (Ball and Possingham 2000).

Both Sites and MARXAN have been used in several different marine planning efforts at regional and smaller scales. In the northern Gulf of Mexico, TNC used Sites to help identify potential conservation areas (Beck and Odaya 2001). The results from using this algorithm were then presented at a scientific workshop where participants were asked to critique the selected areas and identify gaps and problems. The final portfolio of conservation areas integrated results from the area selection algorithm and expert opinion (see Figure 11.3, in color insert). This is a good example of using these algorithms as they were intended—a tool to aid planners and biologists, not a stand-alone approach to selecting conservation areas.

In the Puget Sound/Georgia Straits region in the United States and Canadian Pacific Northwest, TNC and partners are also using Sites to help select marine conservation areas (Ferdaña 2002). In this regional plan, a stepwise analysis was done with Sites to account for differences among the targets in their ecological importance and data quality. Targets that were known to be ecologically important and met high data-quality standards (i.e., the data were comprehensive throughout the region, spatially precise, and recently updated) were run first through the model, and the priority areas that were chosen were then locked in for subsequent runs of the model. This procedure was repeated four times (see Figure 11.4a, in color insert). The final model results were then evaluated by external scientists, and a number of large (or seascape-scale) areas were identified (see Figure 11.4b, in color insert).

There is never just one "optimal" solution (i.e., set of conservation areas) in regional planning, but it is possible to do irreplaceability analyses to identify those areas that must be part of a plan (see Chapter 8). Such analyses have been conducted with data from the Florida Keys (Leslie et al. 2003). Results indicated that a number of areas were irreplaceable in a biodiversity plan; that is, some areas would be necessary in any potential plan to represent the marine diversity in the nearshore areas of the Florida Keys. However, once these core areas were included, there were many options for choosing the remaining conservation areas needed to fulfill the conservation goals. Thus, it would be possible to attempt to choose configurations that could meet biodiversity goals with the greatest benefit and least impact on stakeholder groups. Irreplaceability analyses were also used in California by the scientific advisory panel for the Channel Islands National Marine Sanctuary to identify potential marine protected areas (Airame et al. 2003) and by TNC in the Puget Sound/Georgia Straits region to combine analytic and Delphi results for selecting conservation areas.

One of the key decisions in selecting potential conservation areas is determining their minimum effective size (see below for a discussion of networking multiple areas). Previously, the general advice would have been that marine conservation areas needed to be much larger on average than terrestrial areas, to account for the open and dynamic nature of marine ecosystems and the mobility of marine species. Indeed, it was not even clear that area-based approaches (e.g., marine reserves) would be useful for the conservation of marine diversity. Theoretical analyses suggested that individual conservation areas must exceed the dispersal distance of target species or cover very large sections of the coast (Hastings and Botsford 1999, Botsford et al. 2001). For many targets, these theoretical results would require very large conservation areas, larger than most of the marine reserves in place at present.

Nonetheless, recent compilations of the empirical evidence from a growing number of marine reserves shows that the present area-based efforts can be surprisingly effective for the conservation of biodiversity (Cote et al. 2001, Halpern 2003). After marine reserves are put in place there are increases in density, biomass, organism size, and diversity (Figure 11.5), and these effects can occur surprisingly quickly (Halpern 2003, Halpern and Warner 2002). Moreover, even the smallest marine reserves

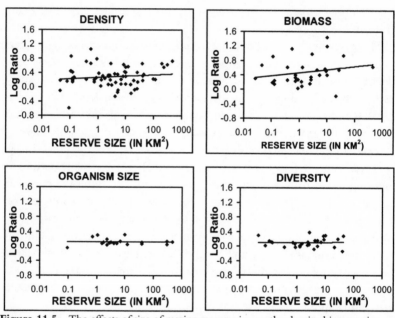

Figure 11.5   The effects of size of marine reserve size on the density, biomass, size, and diversity of invertebrates. Each point represents a paired comparison of the effect of a reserve compared to a control (either the same places before and after reserve designation, or a reserve versus a nearby control site). The difference ratio (*d*) is used for each pair of sites for the biological measure of interest. Data are plotted as the log of the ratio (*d*) versus the log of reserve size. When *d* = 0, the reserves had no effect on the biological measure. When *d* > 0, the reserves had a positive effect on the biological measures. Note that almost all of points are > 0, indicating that reserves had higher density, biomass, size, and diversity of animals than control sites. In all cases except invertebrate biomass, *d* values were significantly different from zero. The slopes of all regression lines are not significantly different from zero, indicating that reserve size did not have a significant proportional impact on the differences between reserves and control sites. (From Halpern 2003. Reprinted by permission of the Ecological Society of America.)

seem highly effective in protecting and enhancing the density, biomass, size, and diversity of marine species (Figure 11.6) (Halpern 2003).

Few scientists would have predicted that some of these smallest reserves would have worked to conserve diversity. Whether or not these reserves can withstand disturbance and ensure viability and integrity is an open question. It seems unlikely, in the absence of other protection for sites in the region (i.e., a network), that these small reserves can be effective in the long term. Part of the effectiveness certainly lies in the fact that even if areas outside marine conservation areas are degraded, they still harbor many of the species and ecosystems of concern. This situation is unlike most terrestrial cases where many historical habitats and ecosystems are now completely lost or uninhabitable.

Designing a Network of Priority Conservation Areas

Single conservation areas are not likely to be effective. Considerations of connectivity are probably more important in marine environments than terrestrial environments because of the mobility of many species (through

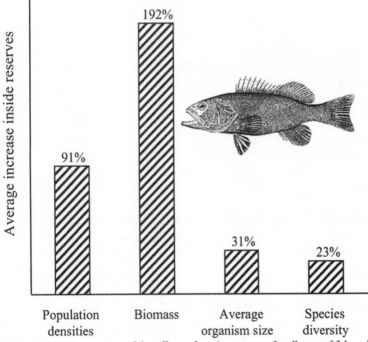

Figure 11.6   A summary of the effects of marine reserves for all taxa of fish and invertebrates.

movement and larval dispersal) and the potential quick spread of marine threats over large areas (e.g., water pollution and invasive species).

The best advice at present in the marine (and terrestrial) environment is for planners to ensure that multiple conservation areas are spread throughout a region, thereby fully representing diversity and guaranteeing that some areas will persist and survive potential catastrophes (e.g., oil spills and hurricanes). Very little is known about spatial variation in diversity in the marine environment; this fact is particularly true for genetic diversity, and it is also relevant at other levels of biological diversity. Classification systems are poorly developed in the marine environment, in part because of the lack of information about consistency and variability in assemblages of plants and animals within broad ecosystem types (e.g., seagrasses).

It is often possible to address variability in diversity in a network design by dividing the region into ecologically relevant subregions and setting conservation goals for each subregion. For instance, the northern Gulf of Mexico ecoregion was divided into three separate subregions delineated largely by the flow from the Mississippi River. The central subregion (Galveston Bay, Texas, to Mobile Bay, Alabama) is dominated by substantial freshwater input and high sediment loads. The waters are turbid and sediments are muddy; salt marshes are the dominant nearshore ecosystem. The western and eastern subregions have much less freshwater input and consequently have clearer water, sandier sediments, and seagrasses dominating the nearshore ecosystem. The strong differences in predominant physical regimes are likely to have strong influences on diversity from genes to ecosystems across the Gulf of Mexico. Therefore, planners set goals for all the targets in each subregion.

There is a rapidly developing literature on how to factor dispersal (connectivity) into the design of a network of marine conservation areas. However, at present there are few clear answers (e.g., Roberts 1997, 1998, Swearer et al. 1999, Cowen et al. 2000, Botsford et al. 2001). In the most basic formulation, the concept is to consider how dominant patterns in ocean currents might influence the sources and settling areas for larvae. The general advice would be to spread priority sites at places along these currents and not bunch them at one end or another. Upstream areas will probably be the main sources for larval recruits that will mainly settle in downstream areas. The principal problems are that this concept assumes larvae are passive drifters and that oceanic currents are the primary flows determining larval movements. It is becoming much clearer that larvae have complex behaviors that afford significant control over how and what currents carry them (e.g., Cowen et al. 2000). Moreover, increasing evidence indicates that larvae are entrained in areas possibly by nearshore currents. These nearshore currents are complex and not well studied by oceanographers. Larvae with long larval stages (weeks to months) that could migrate hundreds to thousands of kilometers in some cases may migrate only tens of kilometers (e.g., Swearer et al. 1999). New techniques in tagging (e.g., otolith microchemistry) are helping reveal patterns in the sources and dispersal of larvae.

Some studies look at the dispersal requirements of one or a few species to assess possible designs for networks of marine reserves. The problem— just as for terrestrial environments—is that a design based on the requirements of one or a small assemblage of species will probably be irrelevant for a wide range of species. Sources for recruits of one species may be a sink for recruits of another species. It is not appropriate to design a regional network for biodiversity conservation with the requirements of just a few species in mind.

A more thorough understanding of dispersal distances of many species on a coastline will help planners make better estimates for balancing minimum size of conservation areas and maximum distance between them. For example, an examination of the dispersal distances for many species on the California coast showed different dispersal distance "peaks" or modes for suites of species (Kinlan and Gaines 2003). These patterns in the data can help lead to fairly robust advice directions for the design of a network of conservation areas. A suite of species, for instance, has very short dispersal distances (a few meters to a few kilometers), and individual conservation areas should be large enough to encompass the dispersal distances of these species. Another group of species has moderate dispersal distances (tens to hundreds of kilometers); the maximum distance between conservation areas should not exceed the median dispersal distances of these species. These estimates provide useful parameters to consider in planning, but they must be combined with knowledge about the locations of appropriate habitats, as well as other features of the marine environment that affect settlement and movement, to avoid having appropriately spaced but poorly placed conservation areas (e.g., Valles 2001).

In a network of areas, planners should not just consider representing current diversity, but also consider whether they can identify areas that are likely to be resilient or resistant to future disturbance. An example of how to begin to design conservation areas with future threats in mind comes in response to the widespread threat of coral bleaching (Salm et al. 2001). Coral bleaching occurs when corals are stressed and the symbiotic zooxanthalea, which provide most of the coral's color, are expelled from the tissues of the coral.

The coral frequently dies. In recent years, the number and extent of coral bleaching events have increased dramatically (Hoegh-Guldberg 1999, Goreau et al. 2000, Glynn et al. 2001). Some large areas have lost more than 90% of their living coral cover. The predominant stress underlying the increase in bleaching events is elevated water temperatures usually associated with El Niño events, which have been increasing in frequency and intensity. Even in massive bleaching events, some reefs remain unbleached and/or recover quickly. It may be possible to predict these patterns in coral resilience and resistance. For example, reefs that are near strong upwelling currents (i.e., cooler water), or that are partially shaded by cliffs, are more likely to survive when sea temperatures are elevated in nearby environments. If areas are particularly resilient or resistant, they can provide potential sources of new recruits to replenish impacted areas in the future.

While planners should weigh all of these potential considerations for designing a network, they should first base their designs on the representation and conservation of the known patterns and distribution of diversity (see Chapter 8). Only after they are certain a set of areas can conserve present patterns of diversity should planners address these issues of connected networks and future threats. Issues of connectivity are a hot topic in marine research, so more rigorous guidance will probably be available in the future.

Connecting Terrestrial, Freshwater, and Marine Priority Areas

Planning in coastal regions should be closely coordinated among terrestrial, freshwater, and marine environments. In coastal regions, it is not ecologically sensible to overlook information from any of the three environments or to conduct wholly separate plans in each one. Obviously, planners cannot focus on only one environment and consider certain ecosystems, such as salt marshes and mangroves, to be wholly terrestrial or marine; likewise, they cannot include targets such as salmon and sturgeon and consider only the relatively short freshwater phase of their life history (Kareiva et al. 2000). There are important connections between these environments in targets (e.g., seabirds), in threats (e.g., nutrients, oil spills, urban sprawl), and in the strategies to address them. Nonetheless, short-sightedness in planning separately in each environment is common, making it difficult to set priorities and coordinate efforts across environments. The coastal zone environment has often been divided separately by academic, governmental, and environmental organizations. Government has often had to devise whole new programs, with the difficult task of linking agencies with separate mandates for environments in Integrated Coastal Zone Management Programs.

Many areas with high estuarine and nearshore marine diversity and productivity occur in areas where uplands are intact and diverse. This correlation appears to occur because many stresses in coastal waters arise upstream. Even if there are not strong direct connections, many places appear to contain correlations between terrestrial and marine hotspots in biodiversity (Roberts et al. 2002). Planners should consider whether efforts can be colocated across environments. This efficiency should make economic sense (e.g., having one office instead of two) and enhance the potential for effective partnerships between groups and agencies with overlapping mandates and priorities in the coastal zone.

Given that few integrated regional plans exist, there is little methodology for how best to incorporate connectivity in coastal planning. The Nature Conservancy has been developing an integrated plan across all three environments in the Puget Sound/Georgia Straits region in the United States and Canada (see Figure 11.7, in color insert). Although it may seem elegant, it is not effective to simply lay a grid-based information system across all the terrestrial, freshwater, and marine environments of a region and run an area selection algorithm. The first problem encountered was that the data were quite different in structure and form in the three different environments (Ferdaña 2002), which resulted in a high degree of error in data transformation and significant information loss. In addition, the areas selected by the algorithm were biased toward places on the coastline because these areas included targets from all three environments. An answer may lie in a watershed-based selection algorithm. Planners can first examine the three environments separately to develop potential conservation areas, then determine if any areas can be moved to be aligned within the same watersheds. Extra weight should be added if there are known connections among environments (e.g., anadromous species targets).

Key Points for Marine Regional Planning

1. Identify conservation targets quickly, focus on ecosystems, and only use species as necessary. Give consideration to aggregation sites and convergences (upwelling, retention zones) as targets in the marine environment.

2. A marine regional plan is not necessarily just a Marine Protected Areas (MPA) plan. A marine regional plan should first identify areas critical for conservation. Appropriate strategies for conservation should then be identified. MPAs are one of the possible strategies.

3. Be clear about the objectives of your plan from the beginning. Two common objectives in marine regional plans are biodiversity representation and conservation, and fisheries sustainability. These objectives will require sometimes substantial differences in the selection of targets, goals, and conservation areas.

4. If the intended use of a plan is to represent and conserve biodiversity, then fished species should be included as targets only if they are imperiled or have declined to such a degree that ecosystem integrity is compromised.

5. Identifying goals is one of the most difficult tasks in regional planning. Current estimates for goals for biodiversity conservation usually vary from 10–40% of the planning area or the present distribution of the targets. Unfortunately, few plans at present have explicit goals. When goals are explicitly identified in plans, they usually vary from 20–30% of the current distribution of the targets. Find as much historical data as possible on conservation targets to help set goals.

6. Involve partners early in marine regional planning efforts. They can provide critical input, and their involvement in planning is important if they are expected to play a role in the conservation of any priority areas.

7. In selecting priority marine conservation areas, give more attention to finding areas of low conservation cost ("low-hanging fruit"). The most well known and contentious areas ("the battle zones") are not necessarily the best areas for biodiversity conservation.

8. Make sure to represent biodiversity based on current distributions first. Then consider other factors that might affect the future distribution of targets (e.g., climate change).

9. Consider connectivity and networking among conservation areas to the extent possible. If dispersal distances for some targets are known, consider whether individual conservation areas are large enough to encompass these dispersal distances or the areas are not too far apart such that they exceed these dispersal distances. Consider the linkages among terrestrial, freshwater, and marine environments in targets and threats, and combine priority areas across environments whenever possible.

CHAPTER TWELVE

Adapting Ecoregional Plans to Anticipate the Impact of Climate Change

EARL C. SAXON

*Economically we are living on our capital;
biologically we are changing radically the
complexion of our share in the carbon cycle.*

A.J. LOTKA (1956)

This chapter examines how the practice of conservation planning at regional scales (Groves et al. 2002a) can guide the allocation of conservation funding and management resources to shelter biodiversity from the effects of climate change. Even in the face of dire scenarios, conservation choices informed by sound science (Orians 1993b, Hannah et al. 2002, Peterson et al. 2002) may anticipate the inevitable, thereby shortening the duration and attenuating the severity of the impact of climate change on biodiversity.

Considerable uncertainly exists as to both the pace and the severity of change indicated by current climate change models for any given locality. It is tempting to ignore warnings that are both dire and vague, but conservation biologists cope with similarly pervasive threats to biodiversity, such as habitat loss and invasive species. In fact, global circulation models diverge more because of their explicit social and economic assumptions than because of different treatments of climate parameters. Where opinions differ, conservation biologists might want to consider less rosy scenarios than those leading global policy makers presently prefer.

Though varying in degree, both models and recent observations are consistent in reporting warmer nights and winters and more climate variability. Extreme fluctuations in climate are inevitable (Committee on Abrupt Climate Change 2002) due to the interaction of climate drivers operating in opposite directions and on different time scales. On the one hand, the present interglacial period could end any time in coming centuries (Schneider and Londer 1984, Pielou 1991). On the other hand, the warming effects of past greenhouse gas emissions will be progressively evident in coming decades. Energy-use and land-use choices already made (Revelle and Suess 1957) ensure that excessive emissions will continue—and that climate instability will push many ecosystems well beyond their natural ability to adapt (Intergovernmental Panel on Climate Change Working Group II 2001).

Climates can shift dramatically and quickly, as some components of the planet's atmosphere–biosphere–geosphere system are delicately balanced. Relatively small changes in Earth's ice masses or its oceanic currents could trigger massive changes in sea level and climate over large areas. Fires following drought in tropical and boreal forests could release enough greenhouse gas from biomass and peat to trigger a runaway greenhouse effect (Met Office 2000).

Change need not be catastrophic or global to be beyond the capacity of "ecosystems to adapt naturally" (United Nations Framework Convention on Climate Change 1992). A host of self-reinforcing climate-mediated threats can be identified at the ecoregion level. Examples are sea-level rise interacting with coastal drainage works, shifting tropical convergence zones interacting with forest clearing, and the melt of permafrost interacting with the effects of infrastructure and trampling in the wet tundra. Each of these processes leads to increased moisture stress on the local biota.

Given that climate shifts will accelerate and not all species will thrive where they currently do, some redistribution of plants and animals is a certainty. However, the patterns of this redistribution will be species-specific and highly unpredictable, so the ecoregion planner's job is not to prepare for some ill-defined, distant, theoretical equilibrium conditions, but to identify potential short-term refugia. These will comprise some existing conservation areas and their suitable adjacent "up-slope" habitats (cooler, higher above sea level, etc.).

Conservation planning can identify refugia where species and ecosystems are most likely to persist for decades to centuries in an unstable and unpredictable environment. Properly implemented, conservation planning might also be a tool for minimizing the emission of greenhouse gases from natural landscapes, thus potentially slowing the pace of climate change to a rate nearer the natural adaptive capacity of species and ecosystems.

I begin by reviewing how ecoregion boundaries and characteristics might change (Step 0) and then explore how the prospect of climate change should influence our thinking and approach to the seven steps in conservation planning. The presentation follows that of Groves et al. (2002), in which the order of steps 3 and 4 differs from that in Part II of this book.

Step 0. Redefining Ecoregions

Ecoregion-based conservation planning is grounded in the assumption that biodiversity needs to be conserved in each environmental setting in which it occurs. It is sound ecological practice to consider both abiotic and biotic factors when describing an environmental setting. But we cannot assume that the current relationships among vegetation patterns and climates either are in equilibrium or will be maintained—in fact, that is precisely what we most want to be able to model. To avoid circular reasoning, ecoregions designed for conservation planning in anticipation of rapid climate change must be defined exclusively on abiotic factors (some climate-dependent and others independent of climate), without any reference to current vegetation or any other biotic patterns. The current location of ecoregions, thus defined, can be readily compared with their modeled future locations, and the persistence or disappearance of environmental conditions at conservation areas within ecoregions can be assessed.

Some of the abiotic characteristics of an ecoregion are climate-dependent, others not. Climate-independent abiotic characteristics (e.g., latitude, continental position, incident solar radiation, day length, landform, topography, geology, and soils) remain stable on time scales of interest to conservation planners, but climate-dependent abiotic characteristics do not. Climate-dependent characteristics of an ecoregion include averages and variability of temperature, precipitation, diurnal and seasonal temperature range, actual and potential evapo-transpiration, and growing season length, as well as the severity, extent, and intensity of extreme disturbance events. All of these characteristics will be dramatically affected by the enhanced greenhouse effect on time scales ranging from decades to a few centuries.

Interaction among the climate-independent and climate-dependent abiotic characteristics of an ecoregion means that areas with currently uniform attributes will both break up and coalesce over time in response to changing climate-dependent characteristics (Bailey 1996), but any displacement of their boundaries will be constrained by unchanging climate-independent factors. Consequently, we cannot assume that ecoregions will be more stable at their centers than on their periphery or that species will readily find the combinations of factors with which they are familiar.

Even where climate change will alter local weather patterns, the effects on conservation targets may vary significantly. For example, if the average precipitation in a montane area declines, glaciers and cloud forests may disappear, fire hazards increase, and rain shadows intensify, but communities sheltered in rocky gorges kept moist by torrent streams may change hardly at all. Places that change least or change slowest have the greatest potential to serve as refugia.

The basic information for anticipating the impact of climate change is an understanding of how a current ecoregion's boundaries may shift and its internal conditions may change. The conservation planner needs to know whether the current conditions in one place will shift toward those now characteristic of a near or distant ecoregion or toward those not presently found anywhere. Similar questions may be asked about the current and future conditions for any planning area equivalent in size to ecoregions and similarly defined.

To answer these questions, Hargrove and Hoffman (1999, 2000) derived current and future abiotic ecoregions for the conterminous 48 United States. First, the area was divided into millions of 1-km^2 (247.1-acre) pixels. Each pixel was given three sets of environmental attributes: (1) climate-independent data (soil nutrients, soil water holding properties, solar irradiance, elevation); (2) current climate-dependent data (temperature and precipitation by months and growing season); and (3) year 2100 climate-dependent data derived by two Global Circulation Models. A massive computer analysis was undertaken to cluster all the current and modeled future combinations of these factors into 100 equally distinct habitat classes and to assign every pixel to a class. Each set of pixels with similar characteristics forms a distinct and largely contiguous ecoregion. The current and modeled future location of each ecoregion can be readily compared.

Hargrove and Hoffman's analysis has significant implications for conservation. For example, almost all the ecoregions currently identified within the conterminous United States persist somewhere, but some ecoregions shrink and others grow dramatically (see Figure 12.1, in color insert). Many places experience a shift in environmental conditions away from those that characteristize their current ecoregion to conditions that presently characterize a different ecoregion.

It should be understood that a long-range prediction of little change is inherently much more reliable than a prediction of change to a different specific set of conditions, because all the variables may not vary synchronously. Also, any areas subject to significant rapid environmental shifts may experience outbreaks of disease, pests, fire, or erosion, such that they end up being unable to support the entire biota currently associated with that environment.

Nonetheless, defining current and future ecoregions in this manner enables planners to answer two questions objectively for any given location (see Figure 12.2): (1) How distant will the nearest future location with the same environment be? (2) How different is the modeled future local environment from the present one?

Most of the literature on adaptation to climate change considers only the first question. Simplistic corridor strategies are proposed to directly link current occurrences with modeled future ones. Such strategies frequently ignore the constraints imposed on current biotic distribution patterns by climate-invariant abiotic characteristics, such as habitat fragmentation and soil parent material, leading to exaggerated expectations of future distribution patterns. In addition, the timely occurrence and persistence of suitably contiguous step-stone habitats is widely assumed, but rarely tested.

In short, designing a successful conservation adaptation strategy for a species' hypothetical long-term movement is impossible, because it requires a great deal more knowledge than we currently have or are ever likely to have. The current relationship between a species' actual and potential environmental niche, its physiological and behavioral adaptability, its ability to disperse, and its capacity to compete with other taxa and colonize unfamiliar habitats are poorly understood, even without considering climate change. Misjudging any of these factors could result in costly, irreversible, and possibly fatal consequences for the conservation targets.

Answers to the second question enable conservationists to develop and implement conservation strategies for all areas with natural land cover, or a reasonable prospect of restoring such cover, because all future natural areas (on a time scale of decades to a century or two) must be some subset of the current ones. These strategies can take into account the near-term implications of long-term trends and the need to be suitably prepared for increased climate variability. The absence of regional-scale climate models previously hampered such an approach, but climate modelers are moving from global to regional models, just as conservation planners are shifting their thinking from sites to regional strategies. The scales of interest in the two disciplines are rapidly converging.

Figure 12.2 Severity of the differences between current (Figure 12.1a, in color insert) and modeled future (Figure 12.1b, in color insert) abiotic ecoregions in the conterminous United States. Areas that remain assigned to the same ecoregion occupy 55% of the conterminous United States (white). The areas that change the most (10% of the conterminous United States) are shaded black. Other areas where environmental conditions change to those found outside their current ecoregion are shaded gray.

Taken together, the answers to both questions enable conservation strategists to identify those current conservation areas that will change least in environmental space and those current ecoregions that will be least fragmented and displaced in geographic space. These are the least at-risk places for conservation effort—places where nature reserves are most likely to maintain their current complement of species and ecoregions to keep their current complement of ecosystems. At the same time, we can identify the most at-risk conservation areas and the most displacement-prone or fragmentation-prone ecoregions. These are the places where we most need to prepare for increases in the severity and frequency of extreme disturbance events, biotic invasions, and other nonequilibrium outcomes. Figure 12.2 provides examples of both the least and the most at-risk areas.

With equal attention to present and future conditions, conservation efforts can attempt to nurture the current biota of an ecoregion (whose needs are known), while learning to recognize and accommodate new arrivals (whose needs are not known). Managers will be able to respond progressively to future data on species' adaptability, niche changes, dispersal ability, and interaction with other taxa as climate conditions change. For example, proactive management of fuel loads in fire-prone forests and stream flows in flood-prone streams can be modified when local responses to changing environmental conditions and changing species assemblages are observed. Most importantly, mistaken refuge management strategies based on inaccurate estimates of climate change at current conservation areas are likely to be much less costly, foreclosing fewer alternatives, and much more readily changed than failed corridor strategies. Other chapters (8, 10, 11) emphasize the cautious use of linkages and corridors in the design of a true network of conservation areas. The refuge management strategies outlined here are not mutually exclusive of a corridor or linkage strategy. In fact, both are likely to benefit biota in the face of climate change, but the refuge strategy may have a higher likelihood of success in the face of specific threats associated with climate change.

Step 1: Identifying Conservation Targets

It is impossible to plan individually for the long-term conservation of each and every element of biodiversity in an ecoregion. Consequently, explicit conservation targets have to be selected during the planning process. These may be elements of biodiversity considered important in their own right (e.g., imperiled, endemic or globally rare taxa, narrow habitat specialists, and small patch communities). In addition, they may be those elements whose direct conservation is thought most likely to indirectly protect many co-occurring elements (e.g., top carnivores, migratory species, matrix-forming communities, and freshwater systems). Chapter 4 provides a more in-depth discussion of conservation targets and their role as filters and surrogates (Hunter et al. 1988).

While co-occurrences of many elements of biodiversity are observed today, it is impossible to predict whether these patterns are immune to the various impacts of climate change. Planning targets and many of their currently associated nontarget elements may respond differently to various changes in climate, so that the predictability of many species associations will be lost. Many coarse-scale targets are forest communities whose long-lived dominant canopy species could persist for centuries, while short-lived understory species composition could turn over relatively quickly. Predators with better dispersal abilities than their prey may alter their distribution patterns abruptly, with cascading consequences for both the food chains they invade and those they abandon. Similarly, many subtle ecosystem relationships could unravel if the timing of insect life cycle stages tracks the growing season of their larval host plants but diverges from the flowering season of the plants they pollinate (Parmesan et al. 1999). This does not detract from the validity of, say, top carnivores or wetlands as sound conservation targets in their own right, but we can no longer take for granted the collateral benefits of their conservation for the conservation of other elements.

We need a strategy for assembling sets of direct planning targets without assuming that the current nested relationships among ecosystems, communities, and species will persist. Direct targets will still function as surrogates for unknown indirect targets, but those collateral beneficiaries will change over time. To ensure that the indirect targets are as numerous, diverse, and representative as possible, separate sets of direct target lists should be developed for ecosystems, communities, and species. Each set should include sufficient direct targets to encompass the full range of abiotic environments in the ecoregion. The elements of biodiversity—not the abiotic environments—remain the conservation targets, and direct target sets could have distinctly different members in environmentally similar, but biogeographically distinct, ecoregions whose biotas fill their environmental space in distinct ways. A portfolio of conservation areas that meets the current needs of conservation targets in every Ecological Land Unit (see Chapter 4) where each now occurs is expected to remain representative of the ecoregion's biota, even as the locations of and interactions among ecosystems, communities, and species change.

For example, the target list for The Nature Conservancy's Alaska and the Yukon Arctic Coast Ecoregion Plan, now in preparation, includes landscape-scale ecosystem targets (e.g., shallow lake complexes, fossil dune fields, and mountain passes), as well as community targets (e.g., riparian shrub lands, ice mounds, and warm freshwater springs). It also includes ecologically diverse species targets, such as polar bears (*Ursus maritimus*), caribou (*Rangifer tarandus*), wolverine (*Gulo gulo*), voles (*Muridae*), and migratory birds. This ecoregion plan does not rely on elements at one level of ecological organization to act as surrogates for others, nor does it attempt to identify which nontarget elements of biodiversity will be free riders in a conservation portfolio built for the specified targets.

Instead, each target set (ecosystem, community, and species) is designed to independently sample the ecoregion's whole current environmental space comprehensively and redundantly. When climate changes occur, losses of target occurrences from some conservation areas are expected to be offset to some extent by gains at others. When some environments disappear altogether (e.g., coastal summer ice pack) and some new ones appear (e.g., an active soil layer where continuous permafrost is presently found), new communities will develop (both simplified by losses and enriched by arriving species). A conservation portfolio built for targets that span the full range of current environments would automatically encompass the widest possible range of future environments. In other words, at least some occurrences of "new" habitats are bound to be within existing conservation areas.

Although definitions of biodiversity often mention the maintenance of ecosystem structures (see Chapter 1) as a desirable goal, these structures are rarely explicit targets. The release of carbon dioxide and methane from standing biomass and decomposing organic matter is a serious source of positive feedback, accelerating the enhanced greenhouse effect. Accordingly, we must consider a new type of conservation target in conservation planning—ecosystem states, such as old-growth forests and peat soils, that sequester substantial amounts of carbon are global conservation targets independent of their biodiversity significance in a particular ecoregion. All conservation portfolios must give priority to such occurrences. Designating sequestered carbon as a conservation target and reducing emissions from land use and land-use change are modest steps toward mitigating the likelihood of catastrophic change.

Step 2: Collecting Information and Identifying Information Gaps

When collating data on the distribution of conservation target occurrences, it is critical to check systematically whether the absence of recorded occurrences in some parts of the ecoregion is due to the absence of targets or simply a failure to look for them there. This step may require additional inventory or some sort of rapid ecological assessment (see Chapter 5). It is a particularly important step when preparing for climate change, because the eventual portfolio of conservation areas must include occurrences across the full range of the ecoregion's climate gradients. The best known, largest, or least disturbed occurrences of a conservation target may turn out to be located in the most climate-vulnerable landscapes of an ecoregion.

It is equally important to review the variation in abundance and condition of targets within their known occurrences, for two reasons. First, existing data may be vague—implying that a target occurs everywhere within an area when, in fact, it has been extirpated from large parts of its former range or naturally occurs only in a few small patch habitats. Second, knowledge about the distribution of the target at different altitudes, on different aspects, on different geological substrates, and so on may prove critical for identifying which occurrences will be naturally somewhat protected from the impacts of climate change and which most vulnerable.

Unfortunately, the climate change literature will not provide conservation planners with precise and convenient guidelines for the timing and location of anticipated climate changes. Nor will conservation planners and biologists find it easy to evaluate the assumptions and limitations of multiple models of global or regional climate parameters. But all is not lost. Often, experts familiar with the ecology of the selected conservation targets will be able to identify critical climate-related thresholds, such as growing degree-days, flood frequency, and evapo-transpiration stress. The planning team can then approach the climate modeling community with specific questions and seek helpful answers on the probability of, say, fire risk exceeding a given level or increased frequency of droughts. Most importantly, since the planner is taking a long-term view, the directional trend in climate-related parameters is much more important than the rate of change. Regardless of whether winter snowpack will disappear from a catchment in 50, 100, or 200 years, its consequences for conserving freshwater habitats should be considered now.

Step 3: Setting Conservation Goals

How much land and water under conservation management is required to ensure the persistence of an ecoregion's ecosystems, communities, and species in an era of global change? Since the impacts of climate change are additional to, and synergistic with, other threatening processes, the simple answer has to be, "More numerous and more extensive than required without considering climate change." How much more? In those regions where we predict a high degree of environmental change and for those targets we expect to be particularly vulnerable, the default assumption should be that all current viable occurrences are needed. The burden of proof should fall on those who would argue that particular occurrences are superfluous (Elton 1958; see Chapter 6, principle 10).

The discussion of species-area curves in Chapter 6 concludes that we need to conserve at least 30–40% of the area of each ecosystem if we are going to conserve 80–90% of its species. That might be true if climate change were unlikely to change the shape of the species-area curve, if one could conserve a random sample of the historic range of occurrences, and if one were satisfied with conserving the species least vulnerable to extinction. Because the most extinction-prone species and the ones most vulnerable to the impact of climate change may well be among the priority targets in a particular planning region, a single numerical goal for all conservation targets is inappropriate. I suggest two alternative approaches.

First, in selecting among substantially intact occurrences, conservation planning should be played as a zero-sum game, assuming the risk to conservation outcomes increases if the extent and condition of any intact occurrence are degraded. Second, a more nuanced approach is appropriate for the numerous occurrences that are already degraded to some degree. Portfolio goals should not be set as a fixed percentage of the current number, size, and distribution of degraded conservation target occurrences. To attenuate the ecoregion's extinction trajectory, the entire set of degraded occurrences needs to be examined and a subset selected that most effectively reduces the net loss over time that would otherwise occur. If, say, 20 occurrences are found with varying degrees of fragmentation and exposure to current threats, then giving more protection to some and less protection to others improves the likelihood that some will survive. In fact, when facing tough regional planning choices, it may be vital to make strategic tradeoffs. Higher conservation status should be secured for those degraded lands and waters that are most vital in the long term, while a gradual erosion of conservation values should be accepted where their protection would require the heroic investment of limited resources.

Finally, conservation planners need to be aware that simply protecting multiple occurrences of a target does not automatically guarantee target persistence. If only current threatening processes are considered, it may be reasonable to assume that multiple occurrences will not suffer the same fate at the same time. Unfortunately, if the threat comes from climate change, many occurrences may share a common fate (see Chapter 6, principle 2).

Step 4: Assessing Existing Conservation Areas for Their Biodiversity Values

The next step in the conservation planning process is to assess whether the previously identified targets occur, and are appropriately managed, in existing conservation areas. These are the places where, in theory, conservation targets are best protected from current threatening processes. Consequently, conservation targets in well-designed, well-managed conservation areas should be safer from the synergistic consequences of climate change interacting with current threats.

Planners must be sure to distinguish between the administrative boundaries of conservation areas and the extent of their intact landscapes (which may be larger or smaller than the conservation area) when considering whether a conservation area currently supports a viable target occurrence. It is misleading to assume that a conservation area adequately meets the needs of all targets that happen to occur there, especially wide-ranging species, migratory species, and species with a minimum viable population size exceeding the carrying capacity of the conservation area. Annual ranges, migratory patterns, and carrying capacity are all likely to be affected by climate change, so even the long-term effectiveness of large intact conservation areas cannot be assumed.

If the persistence of a conservation target in a conservation area already depends on active management (e.g., border patrols, fire regime modification, or water flow controls, etc.), it may be a much less cost-effective component of the conservation portfolio than its administrative status would suggest. Consequently, similar assessments of the current protection status of conservation targets should be carried out on lands not primarily managed for conservation, such as dam catchments, military reservations, and managed forests, to identify areas where some degree of target conservation is a low-cost collateral benefit. This step is particularly important in many ecoregions where the existing conservation areas are small and/or limited to agriculturally marginal habitats and, consequently, unrepresentative of the ecoregion's environmental gradients, such as geological substrates and moisture regimes.

Step 5: Evaluating the Ability of Conservation Targets to Persist

The largest, least threatened, and least degraded occurrences of a particular target have, by definition, been the most persistent and required the least conservation management to date. It seems reasonable to assume that they have the highest likelihood of adapting naturally to the various effects of climate change and to persist, all else being equal. However, all else may not be equal.

Disturbance regimes, such as tree falls, fluctuating predator/prey cycles, and occasional hot fires, that have maintained biodiversity over long periods (Connell 1978) may soon become less dependable. The severity and frequency of extreme events are likely to increase, favoring well-buffered occurrences and those free from synergistic threats such as disease, invasive species, fragmentation, and altered fire regimes.

Models linking species distribution to future climate variables suggest where, if not when, the limits of a species' current environmental envelope are likely to be exceeded. Remnant occurrences isolated at the driest/warmest edge of their natural range appear particularly vulnerable to global warming and should be monitored to determine the validity of their presumed viability thresholds. In the Alaska and Yukon Arctic Coast Ecoregion circumpolar targets (e.g., bears, caribou) occur today at the warmest edge of their historic distribution, while cold-temperate targets such as wolves (*Canis lupus*) (often extirpated from the more mesic parts of their range) are at the coldest edge. The former are at much greater risk from climate instability.

Whole populations that have already lost much of their genetic variability are much less likely to be able to adapt to climate change than polymorphic populations. Patterns of shared vulnerability and, therefore, shared fates have already been identified in salmon populations (Ruckelshaus et al. 2002). For the areas at greatest risk from rapid climate change, viability assessment should place a premium on genetic and life history variety (see chapter 7). Various authors have identified attributes that make occurrences less vulnerable to the local impacts of climate change (Table 12.1), even though such occurrences may not currently be the most viable ones.

Step 6: Assembling a Portfolio of Conservation Areas

Quantitative, target-based conservation area selection directs planners to identify multiple viable occurrences of conservation targets spanning an ecoregion's environmental gradients. Area selection algorithms (see Chapter 8) can then help planners assemble a portfolio of conservation areas that incorporates multiple targets in the least amount of area and the largest contiguous patches, while minimizing land-use conflicts. This efficient approach assumes that many target occurrences will continue to thrive in semi-natural landscapes outside conservation areas, a dubious assumption in the face of climate change.

Table 12.1 *Criteria Affecting the Vulnerability of Target Occurrences to the Effects of Rapid Climate Change*

Increased Risk	Decreased Risk	Reference(s)
An occurrence on the equatorial, lowland, or inland periphery of a target's historic range and physiological limits	An occurrence at the poleward, upland, or maritime periphery of a target's historic range and physiological limits, noting that migratory species have multiple requirements	Peters and Darling 1985, Peters 1992. See also data from Channell and Lomolino 2000.
An exposed or management-dependent occurrence (forest edge ecosystem, small fire-dependent community, heavily exploited species)	A naturally buffered occurrence (forest interior ecosystem, community in an intact catchment, species capable of rebounding from a population crash)	Forman 1997, Noss 2001
An isolated occurrence	An occurrence with functional connectivity to other occurrences	Markham and Malcolm 1966
A genetically impoverished occurrence	A genetically heterogeneous occurrence	Comes and Kadereit 1998, cited by Schlesinger et al. 2001
A topo-edaphically homogeneous occurrence	A topo-edaphically heterogeneous occurrence	Peters 1992
An occurrence with the largest members of each feeding guild, poorly dispersing species, low reproductive rate species, and species characteristic of climax communities	An occurrence without extinction-prone species	Diamond 1976, Terborgh 1976
An occurrence on coastal wetlands unable to migrate inland	An occurrence on coastal wetlands adjacent to low-lying natural areas where natural succession processes are unimpeded	Titus 1998
An occurrence on steep upper-mountain slopes where upward dispersal is limited by slow soil formation, high ultraviolet light, strong diurnal temperature range, and shrinking horizontal area	An occurrence on gentle lower-mountain slopes	Halpin 1997
An occurrence smaller than the target's minimum dynamic area	An occurrence large enough to accommodate more frequent, severe, or extensive disturbances than have historically occurred	Pickett and Thompson 1978, Forman 1997

Developing a first-iteration portfolio that represents all viable occurrences of targets in conservation areas and then assessing the consequences of excluding some areas enable planners to consider whether a portfolio will be adequate as the landscape context deteriorates. The first-iteration portfolio of conservation areas is simply the sum of all lands and waters necessary to conserve all occurrences of conservation targets currently thought to have any reasonable chance of persisting well into the future. If all the occurrences in a particular planning or selection unit (e.g., watershed, landscape element, hexagon; see Chapter 8) are of marginal viability and integrity, then that planning or selection unit is excluded from further consideration.

Planners should be very cautious about "writing off" even the most marginal occurrences as superfluous, given an uncertain climatic future. Since degraded remnants of natural ecosystems embedded in largely converted landscapes are thought to be less robust than more extensive, undisturbed natural ecosystems, these remnants may offer crucial opportunities to observe early evidence of the adaptive ability of remnant natural areas. Also, planners should not assume that errors of omission are correctable and that an ineffective portfolio can be modified or added to at a later time. In fact, excluding relatively high-cost or apparently inefficient conservation areas (e.g., those in which relatively few targets occur) in the first iteration of a plan may make it impossible to reconsider them as portfolio members later. They could become irreversibly degraded over time or committed to other land uses in the region. Unwittingly, conservation planners may be giving up access to all lands and waters outside the conservation areas identified during the planning process, especially if the construction of the conservation plan is a highly public process, with the participation of many partners.

A parsimonious approach to identifying expendable occurrences is justified by two characteristics of climate change scenarios: uncertainty about the pace and even the direction of climate change and the plausibility of dramatic abrupt changes (Committee on Abrupt Climate Change 2002). Even the error terms in climate change models are not widely understood. When asked by a reporter whether the future might turn out to be only half as bad as the standard model employed by the Intergovernmental Panel on Climate Change, its chairman, Bob Watson replied, "Yes, and it is equally likely to be twice as bad" (R. Watson, personal communication).

Are there really superfluous occurrences of conservation targets? The prospect of synergistic effects between current threatening processes and future climate change impacts suggests a relative ranking of conservation areas in terms of their likely long-term contribution to the biodiversity of an ecoregion. Areas unlikely to be affected by either current threats or climate change rank highest, and areas likely to be adversely affected by both rank lowest. Even unthreatened areas not currently in a natural condition but amenable to rehabilitation rank higher (Figure 12.3).

With this shopping list in mind, the conservation planner may be ready to negotiate a form of explicit portfolio membership for "minor league" target occurrences in semi-natural landscapes outside designated conservation areas. Given the tremendous uncertainty of climate change, the sensible response is to keep as many options as possible open for as long as possible (Hilborn 1987, Morgan and Henrion 1990, Walker et al. 1999). This means recognizing that there will inevitably be losses even in the most carefully designed portfolio, but even in a much-impacted ecoregion, there will also be opportunities to abate threatening processes whose full impacts have yet to occur. Just as some areas managed primarily for biodiversity conservation are not managed exclusively for that purpose, some areas managed primarily for other purposes still have a role in biodiversity conservation (see The Semi-Natural Matrix in Chapter 8).

An effective portfolio of conservation areas requires careful design if its components are to interact and maintain the biodiversity formerly characteristic of an entirely intact landscape. For example, designating the upper reaches of one watershed as a conservation area and the lower reaches of another is a recipe for the eventual loss of freshwater habitats from both. Portfolio designers have long recognized the need to balance isolation

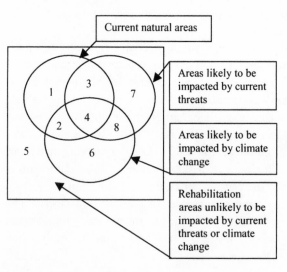

Figure 12.3 Wenn-diagram classifying conservation areas in terms of their likely long-term contribution to the biodiversity of an ecoregion in the face of global climate change. Priorities from 1 (highest) to 8 (most expendable).

with connectivity to ensure the co-occurrence of multiple seral stages and to minimize opportunities for invasive species. The prospect of climate change on a historical time scale makes these goals more difficult and also introduces a daunting new task—avoiding the loss of biodiversity by simultaneously slowing losses and speeding the integration of compatible new community members.

It may be useful to apply different portfolio selection and design criteria for communities that occupy an ecoregion's landscapes and environmental space in different ways, rather than try to find universal design principles for every conservation area, as some authors have proposed (Diamond 1975, Forman 1997). A classification of landscapes based on distinct moisture regimes is suggested, because, for many species, a change in water availability would have greater impact than temperature changes of the order predicted (Peters and Darling 1985, citing Neilson and Wullstein 1983). I suggest three types of ecosystems with distinct moisture regimes:

1. *Chaotic moisture regime.* Ecosystems in which moisture availability is both spatially and temporally unpredictable include deserts, steppe grasslands, and boreal forests. The spatial distribution of successional stages and disturbance states can be stable over large areas and long periods, but is subject to abrupt change. Species richness is low, population numbers may fluctuate dramatically, and long-distance dispersal is an important survival mechanism.

2. *Gradient moisture regime.* Ecosystems in which moisture availability is temporally predictable but strongly patterned spatially are characteristic of riparian habitats, alpine meadows, and tundra zones. The spatial distribution of successional stages and disturbance states is unstable over short distances and tightly linked to seasonal patterns. Species richness is moderate, population numbers fluctuate cyclically, and long dormancy or vegetative propagation is an important survival mechanism.

3. *Saturated moisture regime.* Ecosystems in which moisture availability is never limiting include wetlands and ever-wet forests. The spatial distribution of successional stages is controlled by water tolerance at various life stages, species richness is high, population numbers are stable, and dispersal ability is limited.

Occurrences with such different spatio-temporal patterns, disturbance regimes, and moisture gradients may be expected to differ significantly in their responses to changes in climate variability, the frequency and severity of extreme events, and warmer nights and winters. Consequently, different climate-adaptive rules for selecting conservation areas apply (Tables 12.2, 12.3, and 12.4).

Table 12.2   *Priorities for Selecting Occurrences in Ecosystems with Different Moisture Regimes*

Moisture Regime	Indispensable Occurrences	Most Dispensable Occurrences	Reference(s)
Chaotic	Those occurrences that have all disturbance and successional states	Those occurrences that omit several disturbance and successional states	Anderson et al. 1999, Forman 1997
Gradient	Those occurrences that encompass diverse topo-edaphic conditions	Those occurrences that have uniform topo-edaphic conditions	Markham et al. 1993, Pernetta et al. 1994
Saturated	Those occurrences that have high genetic diversity at species and population levels	Those occurrences that are genetically simplified at species and population levels	Diamond 1975, Peters and Darling 1985, Noss 2001

Table 12.3   *Priorities for Replicate Occurrences in Ecosystems with Different Moisture Regimes*

Moisture Regime	Indispensable Replicates	Most Dispensable Replicates	Reference(s)
Chaotic	Outlier replicates	One of several adjacent replicates	Ruckelshaus et al. 2002
Gradient	Replicate that occupies a distinctive position in several gradients	Replicate that is one of several in the same position on the same gradient	Noss 2001
Saturated	Replicate that occupies a distinctive abiotic environment	Replicate that is one of several in the same abiotic environment	Peters and Darling 1985, Halpin 1997

Table 12.4   *Priorities for Connectivity among Occurrences in Ecosystems with Different Moisture Regimes*

Moisture Regime	Indispensable Connections	Insurmountable Barriers	Reference(s)
Chaotic	Natural linear habitat features (e.g., rivers, passes, ridges)	Artificial linear features (e.g., roads, reservoirs, power lines)	Simberloff et al. 1992, Noss 2001
Gradient	Uninterrupted both along-gradient and across-gradient	Split along a gradient contour (i.e., perpendicular to the gradient)	Peters 1992, Halpin 1997, Walker et al. 1999
Saturated	Low edge/area ratio	Internal gaps and corridors	Diamond 1975, Forman 1997

Step 7: Assessing Threats and Setting Priorities within the Planning Unit

The last step in the conservation planning process is to identify priority conservation areas among the potentially hundreds of conservation areas that most regional or ecoregional planning efforts will identify. A full portfolio of conservation areas will always include sites exposed to a wide range of current and climate-related threats that no single landowner, management agency, or conservation group will have the capacity to address. Priority should be given to places where two conditions are met: where early action can abate those threats that climate change is most likely to amplify (e.g., fire and invasive species) and where conservation targets will not be perpetually dependent on conservation action (Table 12.5). When these conditions cannot be met (e.g., areas with increasingly severe incompatible human land-use pressures, highly fragmented fire-dependent communities, and flow-dependent freshwater ecosystems in regulated rivers), management should focus more on creating opportunities for future elements of biodiversity than on preserving them for current ones.

Assessing each conservation area for the current viability of its target occurrences, future risk factors, and remedial potential is a daunting task. However, the informed allocation of scarce resources among a set of conservation areas intended to survive both present and future threats cannot be made without doing so.

Table 12.5 *Potential for Occurrences of Conservation Targets to Serve as Long-Term Refuges from the Impacts of Climate Change*

High Potential Even If Short-Term Remedial Action Is Required	Low Potential Because Viability Will Depend on Management Interventions, Even If It Does Not Currently Do So	Reference(s)
An occurrence, however disturbed, with the capacity to recover naturally from current extreme events and climate variability	An occurrence vulnerable to current extreme events and climate variability	Burton et al. 1998; Smit and Pilifosova 2001
An occurrence on lands or waters naturally isolated from invasive species	An occurrence exposed to biotic invasions	Committee on Abrupt Climate Change 2002
An occurrence with functional connectivity to other occurrences	An isolated occurrence	McDonald and Brown 1992
An occurrence in which natural fire regimes can eventually be reestablished, even if more intervention is required where fuel loads are abnormally high due to previous fire exclusion regimes	An occurrence in which natural fire regimes will always threaten adjacent land areas	Noss 2001

(table continues)

Table 12.5   *(continued)*

High Potential Even If Short-Term Remedial Action Is Required	Low Potential Because Viability Will Depend on Management Interventions, Even If It Does Not Currently Do So	Reference(s)
An occurrence in which natural food web depth and complexity can eventually be reestablished, even if more intervention is required to manage transitional communities	An occurrence in which natural browsing and predation regimes cannot be reintroduced	Markham et al. 1993; Pernetta et al. 1994
An occurrence where water management infrastructure can be used to reestablish critical hydrological regimes, functional wetlands, and riparian communities	An occurrence in which natural hydrological regimes can not be reintroduced	Markham and Malcolm 1966
An occurrence with an unaltered supply of surface nutrients, soil texture, soil profile, and water table	An occurrence with depleted soil nutrients and degraded soil structure	Scheffer et al. 2001
An occurrence with few threatening processes and ones least likely to be exacerbated by a combination of nitrogen enrichment and increased atmospheric CO_2	An occurrence with multiple synergistic threatening processes	Sala et al. 1999
An occurrence with multiple vegetation canopy layers	An occurrence with simplified vertical structure	Forman 1997

Key Points for Adapting Regional Conservation Plans to Anticipate the Impact of Climate Change

1. When defining planning regions, work on the basis of their uniform abiotic conditions, including current climate. Then use Global Circulation Models to identify potential future locations of those current conditions, to locate potential refugia (areas that change least), and to locate areas where novel communities are most likely to arise (areas that change most).

2. When selecting targets, include sufficient ecosystems, communities, and species to encompass the full range of abiotic environments in the planning region. Do not assume that any particular target will serve as a surrogate or filter for other specific targets, as the distributions of many species will likely be altered in the face of climate change. An additional set of targets not previously considered in the planning process should be identified—targets that sequester substantial amounts of carbon, such as old-growth forests or peat lands.

3. When gathering information on the status and distribution of conservation targets, assess targets across their entire current range and, if possible, their historic range, because the most viable future occurrences may not currently be the largest or least threatened.

4. When setting conservation goals, a more conservative approach is warranted, meaning that more occurrences and more widely distributed occurrences are needed for each conservation target to hedge bets against the impact of climate change.

5. When evaluating the biodiversity value of existing conservation areas, consider the consequences of further fragmentation and loss of habitat inside and outside the conservation area.

6. When evaluating the ability of conservation targets to persist, consider the risk of synergistic effects among climate change and other threatening processes.

7. When assembling a portfolio of conservation areas, take a conservative approach to eliminating only those target occurrences that by any reasonable measure are highly unlikely to persist over the long term or to be feasibly restored to allow such persistence.

8. When assessing threats and setting priorities, protect occurrences that are naturally resilient in the face of extreme events that historically occur and abate existing threats where this will minimize the dependence of conservation areas on long-term human intervention.

PART FOUR

FROM PLANNING
TO PRACTICE

*The final two chapters focus
on the successful implementation of a
regional or ecoregional plan for biodiversity
conservation (Chapter 13) and on improvements that can be made to the science and
application of conservation planning (Chapter 14).*

CHAPTER THIRTEEN

Putting the Pieces Together: Implementing Conservation Plans for Biodiversity at Multiple Scales

Failing to plan is planning to fail.
—ED MCMAHON (2002)

How many plans do you know that have been consigned to the dusty book-shelves of history, rarely resurrected to serve their intended purpose, and remembered primarily by the misery of those who toiled to prepare them and the untold amounts of money spent on the process? This scenario need not play itself out. In this chapter, I discuss factors that make a plan resonate with its intended audience, outline a template for the contents of a regional plan for biodiversity conservation, discuss what we mean by implementation and various approaches for getting there, and provide a number of examples of teams and plans that have successfully made the leap from planning to practice. The chapter concludes with some preliminary figures of the financial cost of conservation at a regional scale, based on several examples from the United States and South Africa, with some global extrapolations.

Elements of a Successful Plan

Having perused many biodiversity conservation plans, I can assure planning teams that there are a few key ingredients to writing and disseminating a plan that a diverse audience will want to read. First, do not title it a plan. The reader is already getting drowsy having just read that fateful word. Call it a biodiversity assessment, conservation blueprint, or conservation priorities for Beringea. Call it anything but a plan. Ensure that the document has an executive summary of substance and not just a couple of pages of rhetoric about the planning process. The summary should focus on the principal findings of the process—the conservation areas that were identified, including a color map of the planning region delineating these important areas. A brief encapsulation of the status of biodiversity in the region, methods used, partners and stakeholders involved, some summary statistics on land ownership and land use in the region, and the major threats to conserving biodiversity, as well as some of the major conservation strategies being considered to implement the plan, are all appropriate content for the executive summary.

Most of the "consumers" of a regional plan for biodiversity have neither the time nor the desire to read a great deal of text, and only a subset of those interested in the plan will want to know the details about the methods that have been used. Liberally using charts, tables, and maps to display results and avoiding lengthy blocks of text in the main body of the document will make it far more readable. Although information on methods is important for the report, the technical details should be placed in an appendix. As much as I do not like to have to keep track of two documents, size does matter. Most readers of these plans find really thick documents a disincentive for reading, so consider placing appendices in a separate volume. In today's era of over-stuffed in-boxes and information overload, a leaner document is much more likely to get read than a 400-page treatise.

Watch out for jargon. It is easy to fall into the trap of writing plan that only staff members of your agency or organization can comprehend. External reviewers can help identify unnecessary jargon. Spell out acronyms the first time they are used in each main section, and explain or define terms that might be confusing or unknown to many readers. Make the data that were used in the plan available to others who may want to better understand them, build upon them, or conduct additional analyses. Post the data on a Web site or an FTP site, or make them available on a CD-ROM. Although your team may be absolutely convinced of the merits of your results and products, scientists and planners from other organizations will want to scrutinize and use the data in their own analysis and planning.

Most readers will focus their attention on the conservation areas or potential conservation areas that have been identified in the planning process. Summary descriptions of these areas are critical and should include information on their locations, what conservation targets they harbor, general threats to the area, and general land ownership (private, federal, or state government agency; conservation organization). Finally, consider making the plan available in a variety of formats. For example, post the plan in a downloadable version on your organization's Web site, make CD-ROM versions available, and for those who are not as digitally connected, prepare printed versions as well. A color brochure that describes the plan's results in a summary fashion can also be a useful vehicle for communicating with many partners, stakeholders, and less technical audiences.

Throughout this book, I have discussed the major steps involved in developing a regional conservation blueprint for biodiversity. Those steps can be translated directly into sections of a written plan, although there are additional components of the plan to be considered for inclusion as well. Below is a summary list of the key components of a written plan.

Key Components of a Written Conservation Plan

1. *Executive Summary.*
2. *Introduction to the Ecoregion or Planning Region:* Ecological and environmental descriptions, biodiversity status, land ownership and land-use patterns, threats to biodiversity, existing conservation areas, human population status and trends.
3. *Overall Purpose, Goals, and Objectives:* A brief overview of the conservation problem(s) the plan is attempting to address, and the overall vision and goals for what the plan hopes to achieve.
4. *Overview of Methods:* A summary of the overall methods, with details in an appendix.
5. *Sources of Information, Data Collection, and Data Management:* A summary of the types of data used in the project (e.g., remote sensing, expert opinion, governmental databases), their origins, and how the information is being managed and accessed (GIS, database software, metadata).
6. *Partners and Stakeholders:* A summary of the organizations and agencies involved in the planning process and the nature of that involvement.
7. *Identification of Conservation Targets:* The elements of biodiversity or surrogates for them upon which the plan is based, the rationale for their selection, and assumptions surrounding their use.
8. *Setting Conservation Goals:* A summary of the target-based goals for each of the conservation targets or groups of targets, and the reasoning and assumptions behind them.
9. *Gap Analysis of Existing Conservation Areas:* A description of the management status of existing conservation areas in the planning region and what conservation targets are found within them.

10. *Assessment of Population Viability and Ecological Integrity:* A brief summary of the methods used to assess viability and integrity, and the outcome of applying these methods to populations and occurrences of target species, communities, and ecosystems in the planning region.

11. *Selection and Design of a Network of Conservation Areas:* Criteria used to select conservation areas, a consideration of linkages and network design, a map and list of selected areas.

12. *Assessment of Threats to Conservation Areas:* Results of threats assessments, including an indication of the scale at which the threat occurs (e.g., single conservation area, multiple conservation areas, entire planning region).

13. *Setting Priorities for Action among Conservation Areas:* Criteria used to set priorities for action among areas in the conservation network, methods for applying criteria, the rationale and assumptions used in the methods, a map and table reporting the results of priority setting.

14. *Development of Conservation Strategies:* A brief introduction to how this planning process will lead into the next phase of identifying conservation strategies at different scales, to help achieve the goals of the plan (i.e., implementation).

15. *Data Gaps and Future Research/Planning Needs:* Identification of critical data gaps in the planning process (e.g., aquatic ecosystem data, human threat information), planning needs in future versions of plan (e.g., incorporation of socioeconomic data), assumptions to be tested.

16. *Conclusions, Literature Cited, and Appendices.*

Having a technically well-developed plan and a well-written plan are critical steps in the conservation planning process. But in and of themselves, neither of these steps is any assurance that the plan will result in achieving actual conservation benefits. In the following sections of this chapter, I begin to explore different approaches to actually implementing the plan. By implementation, I mean taking actions that result in abating threats to conservation targets at real places—be they terrestrial, freshwater, or marine—and maintaining or improving the condition and quality of target occurrences of species, communities, and ecosystems.

From Planning to Places: Noah's Second Job

In Chapter 1, I introduced the notion, conceptually developed by The Nature Conservancy, that actually achieving on-the-ground conservation involved a four-part process, referred to by The Nature Conservancy as the conservation process or conservation approach (Figure 13.1). I referred to the first component of that overall process as "setting priorities," which has been the focus of this book. It essentially answers the "where" question in conservation biology (Redford et al., in press). In other words, where do we, as conservation biologists and planners, need to be working to achieve biodiversity conservation.

The second component of the conservation process answers the "how" question. It is known as the "developing strategies" component. This component is best characterized as a set of proposed actions taken to abate current or future threats to conservation targets at a variety of spatial scales within the planning region. What scale those actions are proposed for depends upon the scale of both the targets themselves and the threats to them. For example, in some planning regions or ecoregions, only one or two areas may be threatened by a specific type of urban or rural residential development, while another threat, such as the invasion of an introduced aquatic species, may threaten numerous watersheds throughout the region. These different types of threats clearly call for different types of strategies to be implemented at different spatial and temporal scales.

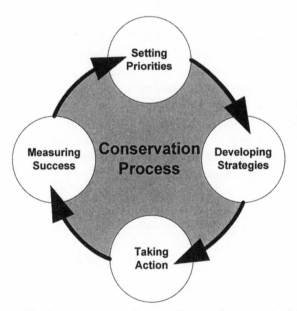

Figure 13.1   The four components of an overall process for conserving biodiversity. The focus of this book has been on the first component—setting priorities—through the development of a scientifically based regional planning process. (From The Nature Conservancy 2001.)

Recall from Chapter 8 the metaphor, first advanced by biologists Mike Scott and Blair Csuti, that Noah really had two jobs—to identify all the important conservation areas and then to individually design the management of each one. A friendly amendment to that metaphor might read: identify the important conservation areas and then develop strategies to abate threats to those areas. Those strategies may be implemented at the scale of the conservation area or at broader spatial scales, depending on the threat. One of the more useful approaches is assessing in greater detail at a finer spatial scale which features of biodiversity (i.e., conservation targets) need to be conserved at each area, what is threatening this biodiversity, and what actions can be taken to abate those threats. This type of detailed planning at the scale of the individual conservation area is referred to as site-based conservation planning, or conservation area planning. Several approaches to conservation area planning by different organizations are described below.

Conservation Area Planning

Several organizations have developed methods for planning the design and management of individual conservation areas, and much has been written on this topic. Probably the best known of these is the Five-S Framework for Site Conservation developed by The Nature Conservancy (Poiani et al. 1998, The Nature Conservancy 2000b). Other methods I will review briefly are the more recently advanced landscape species approach of the Wildlife Conservation Society (Sanderson et al. 2002a) and the landscape approach to forest conservation advanced jointly by the IUCN and the WWF (Maginnis et al. 2001). All of these methods are built upon and have benefited from the considerable knowledge that has accumulated from the design and management of parks, reserves, and protected areas worldwide (e.g., Theberge 1989, Peres and Terborgh 1995).

The Five-S Framework

The Five-S Framework refers to systems, stresses, sources, strategies, and success. This alliteration can be further elaborated as follows (Figure 13.2):

Systems: The conservation targets of Chapter 4, the features intended for conservation at any individual conservation area.

Stresses: The forces of degradation and impairment that impinge upon targets in the conservation area (e.g., increased sedimentation in stream ecosystems).

Sources: The agents that are generating the stresses (e.g., increased sedimentation could be from several sources such as a road, suburban development, or timber harvest operations).

Strategies: The different types of actions that can be taken to abate threats to conservation targets.

Success: Metrics for assessing how well the strategies are working to abate threats to, and improve the status of, targets.

In practice, many Conservancy staff also include a sixth component, *situation,* referring to the human context in which the conservation area is located—what human communities are involved, what sorts of partners and stakeholders the project is engaged with, what government agencies and conservation organizations exist within the region, and so forth.

In contrast to how conservation targets have been used in this book, conservation targets (i.e., "systems") are deployed somewhat differently in conservation area planning. The primary purpose is helping pinpoint threats and develop strategies. Because it is difficult to effectively address a large number of targets at any one area, most conservation area plans developed by TNC limit the number of conservation targets. These targets, referred to as focal targets, range from imperiled species to different types of communities and ecosystems, but in practice, communities and ecosystems tend to be emphasized by TNC in this process. The selected

Figure 13.2   The Five-S Framework for Site Conservation Planning developed by The Nature Conservancy. The components are systems (conservation targets), stresses (threats), sources (of threats), strategies, and success (measures of). In practice, many conservationists add a sixth component to the framework for situation, the socioeconomic and institutional context within which a conservation project takes place. (From The Nature Conservancy 2000b.)

targets are usually intended to serve as surrogates for a broader set of conservation targets at the site or area.

Stresses and sources are the two major components of threats to biodiversity. The key point of importance with regard to stresses and sources is that they apply to specific conservation targets and not necessarily the entire conservation area. For example, one target at an area might be an endangered fish, which could be threatened by a stress such as altered hydrological flows, which, in turn, could be the result of several sources such as a dam, irrigation diversions, or irrigation return flows. Another target at the same area might be the riparian plant community, which likely would have a different set of stresses such as inappropriate livestock grazing or perhaps recreational disturbances, and a different set of sources. Stresses are ranked in importance according to two factors: *severity,* the level of damage to the target within 10 years under current circumstances; and *scope,* the extent of area where the target occurs that will be affected by the stress. Similarly, sources are also ranked in terms of their contribution to the stress based on two factors: *degree of contribution,* the extent to which a particular source is making a substantial contribution to causing a specific stress; and *irreversibility of stress,* whether or not the particular source is producing a stress that is irreversible or reversible.

Strategies are also ranked and evaluated in TNC's Five-S Framework based on three criteria:

1. *Benefits.* Three factors are considered in assessing benefits: the degree of threat abatement by the strategy, the reduction of persistent stresses, and those strategies that are of highest leverage (little effort accrues big results).
2. *Probability of success.* This is defined by the skills of the staff and the feasibility of the conservation strategy.
3. *Costs of implementation.* These refer to the levels of human and financial investment required by the strategy.

Finally, two measures of conservation success are considered in this overall conservation area planning process: *threat abatement,* the degree to which stresses and sources are being abated; and *biodiversity health* (of conservation targets), a measure of the viability of occurrences of target species, or a measure of the ecological integrity of occurrences of target communities and ecosystems based on size, condition, and landscape context (see Chapter 7).

The Nature Conservancy has developed a user-friendly spreadsheet tool that takes users through application of the Five-S Framework and sorts through the ranking of the various factors and measures outlined above. Through a qualitative ranking system, the program identifies the highest-priority stresses and sources to targets at a conservation area and then helps users identify the most promising strategies to abate those sources and stresses (Figure 13.3). By assessing the current viability or integrity of the conservation targets, the program also helps users measure progress over time in improving the "biodiversity" health of the targets.

TNC's two-volume *Five-S Handbook* for site conservation (now referred to as conservation area planning) provides detailed information on the application of the approach (The Nature Conservancy 2000b). This handbook, along with many additional conservation planning references, including examples of conservation area plans, is available at: www. conserveonline.org (go to Library ➤ TNC Programs ➤ Conservation Science ➤ Conservation Planning ➤ Site Conservation Planning).

The Landscape Species Approach

The Wildlife Conservation Society (WCS) has recently conceptualized an approach to conservation area planning referred to as the landscape species approach (Sanderson et al. 2002a). It is predicated on the basic notion that "landscape species," which use large and ecologically diverse areas, can serve as umbrellas (see Chapter 4) or surrogates for the conservation of other species. A suite of landscape species is used in this conservation planning approach. These species are selected on the basis of five factors (Coppolillo et al., in press):

1. *Area.* Area requirements are scored based on the four criteria of home range, dispersal distance, proportion of target landscape occupied by species, and whether connectivity among habitat patches is necessary.
2. *Heterogeneity of habitats.* This factor is based on the proportion of habitats and geopolitical units within a landscape type that a species uses.
3. *Vulnerability to human disturbance.* This factor considers human threats to target species on the landscape, including the severity of the threats, the time period over which they occur, time to recovery from the threat, and the area of species distribution within the landscape affected by the threats.
4. *Ecological functionality.* This factor is based on the number of different ecological functions carried out by a species and the degree of involvement in each function by a species.

Summary of Threats for All Systems
Lower San Pedro River

Systems and Threats	Riparian Forest Mosaic - mainstem	Mixed Broadleaf Riparian Forest - tributary	Aquatic Community - Mainstem	Aquatic Community - Tributary	Upland Plant Community Mosaic	Critical Threat Rank
Excessive Groundwater Withdrawal	High	-	Very High	-	-	High
Invasive/Alien Species	Medium	-	Very High	Medium	-	High
Incompatible Grazing Practices	High	Medium	Medium	Low	Low	Medium
Extirpation of Beaver	High	-	-	-	-	Medium
Incompatible Primary Home Development	Medium	-	-	-	-	Low
Fire Suppression	Low	-	Low	Low	Low	Low
Incompatible Crop Production Practices	Low	-	Low	-	-	Low
Recreational Vehicles	-	Low	-	-	Low	Low
Incompatible Second Home/Resort Development	-	-	-	-	Low	Low
Incompatible Mining Practices	-	-	-	Low	-	Low
Threat Status for Targets and Site	High	Low	Very High	Low	Low	**High**

Go to Targets Directory — Generate / Refresh Data & Color — View Only Chart

Figure 13.3 A sample output from TNC's tool used in the Five-S Framework for site conservation. This tool qualitatively ranks sources and stresses to determine the most critical threats to conservation targets at an area, as well as the most promising strategies for abating them.

378

5. *Socioeconomic significance.* This factor is based on the positive and negative cultural values of a species, positive or negative economic values, and whether the species is a flagship for conservation (see Chapter 4).

The overall landscape species approach to conservation planning involves several steps usually taken within a GIS environment (Figure 13.4). The steps are defining and mapping the biological landscape of the target species, defining and mapping the human landscape (threats), defining the intersections of the biological and human landscape (known as the conservation landscape), delineating those parts of the conservation landscape that are critical to conservation of targeted species (focal landscapes for conservation action), identifying threats within each focal landscape, developing and implementing conservation strategies and interventions, and monitoring results.

Although it has some similarities with the TNC approach (e.g., the same criteria are used to identify threats), the WCS approach relies more heavily on a mapping of both conservation targets and threats. At the same time, it is somewhat less of a cookbook approach than that of the Five-S Framework, more intensive in its data demands, less dependent upon expert opinion, and focused only on species-level targets.

A useful analysis would be to compare these two approaches at several different conservation areas of varying sizes and biotic composition, where the goal is the overall conservation of biodiversity of these areas. Whether a consistent set of similarities or differences in outputs would result from this comparison test is not clear. However, the scientific and conservation communities need to understand the differences between these tools, the strengths and weaknesses of both, and when it is most appropriate to apply one tool, the other, or a blend of both.

The Landscape Approach to Forest Conservation

The World Wildlife Fund (WWF) International and the World Conservation Union (IUCN) are in the early stages of developing, testing, and implementing a conservation planning approach at the landscape scale (Maginnis et al. 2001). They define a landscape as "a contiguous area intermediate in size between an ecoregion and a site, with a specific set of ecological, cultural, and socioeconomic characteristics distinct from its neighbors" (Figure 13.5). (In this case, site refers to a forest or timber stand that may be managed for production, conservation, or both.) The origins of this landscape approach can be traced to the "ecosystem approach" to con-

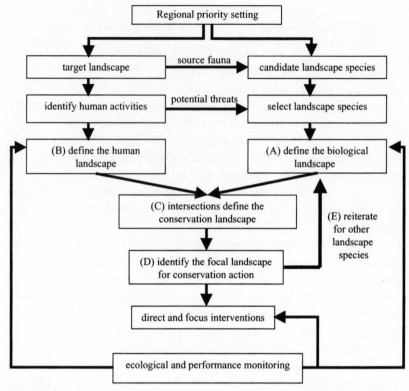

Figure 13.4   A flow chart of the Wildlife Conservation Society's landscape species approach to conservation area planning. Through simultaneous processes, the biological landscape is defined based on the needs of a suite of landscape species, while the human landscape is defined based on threats to these species. The overlay of these two landscapes defines the areas of conservation interest (conservation landscape). By focusing on these areas, conservation strategies and interventions are developed. (Reprinted from *Landscape and Urban Planning,* Volume 58, Sanderson et al., A conceptual model for conservation planning based on landscape species requirements, pages 41–56, copyright © 2002 with permission from Elsevier Science.)

servation enacted by the Convention on Biological Diversity (Fifth Conference of Parties, Decision V/6).

The approach involves eight basic steps: identifying conservation targets and landscapes, conducting stakeholder analysis to determine expectations of others for the conservation landscape, defining the cultural landscapes within the conservation landscape, estimating the performance of the landscape in providing ecological functions, estimating the potential of the landscape to enhance or maintain functions, examining and selecting management options for the landscape, implementing management practices agreed upon by stakeholders, and monitoring and evaluation. The

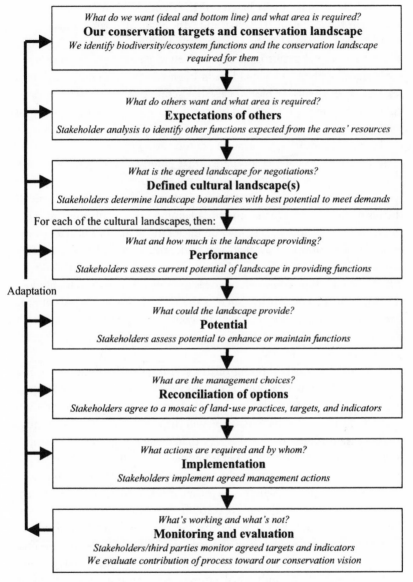

Figure 13.5 A flow diagram for implementing the IUCN-WWF landscape approach. The eight steps involved in this approach to conservation area planning have much in common with similar planning approaches of TNC and the WCS. (From Maginnis et al. 2001. Reprinted by permission of IUCN and WWF International.)

IUCN and WWF International are testing and evaluating the landscape approach. It is conceptually based on experience in Dyfi Valley, Wales; Kinabatangan River, Sabah, Malaysia; and Min Shan, Sichuan, China.

The three approaches to site or conservation area planning outlined above have several points in common. First, each emphasizes landscapes as the scale of application. Second, each begins with an identification of the features or elements of biodiversity (i.e., conservation targets) that are of concern at a particular place and proceeds stepwise to identify strategies and management options for conserving these targets. Third, each approach concludes with monitoring to measure conservation progress.

In practice, the conservation areas that emerge from regional planning come in all shapes and sizes, depending on the methods that were used to select them, the nature of the conservation targets themselves, and the threats to the targets. During the site planning process, planners may need to reconfigure or more accurately delineate the actual boundaries of individual conservation areas that were identified in only general terms during a regional or ecoregional planning effort. The fact that some plans will be prepared for conservation areas intended to conserve highly localized endemic plants in places such as the Cape Floristic Region of South Africa, while others are intended to conserve a population of Siberian tigers, is a clear indication that conservation areas themselves will occur at a wide range of spatial scales.

As I noted at the beginning of this section, the development of individual conservation area plans is but one approach in a suite of strategies that can be deployed in the implementation of a regional or ecoregional conservation plan. In many cases, it will make sense to develop strategies that are appropriate at spatial scales that generally exceed those of the individual conservation area. For example, some threats manifest themselves more or less continuously over vast areas (e.g., acid rain, altered hydrological regimes), whereas in other cases, a particular threat recuts across multiple conservation areas, often involving the same or similar conservation targets. Examples of the latter are fire suppression in many of the coniferous forestlands of the western United States and the bushmeat trade in tropical ecosystems. Below, I examine the development of strategies beyond the individual conservation area in more detail and provide several examples of such strategies. I refer to these strategies collectively as multiarea strategies, with the recognition that their implementation may have benefits that go even beyond the boundaries of conservation areas per se. Of particular interest are strategies that may be implemented at one area but have an impact, either directly or indirectly, at many conservation areas, thereby providing conservationists a bigger bang for the buck.

Multiarea Strategies

In theory, developing and implementing multiarea strategies is a promising avenue of conservation pursuit in that such strategies have the potential to be both efficient and leveraged. They are efficient because a single strategy, if successfully implemented, can reap conservation benefits at several places simultaneously. They are leveraging because the multiarea strategy often originates from the identification and successful implementation of a strategy at a single conservation area that is then applied to multiple areas. In practice, however, successful strategies of this type are often politically difficult to implement and remain few in number.

These strategies need not arise solely from regional or ecoregional planning exercises. For example, The Nature Conservancy has initiated networks of site-based conservation practitioners, organized around a common theme (e.g., wetlands, forestry, fire management, grazing), who work collaboratively with internal and external scientists to develop strategies to abate similar threats that recur across many conservation areas. However, because of the inherent broad overview of biodiversity and threats that occurs during ecoregional and regional planning projects, these planning processes will always represent an important opportunity for the identification of multiarea threats and strategies. Below, I provide examples of multiarea strategies, some of which are successfully being implemented and others that are works in progress.

The Nature Conservancy and the U.S. Department of Defense: Eglin Air Force Base

Eglin Air Force Base (AFB) on the Florida Panhandle was identified as a priority conservation area in The Nature Conservancy's East Gulf Coastal Plain Ecoregional Plan (East Gulf Coastal Plain Core Team 1999). Recognized as a nationally important conservation area for biodiversity, it harbors several endangered species, the best known of which is the red-cockaded woodpecker (*Picoides borealis*), and it contains perhaps the best remaining example of the longleaf pine (*Pinus palustris*) ecosystem, which once occurred throughout the southeastern United States. Following biological surveys conducted by the University of Florida and the Florida Natural Areas Inventory, TNC's Florida program entered into a cooperative agreement with the U.S. Department of Defense (DOD) to develop an Integrated Natural Resources Management Plan for Eglin AFB. With the involvement of more than 100 scientists and managers from different organizations, this plan was completed in 1993.

The success in working at Eglin AFB was leveraged in two distinct ways, both of which had significant positive consequences for biodiversity conservation (J. Hardesty, TNC, personal communication). First, building on the successful program at Eglin and other DOD facilities, TNC worked with the DOD to greatly increase its Legacy Cultural and Natural Resources Program, especially for projects involving integrated natural resource management planning, ecosystem management, ecological monitoring and research, and adaptive management programs. As a result, the Legacy Program has provided more than US$150 million for biological surveys, cooperative management projects, and ecosystem management on DOD lands. Several TNC ecoregional planning projects have been funded, in large part, by the DOD Legacy program (see Sonoran Desert ecoregion example in the next section). In addition, Conservancy staff and other scientists wrote a handbook for managing biodiversity on DOD lands, which is now in its third printing and is extensively used across DOD facilities (Leslie et al. 1996).

The second major leverage point was the creation of the Gulf Coastal Plain Ecosystem Partnership, facilitated by TNC and the DOD. This agreement provides for seven different private and public organizations working toward the conservation of 340,000 hectares (840,000+ acres) of longleaf pine in an area from the Gulf of Mexico on the Florida Panhandle to southern Alabama. This landscape represents the largest contiguous block of longleaf pine ecosystem remaining in the region. Landscape-level planning efforts are under way for the partnership and are addressing the most significant threats to the ecosystem, including habitat fragmentation, fire suppression, altered hydrological flow regimes, and unrestricted off-road vehicle use. Several successful conservation projects have been completed, and the partnership has raised significant funding for additional projects.

The remarkable part of this story is that through working together primarily at one large conservation area with one set of threats, TNC and the DOD were able to launch a multiarea initiative that is abating the same threats far beyond the boundaries of Eglin AFB and is significantly contributing to the recovery and restoration of the longleaf pine ecosystem. In addition, these successes have had positive and far-reaching consequences for biodiversity conservation at DOD facilities throughout the United States.

The Northern Tallgrass Prairie Ecoregion

An ecoregional conservation plan for the Northern Tallgrass Prairie in the central United States (The Nature Conservancy 1998) identified fire suppression of remnant patches of tallgrass prairie as one of the primary threats to biodiversity, especially those remnants on state public lands in Minnesota. The limiting factor appeared to be insufficient funding for state natural resource staff to conduct controlled burns on priority conservation areas that had not been burned previously. The planning team engaged TNC's director of government relations in Minnesota to help them develop a strategy for returning fire to these prairie remnants. Government relations staff worked with an outgoing governor, a new governor, and the state legislature to successfully secure funding for the Department of Natural Resources to use prescribed fire in the management of remnant prairie reserves.

The U.S. Army Corps and The Nature Conservancy: Green River, Kentucky

The Green River in Kentucky is one of the highest-priority conservation areas for freshwater biodiversity in the eastern United States (Interior Low Plateau Planning Team 2001). Scientists have hypothesized that populations of native fish and freshwater mussels of this priority stream ecosystem have been negatively affected by altered hydrological flows from an upstream dam operated by the U.S. Army Corp of Engineers. The dam is operated for both recreational lake management and flood control. During autumn, the reservoir is quickly drawn down with large releases of water to provide storage capacity for anticipated winter floods. These large releases of water are suspected to be detrimental to both native mussels and fish and their prey. Scientists from TNC have been working with the Army Corps to shift the timing and duration of these fall releases to mimic some of the river's natural flow patterns, thereby reducing the impact on native aquatic ecosystems (Richter et al., in press).

The successful collaboration between the Army Corps and the Conservancy at Green River has led to the development of a national Memorandum of Understanding (MOU) between the Army Corps of Engineers and TNC. As a result of this MOU, the two parties will soon be working together to improve natural flow regimes and the overall integrity of aquatic ecosystems at many Corps-operated dams across the United States that occur in priority conservation areas as identified through TNC's ecoregional conservation plans.

Wild Meat Use and the Bushmeat Trade

An assessment of biological conservation priorities in the Guinean–Congolian forest region by WWF International (Blom et al. 2001) concluded that the commercial bushmeat trade is the number-one threat to biodiversity in the region and that "so-called traditional hunting is also considered unsustainable in even remote areas." In fact, the use of wild meat and the trade of bushmeat are global problems, viewed by many professionals as a biodiversity crisis (Bennett et al. 2002). As human populations increase, the demand for bushmeat also increases. Increased logging of tropical forests opens up access to the forest for nonlocal hunters and provides a more cost-effective route to markets for trading bushmeat. This overexploitation of wildlife species has caused a number of species to go extinct, has driven other species to ecological extinction, and is currently a serious threat to many species of birds and mammals classified as critically endangered, endangered, or vulnerable by the IUCN Red List (Rao and McGowan 2002, Rosser and Mainka 2002). The overexploitation of wildlife and trade of bushmeat are clearly multiarea threats that repeat themselves in fairly predictable terms in tropical ecosystems worldwide. As a series of articles in a recent issue of *Conservation Biology* (Volume 16 [3], 2002) indicate, this threat to biodiversity conservation has received a great deal of attention, at the scale of the individual conservation area as well as through the policy arena.

Davies (2002) has suggested four actions to regulate bushmeat hunting and trading, all of which can be considered multiarea conservation strategies:

- Address the underlying and direct causes of the loss of bushmeat species, including the regulation of harvest methods, the regulation of levels of take and access to forest areas (direct causes), and social issues, such as poverty, weak local governments, and unfavorable trading terms (underlying causes).
- Revise poverty assessment procedures to help ensure that environmental concerns such as the bushmeat trade are included in planning by international development agencies.
- Improve environmental assessments for policy changes and for infrastructure projects to incorporate the often detrimental effects of development on increases in the bushmeat trade brought about by improved access to forests.
- Develop new systems of governance in rural areas that involve comanagement by local communities, local government, and the private sector. These comanagement practices are thought to be more effective in managing the bushmeat trade.

Additional information concerning bushmeat can be found at the Bushmeat Crisis Task Force Web site: www.bushmeat.org.

The preceding examples of developing strategies to abate threats that transcend any single conservation area are intended to demonstrate that taking such an approach is or has the potential to be powerful, leveraging, and complementary to actions taken at any single conservation area. Although it is far easier to point to success stories at individual places, this is due, in part, to the fact that the community of practitioners focused on biodiversity conservation has placed more emphasis on individual site-based approaches in the past than it has on more broadly based strategies. Recognizing that there is both efficiency and effectiveness to be gained from employing a multitude of approaches, this emphasis is beginning to shift toward a balance of both single-area-based and multiarea-based approaches.

Developing conservation strategies at scales appropriate to targets and threats with an eye to efficiency and leverage are essential steps in moving a plan into the implementation phase. In the next section, I provide several examples of how these steps and others are helping translate regional and ecoregional plans into conservation action in the broader sense.

Turning Conservation Plans into Conservation Progress

Four examples of successful planning efforts for biodiversity conservation using many of the steps articulated in this book are presented next. These plans were developed at a variety of spatial scales, including ecoregions, parts of several ecoregions, and a geopolitical unit (state).

The Sonoran Desert Ecoregional Plan

In 1998, The Nature Conservancy undertook a binational conservation planning effort in the Sonoran Desert ecoregion (Marshall et al. 2000). Its three principal partners were the U.S. Department of Defense, the Sonoran Institute, and IMADES—the Sonora State Institute for Environment and Sustainable Development (Figure 13.6). The DOD, which provided substantial funding for the project, is a major landowner in the region and wanted to learn about the biodiversity importance of their lands in a broader ecoregional context. The Sonoran Institute is a nonprofit organization that works collaboratively with local communities and interests to conserve and restore natural landscapes in western North America. They

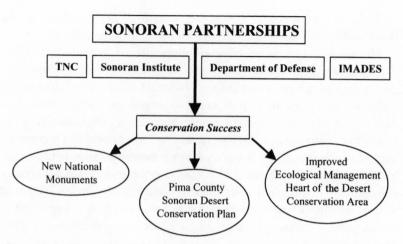

Figure 13.6   The Sonoran Ecoregional Planning Project. This project was conducted by a partnership among four organizations. The partnership has contributed significantly to the successful implementation of the plan, including the creation of new national monuments, the incorporation of plan results into Pima County's Sonoran Desert Conservation Plan, and improved ecological management of key sites through conservation area planning.

were interested in a scientific framework to guide some of their own programmatic activities. Their primary role in the project was outreach to the many stakeholders and interest groups in the region. Technical scientific and information management skills from the Mexican side of the project were provided by IMADES, which also served as the primary liaison with Mexican organizations and institutions.

This capable partnership identified approximately 100 conservation areas covering some 42% of the ecoregion. The planning process was a very public one, with considerable outreach through various media. The results of the planning process were posted in report form in both Spanish and English on a public Web site, and printed copies of the report were provided to over 500 individuals and institutions that cooperated in the planning process. The Nature Conservancy worked closely with a coalition of environmental organizations in the designation of two large conservation areas as National Monuments under the Clinton administration. Scientific information from the planning process helped justify the importance of the designations and contributed to decisions on the boundaries of the monuments. The planning team also worked with Pima County (Arizona) planning staff, who were undertaking the largest county-wide conservation planning effort in the United States at the same time as the ecoregional planning process of the Conservancy. Pima County eventually adopted the conservation areas identified in the Sonoran Ecoregional Plan as the starting point for its own conservation priorities.

The Sonoran ecoregional team has undertaken conservation planning at the highest-priority area in the ecoregion. Working with DOD funding, TNC and six partner organizations are developing a detailed conservation plan to improve the ecological management for the Heart of the Desert conservation area, which includes both public and private land. All told, TNC and conservation partners have already initiated conservation actions on over 3.2 million hectares (8 million acres) of priority conservation lands within the Sonoran ecoregion, within less than 2 years since completion of the plan. The ingredients to success appear to be strong, strategic partnerships, a very public planning process, and thinking and acting upon implementation throughout the planning process, along with no small measure of good timing.

Ecoregional Planning in Yunnan, China

The Yunnan Great Rivers Project is a cooperative venture of The Nature Conservancy and the Yunnan provincial government. The goals of the project are twofold: conservation of biodiversity and the natural and cultural resources of northwestern Yunnan, and compatible and sustainable economic development in the project area.

During 1999–2001, TNC developed a regional plan for biodiversity conservation that covers 15 counties in northwestern Yunnan and includes approximately 66,000 km^2 (25,000 mi^2), an area similar in size to West Virginia (U.S.) or Ireland (The Nature Conservancy of China 2002). Portions of five WWF ecoregions are contained within the project area, although over 95% of the project is located within three of these ecoregions. The regional planning effort largely followed the process that has been outlined in this book. Over 40 academic departments and governmental agencies were engaged in the planning process, three expert workshops were held, and new biological inventory data for two field seasons were collected via Rapid Ecological Assessments (see Chapter 5). Conservation areas were identified separately for the three ecoregions and consisted of both existing protected areas and newly identified areas. The ecoregional analysis identified nearly 1.4 million hectares (3.45 million acres) of new conservation areas across parts of three ecoregions. Threats analyses concluded that four major factors are causing biodiversity loss in the project area:

1. Human population growth, which increases the consumption of natural resources in the region and contributes to illegal hunting.
2. Poverty, which leads to overcollection of natural resources, limited cash to purchase alternative fuel sources and building materials, and increased logging and illegal hunting.
3. Changes in economic market forces in China leading to increases in tourism, which in turn is leading to habitat loss, degradation, and fragmentation due to infrastructure development.
4. Government policies that run counter to conservation, such as a lack of community participation in establishing nature reserves, poorly defined land tenure systems adjacent to nature reserves, and subsidies that encourage the nonsustainable use of natural resources.

The ecoregional plan identified five high-priority areas for conservation action spread across parts of three ecoregions. Program offices are being established at each of these areas, detailed conservation area plans are under development for each, and strategies are being implemented to effectively create new conservation areas. The ecoregional plan has been accepted as one of the Yunnan government's subplans in China's overall 5-year plan. Action items in the plan include expansion and additions to the current protected area system, identification of management needs of conservation areas, strategies for abating threats at these areas, and economic development opportunities.

By any measure, the systematic approach to conservation planning being used by TNC in Yunnan has been tremendously successful to date. As a result, several new conservation areas that are significant in size and biodiversity composition are being created. There is interest from other provinces in China and the national government in expanding the conservation approach that TNC has used. The initial partnership with the Yunnan government, the largely in-country staff used by TNC, the large outreach program to Chinese institutions and experts, and the melding of the conservation project into the larger planning framework of the Yunnan and Chinese national governments all appear to be factors contributing to the success of the conservation effort to date.

Regional Forestry Agreements in Australia

There has long been a conflict in Australia over the use of its forested lands, largely split along the lines of extractive use and conservation. In 1992, the Commonwealth government issued the National Forest Policy

Statement, which provided the basis for the sustainable development of forested lands. Regional Forestry Agreements (RFAs) are viewed as the vehicle for implementing the National Forest Policy Statement in the commercial forested lands of Australia (Davey et al. 2000). These legally binding agreements are a mechanism for balancing conservation and the sustainable use of the natural, cultural, social, and economic values of Australian forests. The legal specifications of these agreements include identifying and implementing a comprehensive, adequate, and representative reserve system for biodiversity (see Chapter 8); providing for the ecological and sustainable management of forests; providing long-term stability for forest industries; and incorporating studies on environmental values, indigenous heritage values, economic values, social values of communities, and principles of sustainable management.

Nine RFAs were in place as of March 2000, covering ten forest regions in four of the six states of Australia, with two additional regions pending. These regional agreements cover a 20-year time frame and currently account for 16% of the 157 million hectares (388 million acres) of forested lands in Australia. The process for RFAs, outlined in Figure 13.7, includes a deferred forest and interim management agreement to avoid foreclosing conservation options in the future. Comprehensive Regional Assessments collected and synthesized vast amounts of existing and new information on environmental (including biodiversity), heritage, economic, and social values and also developed principles of ecologically sustainable forest management. Designing comprehensive, adequate, and representative reserve systems under the auspices of RFAs is a key component of the process and includes standards for placing a minimum of 15% of each forest type (pre-1750 type and extent of forests) in reserves with higher rates for threatened forest types (e.g., old-growth) or types with uncertainty as to original extent. A Scientific Advisory Group was established to advise the Commonwealth on the implementation of the three reserve criteria and other significant scientific issues. More detailed information on the RFA process is available at www.rfa.gov.au.

A reserve selection algorithm developed by Pressey and colleagues (Pressey 1998; see Chapter 8) known as C-Plan was used in interactive negotiations between conservation and timber interests to design a conservation reserve system for four RFA regions in New South Wales. Details of the first round of the negotiation process over the reserve system are provided by Pressey (1998). Since that first round, the same software system with progressive refinements has been used in three subsequent stages of the RFA process in New South Wales. The concept of irreplaceability (see

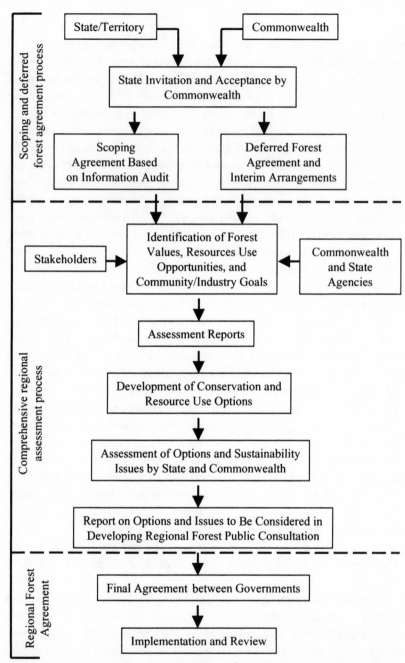

Figure 13.7 A flow chart for the development of a Regional Forest Agreement (RFA) in Australia. These legally binding agreements were put in place by the Commonwealth government to help overcome the impasse between conservationists and timber industries over the future of public commercial timberlands in much of Australia. A consideration of biodiversity values and the design of a network of conservation areas is an integral part of the RFA process. (From Davey et al. 2000.)

Chapter 8) was the cornerstone of the analyses and negotiations for the use of public production forests of eastern New South Wales. The outcome of the negotiation process was nine new national parks and nature reserves, 816,000 hectares (2 million acres) of forest temporarily deferred from logging, new wilderness areas, and agreements on the supply of timber for 5 years.

The conservation planning process used in these RFAs was the first in New South Wales to involve explicit goals and conservation targets to make decisions about a system of reserves (i.e., conservation areas). It was also the first to involve an open negotiation process among a wide array of interest groups. At the same time, some scientists have questioned the scientific credibility and transparency of the RFA process as it has taken place over all regions (Davey et al. 2000). Among other criticisms, the peer review of the scientific process was limited at best, as was publishing findings in peer-reviewed journals, primarily due to the short, politically driven time frame of the negotiations. A retrospective look also suggests that although the planning process had a strong scientific foundation and used the best available scientific information, it did not always result in a balance between sustainable use and conservation.

One of the few quantitative assessments of RFA outcomes was published recently (Pressey et al. 2002). Results showed that for the forest of northeastern New South Wales, the achievement of goals on public lands was reasonably successful, largely due to extensive areas deferred from logging. However, the results also highlighted the strong bias in achieving of goals resulting from the political decision to restrict negotiations to public tenures. Forest types that occurred primarily on private land were not conserved by this process and remain vulnerable to further clearing.

Statewide Conservation Plans in Florida

Although several states have or are developing statewide comprehensive plans for biodiversity conservation (Environmental Law Institute 2001), Florida has a relatively long and rich history in conservation planning. Noss (1987) first presented a conceptual map of a proposed biodiversity reserve network for Florida. Noss and Cooperrider (1994) later published a more detailed map of existing public lands, private nature reserves, private lands that were priorities for conservation management, and other areas of general conservation concern, all based on the findings of a 1991 workshop that involved the participation of 40 biologists from throughout the state (Kautz and Cox 2001). However, many of the areas identified in this map were still too general to establish specific private or public land conservation priorities, some areas already contained considerable amounts of urban or agricultural use, and some important linkage areas were already affected by interstate highways.

Beginning in 1990, the Florida Fish and Wildlife Conservation Commission initiated a project to identify private lands in the state that should be conserved to meet the long-term conservation needs of Florida's biodiversity. Results of this effort were first published in 1994 in a report entitled *Closing the Gaps in Florida's Wildlife Habitat Conservation System* (Cox et al. 1994). The project focused on the population and habitat needs of 54 focal wildlife species, rare plant communities, and 105 globally rare species of plants. Analyses concluded that the system of public land conservation areas in the state would not adequately conserve several key focal species such as black bear (*Ursus americanus*), Florida panther (*Felis concolor*), American swallow-tailed kite (*Elanoides forficatus*), and crested caracara (*Caracara plancus*). Cox and colleagues then designed a system of Strategic Habitat Conservation Areas (SHCAs) that was thought to adequately conserve the conservation targets over the long term. These SHCAs included some 1.95 million hectares (4.82 million acres) of private land that needed conservation management through a variety of tools, including acquisition, easements, and land-use planning.

Concerned that the initial analyses of 54 focal species, rare communities, and rare plants may have overlooked some important components of biodiversity, Cox and Kautz (2000) examined 124 additional species of rare and imperiled wildlife to determine how well existing conservation lands and the SHCAs of the 1994 report performed in conserving habitat for these additional focal species. Their results were encouraging. Only 17 of the 124 species examined were found to need additional habitat conservation efforts, suggesting that the original approach of Cox and colleagues (1994) worked reasonably well in conserving rare and declining species in Florida. No new SHCAs were identified because most of the needs of the 17 species could be met through the protection of select and relatively small parcels of land due to the relatively small home ranges and restricted distributions of the target species.

Since 1994, about 15% of the lands (287,333 hectares, or 710,000 acres) identified in the SHCAs have been acquired through state and federal land acquisition programs. Results of the 1994 and 2000 reports have also been used in land-use planning, environmental impact assessments, development regulations, and private landowner initiatives (Kautz and Cox 2001). Nevertheless, strong lobbying and public relations efforts from agricultural, timber, and development interests have generated opposition to the results of these planning efforts in some forums.

In general, the analyses of Cox, Kautz, and colleagues closely parallel the steps outlined in this book. These researchers identified a set of conservation targets, collected information on them from a wide variety of sources, and set specific goals (10 populations each of 200 breeding individuals; see Chapter 6). Moreover, they used population viability analyses (see Chapter 7) to establish these conservation goals. Various habitat models were then used to identify the SHCAs necessary to conserve the target species. The agency is now in the process of updating these analyses using more recent remote-sensing data, reanalyzing species models to incorporate protected and lost habitats, and identifying new SHCAs as needed (Environmental Law Institute 2001). It expects to publish an update in late 2002.

We have now examined several different cases of implementing the results of regional biodiversity conservation plans. A reasonable question from any conservation practitioner at this point might be: What is all of this going to cost? Recent analyses provide some hints.

Implementing Regional Conservation Plans: The Cost of Conservation

James et al. (2001) estimated the costs of achieving a global set of protected areas based on the World Conservation Union's stated goal of 10% of the land area of each nation or ecosystem set aside for conservation purposes. Their projections were based on five steps:

1. Estimate current expenditures on protected areas worldwide.
2. Estimate funding shortfalls in the current system of protected areas.
3. Project costs of acquiring additional lands to expand the network of protected areas to achieve ecological representation.
4. Calculate the costs of appropriate management of the additional protected areas.
5. Estimate opportunity costs for compensation to local communities of the additional protected areas.

To develop these estimates, they divided the world into ten ecologically and economically similar regions, then used data from within each region to extrapolate projections to the rest of the region. Their analyses suggest that current expenditures on protected areas (as defined by IUCN Categories I–VI; see Chapter 5) total around US$6 billion annually. Improving the management of existing protected areas, expanding the network especially in the IUCN Category I–III protected areas, and accounting for opportunity costs to local communities would increase these costs to approximately US$28 billion annually. Of course, these figures do not account for adequate conservation of freshwater and marine ecosystems, or the conservation of lands outside protected areas that would be necessary to maintain the integrity of protected areas and conserve elements of biodiversity that cannot be conserved in protected areas. By one estimate, James and colleagues place that additional figure at about US$290 billion per year, although this estimate is very crude at this point in time.

In the United States, Shaffer and colleagues (2002) took a somewhat different approach. Based on two statewide plans to conserve biodiversity (Oregon and Florida) and one TNC ecoregional plan (the Central Shortgrass Prairie), they estimated that about 25% of the land in the contiguous United States would be required for a comprehensive national system of habitat conservation areas and that about half of this land is in private ownership. Note that the 25% figure is considerably greater than the 10% IUCN goal mentioned above and, based on many examples presented in this book, far closer to the reality of what will really be needed to achieve biodiversity conservation. They examined three options for conservation on private lands: fee simple acquisition, conservation easements, and land rental agreements. Assuming it would take three to four decades to accomplish this system, they prorated the amount of land protected over 30- and 40-year periods. Adjusting for both inflation and the additional management costs of new lands brought into this system, Shaffer and colleagues estimated the annual costs of the acquisition option to be US$7.7 billion, US$5.4 billion for the conservation easement option, and US$0.3 billion for the land rental option. Spread out over 40 years, however, the costs ranged from US$256.8 billion for conservation easements to US$502 billion for land rental agreements.

These estimates assume that there is no need for increases in the management costs of public lands that would fall in a national system of habitat conservation areas. In addition, they include no estimates of "opportunity costs" such as those of James et al. (2001) and likely fall considerably short of expenditures necessary to conserve both freshwater and nearshore marine diversity. Nevertheless, they do provide for the first time in the United States a ballpark figure (US$5–US$8 billion annually) for what it will take to seriously curb the ever-increasing number of endangered species and slow the loss of biological diversity. As Shaffer and colleagues point out, these annual costs are about one-third to one-fourth of what the United States spends to maintain its national highway system.

In the Agulhas Plain of the Cape Floristic region of South Africa, a 2160-km^2 (834-mi^2) area known globally for its plant endemism, Pence and colleagues (in press) investigated the financial costs of alternative conservation strategies for an expanded reserve system. Traditional reserve approaches were estimated to cost approximately R240M (US$30 million) in acquisition costs, whereas a mixed-management approach involving some acquisitions as well as property rights management agreements could save as much as 80% in acquisition costs. Subsequent management costs (primarily alien plant removal), estimated to cost R401M (approximately US$50 million), could also be substantially reduced with a mixed-management approach and tax rate relief (incentive) to private landowners.

As overwhelming as these figures may first appear, when contrasted against other expenditures that society makes on a regular basis, they seem neither unrealistic nor undoable. The good news about the results of systematic conservation planning for biodiversity and these sorts of financial estimates of the costs of conservation is that for the first time, society is being provided some transparent information and analyses on where to do conservation work and how much it might cost. That, in itself, is a major step forward for the science of conservation planning and for the practice of biodiversity conservation.

CHAPTER FOURTEEN

Conservation Planning at the Crossroads

Well-designed conservation landscapes—with representative systems of conservation areas of sufficient size, condition, and connectivity to maintain even the most sensitive species and ecological processes—are an essential foundation of any conservation strategy.

DAVID OLSON AND COLLEAGUES (2002)

Conservation biologists will remember the 1990s as a time when many organizations and institutions jumped on the conservation planning bandwagon. The global priority-setting exercises of the World Wildlife Fund and Conservation International, the U.S. government's planning efforts in the Pacific Northwest for conservation of the northern spotted owl and other old-growth-dependent species, the plethora of biodiversity conservation plans being prepared under the auspices of the Convention on Biological Diversity, and The Nature Conservancy's initiation of ecoregional conservation plans for all of the United States are but a few examples demonstrating the wave of enthusiasm for conservation planning that both governmental and nongovernmental organizations worldwide are exhibiting. To what can we attribute this enthusiasm?

I suspect the reasons are many and varied. First and foremost, the continued degradation and disappearance of natural habitats, and along with them the extinction of many species at alarming rates, is of great concern to many conservation biologists, conservation practitioners, and the public at large. For some parts of the world, both natural capital and real resources to conserve this capital are running dry. With conservation options being rapidly foreclosed and limited funds to exercise these remaining options, the development of plans that highlight the most important places and strategies for conserving biodiversity is a reasoned and critical response.

In the United States, the legal, policy, and managerial quagmires that have resulted from the listing of some endangered species (which former Secretary of the Interior Bruce Babbitt referred to as "train wrecks") have caused many inside and outside government to think more proactively about species and habitat conservation. One of the responses has been to develop large-scale conservation assessments and plans that consider the needs of many species simultaneously so as to avoid the last-minute "emergency room procedures" that have become far too common with some endangered species. In other cases, organizations like The Nature Conservancy realized that, despite enormous investments in habitat protection, those investments were only scratching the surface of what was really needed to conserve biodiversity and that a different approach that mobilized and catalyzed many interested conservation players was needed.

Finally, there has been an acknowledgment by the scientific, conservation, and donor communities that so much of what had been accomplished in conservation to date has been through ad hoc means and that there are serious biases in the distribution of most conservation efforts worldwide. All of these lines of thinking lead us to believe that a more systematic approach to conservation might bear more fruit.

Much conservation has been accomplished in the last decade or so because of these efforts. Yet with so many planning efforts under way by so many different organizations, it seems appropriate to ask: How can this planning work be improved in terms of its technical side, and how, on the practical side, can planning efforts lead to more effective conservation of the world's diminishing biological resources?

There are several areas in which the science of conservation planning is falling well short of where it needs to be. First, the setting of target-based goals and the assessment of viability and integrity for the long-term persistence of biodiversity are in their infancy in terms of the scientific rigor behind these steps. One could easily argue that for some species and ecosystems in some parts of the world, the setting of conservation goals is as much an art as it is a scientific endeavor. That is more of a criticism about our state of knowledge about biodiversity than it is of our abilities as scientists and conservation practitioners. Second, we are in dire need of information on the ecological variability of ecosystems across the environmental gradients in which they are distributed and how large conservation areas need to be to maintain natural disturbance regimes and ecological functionality. Both types of information are critical to setting informed goals and to assessing which examples of these ecosystems are the best candidates for long-term conservation efforts.

Third, despite two decades of research and scores of books and scientific papers in population viability analyses, there has been little synthetic work to provide conservation planners with even the broadest of generalizations across major taxonomic or life history groups on how many populations are needed for a species to persist and how large those individual populations need to be. Fourth, although progress is being made, conservation planning and action in freshwater and marine ecosystems continue to badly lag behind terrestrial efforts, despite overwhelming evidence, in the case of freshwater biota and ecosystems, that they are among the most imperiled in the world. We need to improve our abilities to estimate which conservation targets and which conservation areas are the most vulnerable to degradation and destruction in the future. Although Bob Pressey and colleagues in Australia have undertaken some promising efforts in this arena, it remains unclear as to whether current or past levels of land clearing in a region, for example, are good predictors of future levels of land clearing.

Some interesting new research strongly hints that the ghosts of land-use past continue to haunt and plague our conservation efforts in ways we are only beginning to understand. Much of the work in conservation planning related to threats and vulnerability is anecdotal in nature, largely based on expert opinion. With the advent of GIS and the increasing availability of digital data from remote sensing and other sources, our ability to predict those areas that will be most threatened in the future needs to be strengthened. At the same time, our capacity to effectively spend scarce conservation money also needs to improve.

The age of digital ecology is upon us. With it comes the ability to amass a variety of different types of information in powerful and compelling ways with maps and analyses. As seductive as these tools may appear for conservation, biologists and planners should avoid falling into the trap of accepting all the conclusions and results of these techniques at face value. Of particular concern is the lack of field verification and accuracy assessments associated with the increasing number of modeling efforts being used to predict the distribution and status of critical conservation targets. Our ability to select conservation areas that are known to or predicted to contain specific biological features in a particular condition is only as good as the information behind these modeling efforts. It is not a trivial point to suggest that the ability of scientists to produce sophisticated models that predict the status and distribution of species and communities has outstripped our willingness and enthusiasm to verify the accuracy, and ultimately, the usefulness, of those models.

On the practical side, a number of actions can be considered to help ensure that these planning efforts lead to as much effective practice of conservation on the ground (or in the water) as possible. For instance, in the opening chapter of this book, I noted that many decisions regarding the conservation of biological diversity are made locally—through village, city, township, watershed, or county governments. Yet it is at this local level that the least amount of information and expertise on biodiversity is available for decision making. If the most extensive and highest-quality examples of a particular species or ecosystem are located within the jurisdiction of a particular local government, then that government, at the very least, needs to be aware of the status and location of these important biological features. In addition, local government officials should have some appreciation for why they should care about these features of biodiversity. In the United States, a consortium of conservation-oriented groups is tackling this thorny problem. Regional and ecoregional conservation planning exercises will identify scores of areas that are critical to the future existence of a region's native species, communities, and ecosystems. To conserve these important places, we must find ways to deliver information to local jurisdictions in a form that is not only usable but compelling as well.

Of course, just knowing that a particular place is important for conservation begs the question of how that area will get conserved. In most cases, regional or ecoregional conservation plans will identify generalized places, areas of biodiversity significance that contain particular biological features (i.e., conservation targets). There may also be some information on what general threats are impinging on those targets. In and of itself, that is not nearly enough information to undertake conservation action. As I mentioned earlier, Scott and Csuti referred to this conundrum as the two jobs of Noah—the selection of important conservation areas and their design and management. I have suggested some friendly amendments to that metaphor, but the point still stands: At the local level, conservation practitioners need more detailed information on the status and distribution of conservation targets, and the threats to them, in order to formulate and implement effective conservation strategies. Inevitably, a closer examination of conservation areas will reveal additional important targets for conservation and additional threats to biodiversity that were not apparent to planners and biologists at the regional planning level. What is needed now are more effective ways for information and data to flow from the results of conservation planning exercises as outlined in this book to on-the-ground natural resource managers and conservation practitioners. And vice versa: As practitioners and managers learn more about particular areas, that information needs to flow back into these larger-scale planning efforts so that they can be updated and revised over time.

Conservation planning, at whatever scale it is conducted, will always make a variety of assumptions about how the natural world is organized and functions. To effectively put these plans into action, others who will want to use them need to understand and appreciate these assumptions. For example, a plan might assume that if a particular group of representative ecosystems is conserved, then the majority of native species in the region will also be conserved. That may or may not be the case, but regardless, planning teams need to explicitly acknowledge the assumptions behind their work. If the team assumes a vegetation map of a region is largely accurate even though it has never been checked for errors, then readers and users of the plan need to appreciate that assumption, because it may well lead to erroneous conclusions about the importance of particular places or actions. Most plans do not adequately acknowledge the underlying assumptions of the planning process, and that is unfortunate. Explicit and clearly stated assumptions will not only improve others' ability to implement and use the results, but they will add to the credibility of the planning process and those involved in developing the plan. We need to be equally honest about what we know and what we do not know.

As I have noted throughout this book, many different planning efforts for biodiversity are being conducted by numerous organizations at a variety of scales. Some of these endeavors have similarities in approach and methodology, while others are unique. Although many scientists and planners may immediately recognize the differences among various efforts and the need to develop priorities for biodiversity conservation at different scales, the interested public at large is unlikely to do so. To them, as well as to many conservation biologists and practitioners who are not directly involved, these efforts may constitute a bewildering array of bureaucratic projects with competing agendas—one organization promoting one set of priority areas, and another institution emphasizing a different set.

There is, as they say, strength in numbers. The community of conservation scientists, planners, and practitioners who care deeply about biodiversity conservation would undoubtedly be far more effective with a more unified and consistent approach to identifying the most important areas for conservation action. From those who are implementing the Convention on Biological Diversity at global and national levels, to state and federal natural resource agencies, to local land trusts and conservation organizations, there is an urgent need to reach agreement on how we should identify the most important remaining natural areas for biodiversity conservation. The Four-R Framework outlined in Chapter 2—representative, resilient, redundant, and restorative—is one reasoned approach that has both scientific and practical underpinnings. There are clearly others.

When President Lyndon Johnson was touring Cape Canaveral during the Apollo space missions, he approached a janitor and asked him what his job was, to which the janitor replied, "To put a man on the moon, sir." That's the sort of collective effort the conservation community needs in setting and acting on priority places for conserving biodiversity. If conservation planning as a discipline is going to be a more effective vehicle for conservation action in the future, the approaches and methods need to be more unified, consistent, and consensus-driven. As practitioners, planners, and biologists, we are in the driver's seat. The natural world waits to see which road we take.

LITERATURE CITED

Abell, R. 2002. Conservation biology for the biodiversity crisis: a freshwater follow-up. *Conservation Biology* 16:1435–1437.

Abell, R. A., D. M. Olson, E. Dinerstein, P. T. Hurley, J. T. Diggs, W. Eichbaum, S. Waltgers, W. Wettengell, T. Allnutt, C. J. Loucks, and P. Hedao. 2000. *Freshwater Ecoregions of North America: A Conservation Assessment*. Washington, DC: Island Press.

Abell, R., M. Theime, E. Dinerstein, and D. Olson. 2002 *A Source Book for Conducting Biological Assessments and Developing Biodiversity Visions for Ecoregion Conservation. Volume II: Freshwater Ecoregions*. Washington, DC: World Wildlife Fund.

Adams, J. S., B. A. Stein, and L. S. Kutner. 2000. Biodiversity: our precious heritage. In B. L. Stein, L. S. Kutner, and J. S. Adams, eds., *Precious Heritage: The Status of Biodiversity in the United States*, pp. 3–18. Oxford, U.K.: Oxford University Press.

African Wildlife Foundation. 2000. Annual Report. Washington, DC: African Wildlife Foundation.

Airame, S., J. E. Dugan, K. D. Lafferty, H. M. Leslie, D. A. McArdle, and R. R. Warner. 2003. Applying ecological criteria to marine reserve design: a case study from the California Channel Islands. *Ecological Applications* 13:S170–S184.

Albert, D. A. 1994. *Regional Landscape Ecosystems of Michigan, Minnesota, and Wisconsin: A Working Map and Classification*. Fourth revision. General Technical Report NC-178, North Central Forest Experiment Station. St Paul, MN: USDA Forest Service.

Alison, G. W., S. D. Gaines, J. Lubchenco, and H. P. Possingham. 2003. Taking the long view of marine reserves: catastrophes and an insurance factor. *Ecological Applications* 13:S58–S24.

Allan, J. D., and A. S. Flecker. 1993. Biodiversity conservation in running waters. *BioScience* 43:32–43.

Allee, R. J., M. N. Dethier, D. Brown, L. A. Deegan, F. G. Ford, T. F. Hourigan, J. Maragos, G. C. Schoch, K. Sealey, R. R. Twilley, M. P. Weinstein, and M. Yoklavich. 2000. *Marine and Estuarine Ecosystem and Habitat Classification. NOAA Technical Memorandum NMFS-F/SPO-43*, Silver Spring, MD: National Oceanographic and Atmospheric Administration.

Allendorf, F. W., and N. Ryman. 2002. The role of genetics in population viability analysis. In S. R. Beissinger and D. R. McCullough, eds., *Population Viability Analysis*, pp. 50–85. Chicago, IL: University of Chicago Press.

Anand, P. 2002. Decision-making when science is ambiguous. *Science* 295:1839.

Andelman, S. J., S. Beissinger, J. Cochrane, L. Gerber, P. Gomez-Priego, C. Groves, J. Haufler, R. Holthausen, D. Lee, L. Maguire, B. Noon, K. Ralls, and H. Regan. 2001. *Scientific Standards for Conducting Viability Assessments under the National Forest Management Act: Report and Recommendations of the NCEAS Working Group.* Santa Barbara, CA: National Center for Ecological Analysis and Synthesis, University of California, Santa Barbara.

Andelman, S. J., and W. F. Fagan. 2000. Umbrellas and flagships: efficient conservation surrogates or expensive mistakes. *Proceedings of the National Academy of Science* 97:5954–5959.

Anderson, J. E. 1991. A conceptual framework for evaluating and quantifying naturalness. *Conservation Biology* 5:347–352.

Anderson, J. R., E. E. Hardy, and J. T. Roach. 1976. *Land Use and Land Cover Classification System for Use with Remote Sensing Data.* Geological Survey Professional Paper 964 (a revision of the land-use classification system as presented in U.S. Geological Circular 671), Washington, DC: U.S. Government Printing Office.

Anderson, M., P. Comer, D. Grossman, C. Groves, K. Poiani, M. Reid, R. Schneider, B. Vickery, and A. Weakley. 1999. *Guidelines for Representing Ecological Communities in Ecoregional Conservation Plans.* Arlington, VA: The Nature Conservancy. Available online: www.conserveonline.org.

Anderson, M. G. 1999. Viability and spatial assessment of ecological communities in the Northern Appalachian Ecoregion. Ph.D. diss., University of New Hampshire, Durham.

Anderson, M. K. 2001. The contribution of ethnobiology to the reconstruction and restoration of historic ecosystems. In D. Egan, E. A. Howell, and C. Meine, eds., *The Historical Ecology Handbook: A Restorationist's Guide to Reference Ecosystems*, pp. 55–72. Washington, DC: Island Press.

Angermeier, P. L. 1995. Ecological attributes of extinction-prone species: loss of freshwater fishes in Virginia. *Conservation Biology* 9:143–158.

———. 2000. The natural imperative for biological conservation. *Conservation Biology* 14:373–381.

Angermeier, P. L., and J. R. Karr. 1994. Biological integrity versus biological diversity as policy directives. *BioScience* 44:690–697.

Angermeier, P. L., and I. J. Schlosser. 1995. Conserving aquatic biodiversity: beyond species and populations. *American Fisheries Society Symposium* 17:911–927.

Angermeier, P. L., and M. R. Winston. 1999. Characterizing fish community diversity across Virginia landscapes: prerequisite for conservation. *Ecological Applications* 9:335–349.

Araújo, M. B., and P. H. Williams. 2001. The bias of complementarity hotspots toward marginal populations. *Conservation Biology* 15:1710–1720.

Askins, R. A. 2001. Sustaining biological diversity in early successional communities: the challenge of managing unpopular habitats. *Wildlife Society Bulletin* 29:407–412.

Austin, M. P., and C. R. Margules. 1986. Assessing representativeness. In M. B. Usher, ed., *Wildlife Conservation Evaluation*, pp. 45–67. London, U.K.: Chapman and Hall.

Austin, M. P., A. O. Nicholls, and C. R. Margules. 1990. Measurement of the realized qualitative niche: environmental niches of five *Eucalyptus* species. *Ecological Monographs* 60:161–177.

Bailey, R. G. 1995. *Description of Ecoregions of the United States.* U.S. Forest Service Miscellaneous Publication Number 1391, Washington, DC: USDA Forest Service.

———. 1996. *Ecosystem Geography.* New York: Springer-Verlag.

————. 1998. *Ecoregions: The Ecosystem Geography of Oceans and Continents.* New York: Springer-Verlag.

Baker, W. L. 1992. The landscape ecology of large disturbances in the design and management of nature reserves. *Landscape Ecology* 7:181–194.

Ball, I., and H. Possingham. 2000. Marxan (v1.2). *Marine Reserve Design Using Spatially Explicit Annealing.* Brisbane, Queensland, AU: University of Queensland. Available online: http://www.ecology.uq.edu.au/marxan.htm.

Balmford, A., and K. J. Gaston. 1999. Why biodiversity surveys are good value. *Nature* 398:204–205.

Balmford, A., G. M. Mace, and J. R. Ginsberg. 1998. The challenges to conservation in a changing world: putting processes on the map. In G. M. Mace, A. Balmford, and J. R. Ginsburg, eds., *Conservation in a Changing World,* pp. 1–28. Cambridge, UK: Cambridge University Press.

Banks, D., M. Williams, J. Pearce, A. Springer, R. Hagenstein, and D. Olson. 2000. *Ecoregion-Based Conservation in the Bering Sea: Identifying Important Areas for Biodiversity Conservation.* Washington, DC: World Wildlife Fund and The Nature Conservancy.

Barber, P. H., S. R. Palumbi, M. V. Erdmann, M. K. Moosa. 2000. A marine Wallace's line? *Nature* 406:692–693.

————. 2002. Sharp genetic breaks among populations of *Haptosquilla pulchella* (Stomatopoda) indicate limits to larval transport: patterns, causes, and consequences. *Molecular Ecology* 11:(4):659–674.

Baydack, R. K., and H. Campa III. 1999. Setting the context. In R. K. Baydack, H. Campa III, and J. B. Haufler, eds., *Practical Approaches to the Conservation of Biological Diversity,* pp. 3–16. Washington, DC: Island Press.

Beck, M. W., K. L. Heck Jr., K. W. Able, D. L. Childers, D. B. Eggleston, B. M. Gillanders, B. Halpern, C. G. Hays, K. Hoshino, T. J. Minello, R. J. Orth, P. F. Sheridan, and M. P. Weinstein. 2001. The identification, conservation and management of estuarine and marine nurseries for fish and invertebrates. *BioScience* 51:633–641.

Beck, M. W., and M. Odaya. 2001. Ecoregional planning in marine environments: identifying priority sites for conservation in the Northern Gulf of Mexico. *Aquatic Conservation* 11:235–242.

Bedward, M. R., L. Pressey, and D. A Keith. 1992. A new approach for selecting fully representative reserve networks: addressing efficiency, reserve design and land suitability with an iterative analysis. *Biological Conservation* 62:115–125.

Beehler, B. M., ed. 1993. *Papua New Guinea Conservation Needs Assessment Volume 2.* Biodiversity Support Program, and Government of Papua New Guinea, Department of Environment and Conservation, Landover, MD: Corporate Press.

Beier, P., and R. F. Noss. 1998. Do habitat corridors provide connectivity? *Conservation Biology* 12:1241–1252.

Beissinger, S. R., and D. R. McCullough, eds. 2002. *Population Viability Analysis.* Chicago, IL: University of Chicago Press.

Beissinger, S. R., and M. I. Westphal. 1998. On the use of demographic models of population viability in endangered species management. *Journal of Wildlife Management* 62:821–841.

Belbin, L. 1993. Environmental representativeness: regional partitioning and reserve selection. *Biological Conservation* 66:223–230.

————. 1995. A multivariate approach to the selection of biological reserves. *Biodiversity and Conservation* 4:951–963.

Belden, Russonello, and Stewart. 2002. *Americans and Biodiversity: New Perspectives in 2002.* The Biodiversity Project, Washington, DC: Belden, Russonello, and Stewart.

Benedict, M. A., and E. T. McMahon. 2002. *Green Infrastructure: Smart Conservation for the 21st Century.* Washington, DC: Sprawl Watch Clearinghouse. Available online: www.sprawlwatch.org.

Bennett, A. F. 1999. *Linkages in the Landscape: The Role of Corridors and Connectivity in Wildlife Conservation.* Gland, Switzerland: IUCN World Conservation Union.

Bennett, E., H. Eves, J. Robinson, and D. Wilkie. 2002. Why is eating bushmeat a biodiversity crisis? *Conservation in Practice* 3:28–29.

Biodiversity Support Program. 1994. *Conserving Biological Diversity in Bulgaria: The National Biological Diversity Conservation Strategy.* Washington, DC: Biodiversity Support Program.

Black, A. E., E. Strand, R. G. Wright, J. M. Scott, P. Morgan, and C. Watson. 1998. Land use history at multiple scales: implications for conservation planning. *Landscape and Urban Planning* 43:49–63.

Block, W. M., M. L. Morrison, J. Verner, and P. N. Manley. 1994. Assessing wildlife-habitat relationship models: a case study with California oak woodlands. *Wildlife Society Bulletin* 22:549–561.

Blom, A., A. K. Tohem, J. D'Amico, D. O'Hara, R, Abell, and D. Olson. 2001. *Assessment of Biological Priorities for Conservation in the Guinean–Congolian Forest Region.* Report of the Guinean–Congolian Forest Region Workshop, Libreville, Gabon: World Wildlife Fund.

Boecklen, W. J., and D. Simberloff. 1986. Area-based extinction models. In D. K. Elliot, ed., *Dynamics of Extinction*, pp. 247–276. New York, NY: John Wiley and Sons.

Boersma, P. D., P. Kareiva, W. F. Fagan, J. A. Clark, and J. M. Hoekstra. 2001. How good are endangered species recovery plans? *BioScience* 51:643–649.

Bojorquez-Tapia, L. A., I. Azuara, E. Ezcurra, and O. Flores-Villela. 1995. Identifying conservation priorities in Mexico through geographic information systems and modeling. *Ecological Applications* 5:215–231.

Boon, P. J. 1992. Essential elements in the case for river conservation. In P. J. Boon, P. Calow, and G. E. Petts, eds., *River Conservation and Management*, pp. 11–33. Manchester, England: John Wiley and Sons.

Bormann, F. H., and G. E. Likens. 1979. Catastrophic disturbance and the steady state in northern hardwood forests. *American Scientist* 67:660–669.

Boston, A. N. 1996. *Volume One: The BioRap Biological Database, BioRap: Rapid Assessment of Biodiversity.* Canberra, AU: The Australian BioRap Consortium.

Botsford, L. W., A. Hastings, and S. D. Gaines. 2001. Dependence of sustainability on the configuration of marine reserves and larval dispersal distance. *Ecology Letters* 4:144–150.

Boyce, M. S. 1992. Population viability analysis. *Annual Review of Ecology and Systematics* 23:481–506.

Boyce, M. S., and L. L. McDonald. 1999. Relating populations to habitat using resource selection functions. *Trends in Ecology and Evolution* 14:268–272.

Braun-Blanquet, J. 1928. *Pflanzensoziologie. Grundzuge der vegetationskunde.* Springer-Verlag, Berlin, Germany. (English translation published by McGraw-Hill, New York, in 1932.)

Briers, R. A. 2002. Incorporating connectivity into reserve selection procedures. *Biological Conservation* 103:77–83.

Brook, B. W., M. A. Burgman, H. R. Ackaya, J. J. O'Grady, and R. Frankham. 2002. Critiques of PVA ask the wrong questions: throwing the heuristic baby out with the numerical bath water. *Conservation Biology* 16:262–263.

Brook, B. W., J. J. O'Grady, A. P. Chapman, M. A. Burgman, H. R. Ackakaya, and R. Frankham. 2000. Predictive accuracy of population viability analysis in conservation biology. *Nature* 404:385–387.

Brunckhorst, D. J. 2000. *Bioregional Planning: Resource Management beyond the New Millennium*. Amsterdam, Netherlands: Harwood Academic Publishers.

Bryer, M. T., K. Maybury, J. S. Adams, and D. H. Grossman. 2000. More than the sum of the parts: diversity and status of ecological systems. In B. A. Stein, L. S. Kutner, and J. S. Adams, eds., *Precious Heritage: The Status of Biodiversity in the United States,* pp. 201–238. Oxford, UK: Oxford University Press.

Burnett, M. R., P. V. August, J. H. Brown Jr., and K. T. Killingbeck. 1998. The influence of geomorphological heterogeneity on biodiversity I. A patch-scale perspective. *Conservation Biology* 12:363–370.

Burton, I., J. B. Smith, and S. Lenhart, 1998. Adaptation to climate change: theory and assessment. In J. F. Feenstra, I. Burton, J. B. Smith, and R. S. J. Tol, eds., *Handbook on Methods for Climate Change Impact Assessment and Adaptation Strategies.* Version 2.0. Amsterdam: Institute for Environmental Studies and United Nations Environment Program, Chapter 5. Available online: http://www.vu.nl/english/o_o/instituten/IVM/research/climatechange/fb_handbook.htm.

Busch, W. D. N., and P. G. Sly, eds. 1992. *The Development of an Aquatic Habitat Classification System for Lakes.* Boca Raton, FL: CRC Press.

Cabeza, M., and A. Moilanen. 2001. Design of reserve networks and the persistence of biodiversity. *Trends in Ecology and Evolution* 16:242–248.

Caicco, S. L., J. M. Scott, B. Butterfield, and B. Csuti. 1995. A gap analysis of the management status of the vegetation of Idaho (U.S.A.). *Conservation Biology* 9:498–511.

California Resources Agency. 1997. An introduction to NCCP. Sacramento, CA: California Resources Agency.

————. 2001. *First Draft Report on the Methodology to Identify State Conservation Priorities.* California Continuing Resources Investment Strategy Project (CCRISP), Sacramento, CA. Available online: http://legacy.ca.gov/documents.html.

————. 2002. *Resource Status Assessment and Trends Methodology.* California Legacy Project, Sacramento, CA: California Resources Agency. Available online: http://legacy.ca.gov/pub_docs/Natural_Resource_Health_and_Condition_Methodology_Report_FINAL.pdf.

Callicott, J. B. 1997. Conservation values and ethics. In G. K. Meffe, C. R. Carroll, and contributors, *Principles of Conservation Biology,* 2nd ed., pp. 29–56. Sunderland, MA: Sinaeur Associates, Inc.

Callicott, J. B., L. B. Crowder, and K. Mumford. 1999. Current normative concepts in conservation. *Conservation Biology* 13:22–35.

Cape Action Plan for the Environment (CAPE) Team. 2000. *Cape Action Plan for the Environment: A Biodiversity Strategy and Action Plan for the Cape Floral Kingdom.* World Wide Fund for Nature–South Africa. Available online: http://www.panda.org.za/publications.php.

Carlton, J. T. 1998. Apostrophe to the Ocean. *Conservation Biology* 12:1165–1167.

Carlton, J. T., J. B. Geller, M. L. Reaka-Kudla, and E. A. Norse. 1999. Historical extinctions in the sea. *Annual Review of Ecology and Systematics* 30:515–538.

Caro, T. 2000. Focal species. *Conservation Biology* 14:1569–1570.

———. 2002. Focal and surrogate species: getting the language right. *Conservation Biology* 16:285–287.

Caro, T. M., and G. O'Doherty. 1999. On the use of surrogate species in conservation biology. *Conservation Biology* 13:805–814.

Carroll, C., R. F. Noss, and P. C. Pacquet. 2001. Carnivores as focal species for conservation planning in the Rocky Mountain Region. *Ecological Applications* 11:961–980.

Carroll, C., M. K. Philips, N. H. Schumaker, and D. W. Smith. 2003. Impacts of landscape change on wolf restoration success: planning a reintroduction program using static and dynamic spatial modeling. *Conservation Biology* 17:536–548.

Carter, M. F., W. C. Hunter, D. N. Pashley, and K. V. Rosenberg. 2000. Setting conservation priorities for landbirds in the United States: the partners in Flight Approach. *Auk* 117:541–548.

Casey, J. M., and R. A. Myers. 1998. Near extinction of a large widely distributed fish. *Science* 281:690–692.

Cassidy, K. M., C. E. Grue, M. R. Smith, R. E. Johnson, K. M. Dvornich, K. R. McAllister, P. W. Mattocks Jr., J. E. Cassidy, and K. B. Aubry. 2001. Using current protection status to assess conservation priorities. *Biological Conservation* 97:1–20.

Ceballos, G., P. Rodriguez, and R. A. Medellin. 1998. Assessing conservation priorities in megadiverse Mexico: mammalian diversity, endemicity, and endangerment. *Ecological Applications* 8:8–17.

Central Appalachians Ecoregion Team. 2001. *Central Appalachian Forest Ecoregional Plan.* Arlington, VA: The Nature Conservancy.

Central Shortgrass Prairie Ecoregional Planning Team. 1998. *Ecoregion-Based Conservation in the Central Shortgrass Prairie.* Arlington, VA: The Nature Conservancy. Available online: www.conserveonline.org.

Central Tallgrass Prairie Ecoregional Planning Team. 2000. *Conservation in a Highly Fragmented Landscape: The Central Tallgrass Prairie Ecoregional Conservation Plan.* Arlington, VA: The Nature Conservancy. Available online: www.conserveonline.org.

Channell, R., and M. V. Lomolino. 2000. Dynamic biogeography and conservation of endangered species. *Nature* 403:84–86.

Chaplin, S. J., R. A. Gerrard, H. M. Watson, L. L. Master, and S. R. Flack. 2000. The geography of imperilment: targeting conservation toward critical biodiversity areas. In B. A. Stein, L. S. Kutner, and J. S. Adams, eds., *Precious Heritage: The Status of Biodiversity in the United States,* pp. 159–200. Oxford, UK: Oxford University Press.

Chapman, S. S., J. M. Omernik, J. A. Freeouf, D. G. Huggins, et al. 2001. Ecoregions of Nebraska and Kansas. (Color poster with map, descriptive text, summary tables, and photographs): Reston, VA: U.S. Geological Survey.

Clark, J. R. 1982. Assessing the national estuarine research reserve system: Action summary. In *Estuarine Sanctuary Program: A Summary Report to the Office of Coastal Zone Management,* pp. 1–17. Washington, DC: National Oceanographic and Atmospheric Administration.

Clark, T. W. 1997. *Averting Extinction: Reconstructing Endangered Species Recovery.* New Haven, CT: Yale University Press.

———. 1998. Conservation biologists in the policy process: learning to be practical and effective. In G. K. Meffe, C. R. Carroll, and contributors, *Principles of Conservation Biology,* 2nd ed., pp. 575–598. Sunderland, MA: Sinauer Associates, Inc.

————. 2002. *The Policy Process: A Practical Guide for Natural Resource Professionals.* New Haven, CT: Yale University Press.

Cleaves, D. A. 1994. *Assessing Uncertainty in Expert Judgments about Natural Resources.* General Technical Report SO-110, Southern Forest Experiment Station. New Orleans, LA: USDA Forest Service.

Cleland, D. T., J. B. Hart, G. E. Host, K. S. Pregitzer, and C. W. Ramm. 1994. *Ecological Classification and Inventory System of the Huron-Manistee National Forest.* Milwaukee, WI: USDA Forest Service, Region 9.

Clevenger, A. P., J. Wierzchowski, B. Chruszcz, and K. Gunson. 2002. GIS-generated, expert-based models for identifying wildlife habitat linkages and planning mitigation passages. *Conservation Biology* 16:503–514.

Cofré, H., and P. A. Marquet. 1999. Conservation status, rarity, and geographic priorities for conservation of Chilean mammals: an assessment. *Biological Conservation* 88:53–68.

Collier, M. P., R. H. Webb, and E. D. Andrews. 1997. Experimental flooding in the Grand Canyon. *Scientific American* 276:82–89.

Comes, H. P., and J. W. Kadereit. 1998. The effect of Quaternary climatic changes on plant distribution and evolution. *Trends in Plant Science Reviews* 3:432–438.

Committee on Abrupt Climate Change. 2002. *Abrupt Climate Change: Inevitable Surprises.* Washington, DC: National Academy of Sciences. Available online: http://www.nap.edu.

Committee of Scientists. 1999. *Sustaining the People's Lands: Recommendations for Stewardship of National Forests and Grasslands into the Next Century.* Washington, DC: U.S. Department of Agriculture.

Commonwealth of Australia. 1999. *Australian Guidelines for Establishing the National Reserve System.* Canberra, AU: Environment Australia.

Connell, J. H. (1978) Diversity in tropical rain forests and coral reefs. *Science* 199:1304–1310.

Conner, R. N. 1988. Wildlife populations: minimally viable or ecologically functional? *Wildlife Society Bulletin* 16:80–84.

Conroy, M. J., and B. R. Noon. 1996. Mapping of species richness for conservation of biological diversity: conceptual and methodological issues. *Ecological Applications* 6:763–773.

Conservation International. 1999. *The Irian Jaya Biodiversity Conservation Priority-Setting Workshop.* Washington, DC: Conservation International.

Cooperrider, A. Y., R. J. Boyd, and H. R. Stuart. 1986. *Inventory and Monitoring of Wildlife Habitat.* Washington, DC: U.S. Government Printing Office.

Coppolillo, P., H. Gomez, F. Maisels, and R. Wallace. In press. Building a better umbrella: selecting suites of landscape species as a basis for site-based conservation. *Biological Conservation.*

Corsi, F., J. De Leeuw, and A. Skidmore. 2000. Modeling species distribution with GIS. In L. Boitani and T. K. Fuller, eds., *Research Techniques in Animal Ecology: Controversies and Consequences,* pp. 389–434. New York: Columbia University Press.

Costanza, R., R. d'Agre, R. de Groot, S. Farber, M. Grasso, B. Hannon, K. Limburg, S. Naeem, R. V. O'Neill, J. Paruelo, R. G. Raskin, P. Sutton, and M. van den Belt. 1997. The value of the world's ecosystem services and natural capital. *Nature* 387:253–260.

Cote, I. M., I. Mosqueira, and J. D. Reynolds. 2001. Effects of marine reserve characteristics on the protection of fish populations: a meta-analysis. *Journal of Fish Biology* 59:178–189.

Covey, S. 1990. *The Seven Habits of Highly Effective People: Powerful Lessons in Personal Change.* New York: Simon and Schuster.

Cowardin, L. M., V. Carter, F. C. Golet, and E. T. LaRoe. 1979. *Classification of the Wetlands and Deepwater Habitats of the United States.* Washington, DC: U.S. Fish and Wildlife Service.

Cowen, R. K., K. M. M. Lwiza, S. Sponaugle, C. B. Paris, and D. B. Olson. 2000. Connectivity of marine populations: open or closed? *Science* 287:857–859.

Cowling, R. M. 1999. Planning for persistence—systematic reserve design in southern Africa's Succulent Karoo desert. *Parks* 9:17–30.

Cowling, R. M., and C. E. Heijnis. 2001. The identification of Broad Habitat Units as biodiversity entities for systematic conservation planning in the Cape Floristic Region. *South African Journal of Botany* 67:15–38.

Cowling, R. M., R. L. Pressey, A. T. Lombard, P. G. Desmet, and A.G. Ellis. 1999. From representation to persistence: requirements for a sustainable system of conservation areas in the species-rich mediterranean-climate desert of southern Africa. *Diversity and Distributions* 5:51–71.

Cox, J., R. Kautz, M. MacLaughlin, and T. Gilbert. 1994. *Closing the Gaps in Florida's Wildlife Habitat Conservation System.* Tallahassee, FL: Florida Game and Freshwater Fish Commission.

Cox, J. A., and R. S. Kautz. 2000. *Habitat Conservation Needs of Rare and Imperiled Wildlife in Florida.* Office of Environmental Services, Tallahassee, FL: Florida Fish and Wildlife Conservation Commission.

Craighead, J. J., J. S. Sumner, and J. A. Mitchell. 1995. *The Grizzly Bears of Yellowstone: Their Ecology in the Yellowstone Ecosystem, 1959–1992.* Washington, DC: Island Press.

Crist, P., and R. Deitner. 1998. Assessing land cover map accuracy. In *A Handbook for Conducting Gap Analysis.* National Gap Analysis Program, Moscow, ID: U.S. Geological Survey. Available online: www.gap.uidaho.edu/handbook.

Crist, P. J., T. C. Edwards, C. G. Homer, S. D. Bassett, B. C. Thompson, and E. Brackney 1998. Mapping and categorizing land stewardship. In *A Handbook for Conducting Gap Analysis.* National Gap Analysis Program, Moscow, ID: U.S. Geological Survey. Available online: www.gap.uidaho.edu/handbook.

Croker, R. 1991. *Pioneer Ecologist: The Life and Work of Victor Earnest Shelford, 1877–1968.* Washington, DC: Smithsonian Institute Press.

Crouse, D. T., L. B. Crowder, and H. Caswell. 1987. A stage-based population model for loggerhead sea turtles and implications for conservation. *Ecology* 68:4112–1423.

Csuti, B., S. Polasky, P. H. Williams, R. L. Pressey, J. D. Camm, M. Kershaw, A. R. Kiester, B. Downs, R. Hamilton, M. Huso, and K. Sahr. 1997. A comparison of reserve selection algorithms using data on terrestrial vertebrates in Oregon. *Biological Conservation* 80:83–97.

Cupp, C. E. 1989. *Valley Segment Type Classification for Forested Lands of Washington.* Report TFW-AM-89-001, Olympia, WA: Washington Forest Protection Association.

Czech, B., P. R. Krausman, and P. K. Devers. 2000. Economic associations among causes of species endangerment in the United States. *BioScience* 50:593–601.

Da Fonseca, G. A. B., A. Balmford, C. Bibby, L. Boitani, F. Corsi, T. Brooks, C. Gason, S. Olivieri, R. A. Mittermeier, N. Burgess, E. Dinerstein, D. Olson, L. Hannah, J. Lovett, D. Moyer, C. Rahbek, S. Stuart, and P. Williams. 2000. Following Africa's lead in setting priorities. *Nature* 405:393–394.

Daily, G. C., ed. 1997. *Nature's Services: Societal Dependence on Natural Ecosystems.* Washington, DC: Island Press.

Dale, V. H., S. Brown, R. A. Haeuber, N. T. Hobbs, N. Huntly, R. J. Naiman, W. E. Riebsame, M. G. Turner, and T. J. Valone. 2000. Ecological principles and guidelines for managing the use of land. *Ecological Applications* 10:639–670.

Dasmann, R. F. 1973. *A System for Defining and Classifying Natural Regions for Purposes of Conservation.* IUCN Occasional Paper No. 7, Morges, Switzerland: International Union for Conservation of Nature and Natural Resources.

———. 1974. *Biotic Provinces of the World: Further Development of System for Defining and Classifying Natural Regions for Purposes of Conservation.* IUCN Occasional Paper No. 9, Morges, Switzerland: International Union for Conservation of Nature and Natural Resources.

Daubenmire, R. F. 1952. Forest vegetation of northern Idaho and adjacent Washington, and its bearing on concepts of vegetation classification. *Ecological Monographs* 22:301–330.

Davey, A. 1998. *National System Planning for Protected Areas.* Best Practice Protected Areas Guidelines Series No. 1, World Commission on Protected Areas, Gland, Switzerland: World Conservation Union (IUCN).

Davey, S. M., J. Hoare, and K. Rumba. 2000. Science and its role in Australian Regional Forest Agreements. Paper presented at Society of American Foresters 2000 National Convention, November 16–20, Washington, DC.

Davies, C. E., and D. Moss. 1999. EUNIS Habitat Classification. Final Report to the European Topic Centre on Nature Conservation, European Environment Agency. Copenhagen, Denmark: European Environment Agency.

Davies, G. 2002. Bushmeat and international development. *Conservation Biology* 16:587–589.

Davis, F. W., P. A. Stine, D. M. Stoms, M. I. Borchert, and A. D. Hollander. 1995. Gap analysis of the actual vegetation of California: I. The southwestern region. *Madroño* 42:40–78.

Davis, F. W., D. M. Stoms, and S. Andelman. 1999. Systematic reserve selection in the USA: an example from the Columbia Plateau ecoregion. *Parks* 9:31–41.

Davis, W. S., and T. P. Simon, eds. 1999. *Biological Assessment and Criteria: Tools for Water Resource Planning and Decision Making.* Boca Raton, FL: CRC Press.

Day, J., and J. C. Roff. 2000. Planning for representative marine protected areas. Toronto: World Wildlife Fund Canada.

Defenders of Wildlife. 1998. *Oregon's Living Landscape: Strategies and Opportunities to Conserve Biodiversity.* Washington, DC: Defenders of Wildlife.

DellaSala, D. A., N. L. Staus, J. R. Strittholt, A. Hackman, and A. Iacobelli. 2001. An updated protected areas database for the United States and Canada. *Natural Areas Journal* 21:124–135.

DellaSala, D. A., J. R. Strittholt, R. F. Noss, and D. M. Olson. 1996. A critical role for core serves in managing Inland Northwest landscapes for natural resources and biodiversity. *Wildlife Society Bulletin* 24:209–221.

Department of Primary Industries Water and Environment. 2001. Tasmanian Marine Protected Areas Strategy. Hobart, Tasmania: Crown in Right of the State.

Dethier, M. N. 1992. Classifying marine and estuarine natural communities: an alternative to the Cowardin system. *Natural Areas Journal* 12:90–100.

DeVelice, R. L., and J. R. Martin. 2001. Assessing the extent to which roadless areas complement the conservation of biological diversity. *Ecological Applications* 11:1008–1018.

Diamond, J. M. 1975. The island dilemma: lessons of modern biogeographic studies for the design of natural preserves. *Biological Conservation* 7:129–146.

———. 1976. Island biogeography and conservation: strategy and limitations. *Science* 193:1027–1029.

Dinerstein, E., D. M. Olson, D. H. Graham, A. L. Webster, S. A. Pimm, M. P. Bookbinder, and G. Ledec. 1995. *A Conservation Assessment of the Terrestrial Ecoregions of Latin America and the Caribbean.* Washington, DC: World Wildlife Fund and the World Bank.

Dinerstein, E., G. Powell, D. Olson, E. Wikramamayake, R. Abell, C. Loucks, E. Underwood, T. Allnut, W. Wettengull, T. Ricketts, H. Strand, S. O'Connor, N. Burgess, and M. Mobley. 2000. *A Workbook for Conducting Biological Assessments and Developing Biodiversity Visions for Ecoregion-Based Conservation.* Washington, DC: World Wildlife Fund.

Dobson, A., K. Ralls, M. Foster, M. E. Soulé, D. Simberloff, D. Doak, J. A. Estes, L. S. Mills, D. Mattson, R. Dirzo, H. Arita, S. Ryan, E. A. Norse, R. F. Noss, and D. Johns. 1999. Corridors: reconnecting fragmented landscapes. In M. E. Soulé and J. Terborgh, eds., *Continental Conservation: Scientific Foundations of Regional Reserve Networks,* pp. 129–170. Washington, DC: The Wildlands Project and Island Press.

Dobson, A. P. 1996. *Conservation and Biodiversity.* New York: Scientific American Library.

Dobson, A. P., J. P. Rodriguez, W. M. Roberts, and D. S. Wilcove. 1997. Geographic distribution of endangered species in the United States. *Science* 275:550–553.

Dobson, A. P., J. P. Rodriguez, and W. M. Roberts. 2001. Synoptic tinkering: integrating strategies for large-scale conservation. *Ecological Applications* 11:1019–1026.

Doppelt, B., M. Sculock, C. Frissell, and J. Karr. 1993. *Entering the Watershed.* The Pacific Rivers Council, Washington, DC: Island Press.

Driscoll, R. S., D. L. Merkel, D. L. Radloff, D. E. Snyder, and J. S. Hagihara. 1984. *An Ecological Land Classification Framework for the United States.* Miscellaneous Publication 1439, Washington, DC: USDA Forest Service.

Duker, L., and L., Borre. 2001. *Biodiversity Conservation of the World's Lakes: A Preliminary Framework for Identifying Priorities.* LakeNet Report Series, Number 2. Annapolis, MD: Monitoring International.

Eardley, K. A. 1999. *A Foundation for Conservation in the Riverina Bioregion.* Hurtsville, New South Wales, AU: New South Wales National Parks and Wildlife Service.

East Gulf Coastal Plain Core Team. 1999. *East Gulf Coastal Plain Ecoregional Plan.* Arlington, VA: The Nature Conservancy.

Ecological Stratification Working Group (ESWG). 1995. *A National Ecological Framework for Canada.* Ottawa, Canada: Agriculture and Agri Food Canada, Research Branch, Centre for Land and Biological Resources Research and Environment Canada, State of the Environment Directorate, Ecozone Analysis Branch.

ECOMAP. 1993. *National Hierarchical Framework of Ecological Units.* Unpublished paper, Washington, DC: USDA Forest Service.

Edgar, G. J., N. S. Barrett, D. J. Graddon, and P. R. Last. 2000. The conservation significance of estuaries: a classification of Tasmanian estuaries using ecological, physical, and demographic attributes as a case study. *Biological Conservation* 92:383–397.

Edmonds, M. 2001. The pleasures and pitfalls of written records. In D. Egan and E. A. Howell, eds., *The Historical Ecology Handbook: A Restorationist's Guide to Reference Ecosystems,* pp. 73–100. Washington, DC: Island Press.

Ehrenfeld, D. W. 1981. *The Arrogance of Humanism*. New York, NY: Oxford University Press.

Ellner, S. P., J. Fieberg, D. Ludwig, and C. Wilcox. 2002. Precision of population viability analysis. *Conservation Biology* 16:258–261.

Elton, C. S., 1958. *The Ecology of Invasions by Animals and Plants*. Chicago, IL: University of Chicago Press.

English Nature. 2001. *Biodiversity Action Plans: From Rio—The Background*. Peterborough, UK: English Nature. Available online: www.english-nature.org.uk/baps.rio/htm.

Engstrom, R. T., S. Gilbert, M. L. Hunter Jr., D. Merriwether, G. J. Nowacki, and P. Spencer. 1999. Practical applications of disturbance ecology to natural resource management. In R.C. Szaro, N. C. Johnson, W. T. Sexton, and A. J. Malk, eds., *Ecological Stewardship: A Common Reference for Ecosystem Management*, pp. 313–330. Oxford, UK: Elsevier Science, Ltd.

Environment Canada and the U.S. Environmental Protection Agency. 2001. *State of the Great Lakes*. Report No. EPA 905-R-01-003, Washington, DC: Environmental Protection Agency.

Environmental Law Institute. 2001. *Status of the States: Innovative State Strategies for Biodiversity Conservation*. A Report on the First State Biodiversity Symposium. Washington, DC: Environmental Law Institute.

Environmental Protection Agency. 2000a. *Environmental Monitoring and Assessment Program-Surface Waters: Field Operations and Methods for Measuring the Ecological Condition of Non-Wadeable Rivers and Streams*. Report No. EPA/620/R-00/007, Washington, DC: Environmental Protection Agency.

Environmental Protection Agency. 2000b. *Mid-Atlantic Highlands Stream Assessment*. Report No. EPA/903/12-00/015, Washington, DC: Environmental Protection Agency.

Estes, J. A., D. O. Duggins, and G. B. Rathbun. 1998. The ecology of extinctions in kelp forest communities. *Conservation Biology* 3:252–264.

Etnier, D. A., and W. C. Starnes. 1993. *The Fishes of Tennessee*. Knoxville, TN: University of Tennessee Press.

European Commission. 2000. *Natura 2000: Managing Our Heritage*. Luxembourg, Belgium: European Commission.

Everett, R. L., and J. F. Lehmkuhl. 1996. An emphasis use approach to conserving biodiversity. *Wildlife Society Bulletin* 24:192–199.

———. 1997. A forum for presenting alternative viewpoints on the role of reserves in conservation biology? A reply to Noss (1996). *Wildlife Society Bulletin* 25:575–577.

Faaborg, J., M. Brittingham, T. Donovan, and J. Blake. 1993. Habitat fragmentation in the temperate zone: a perspective for managers. In D. M. Finch and P. W. Stangel, eds., *Status and Management of Neotropical Migratory Birds*, pp. 331–338. General Technical Report RM-23, Rocky Mountain Forest and Range Experiment Station. Fort Collins, CO: USDA Forest Service.

Fagan, W. F., E. Meir, J. Prendergast, A. Folarin, and P. Kareiva. 2001. Characterizing population vulnerability for 758 species. *Ecology Letters* 4:132–138.

Fahrig, L. 1997. Relative effects of habitat loss and fragmentation on population extinction. *Journal of Wildlife Management* 61:603–610.

———. 2001. How much habitat is enough? *Biological Conservation* 100:65–74.

———. 2002. Effect of habitat fragmentation on the extinction threshold: a synthesis. *Ecological Applications* 12:346–353.

Fairbanks, D. H. K., and G. A. Benn. 2000. Identifying regional landscapes for conservation planning: a case study from KwaZulu-Natal, South Africa. *Landscape and Urban Planning* 50:237–257.

Fairbanks, D. H. K., B. Reyers, and A. S. van Jaarsveld. 2001. Species and environment representation: selecting reserves for the retention of avian diversity in KwaZulu-Natal, South Africa. *Biological Conservation* 98:365–379.

Faith, D. P., and A. O. Nicholls, eds. 1996. *Volume Three: Tools for Assessing Biodiversity in Priority Areas, BioRap: Rapid Assessment of Biodiversity.* Canberra, AU: The Australian BioRap Consortium.

Faith, D. P., and P. A. Walker. 1996. Environmental diversity: on the best possible use of surrogate data for assessing the relative biodiversity of sets of areas. *Biodiversity and Conservation* 5:399–415.

Fausch, K. D., C. T. Torgersen, C. V. Baxter, and H. W. Li. 2002. Landscapes to riverscapes: bridging the gap between research and conservation of stream fishes. *BioScience* 56:483–498.

Fearnside, P. M., and J. Ferraz. 1995. A conservation gap analysis of Brazil's Amazonian vegetation. *Conservation Biology* 9:1134–1147.

Ferdaña, Z. A. 2002. Approaches to integrating a marine GIS into The Nature Conservancy's ecoregional planning process. In J. Breman, ed., *Marine Geography: GIS for the Oceans and Seas,* pp. 151–158. Redlands, WA: ESRI.

Ferrier, S. 1997. Biodiversity data for reserve selection: making best use of incomplete data. In J. J. Pigram and R. C. Sundell, eds., *National Parks and Protected Areas: Selection, Delimitation, and Management,* pp. 315–329. Armidale, New South Wales, AU: Centre for Water Policy Research, University of New England.

Ferrier, S., R. L. Pressey, and T. W. Barrett. 2000. A new predictor of the irreplaceability of areas for achieving a conservation goal, its application to real-world planning, and a research agenda for further refining. *Biological Conservation* 93:303–325.

Ferrier, S., and G. Watson. 1997. *An Evaluation of the Effectiveness of Environmental Surrogates and Modelling Techniques in Predicting the Distribution of Biological Diversity.* Canberra, AU: Environment Australia.

Fisher, R. N., and H. B. Shaffer. 1996. The decline of amphibians in California's Great Central Valley. *Conservation Biology* 10:1387–1397.

Fleishman, E., D. D. Murphy, and P. F. Brussard. 2000. A new method for the selection of umbrella species for conservation planning. *Ecological Applications* 10:569–579.

Flink, C. A., and R. M. Searns. 1993. *Greenways: A Guide to Planning, Design, and Development.* Washington, DC: Island Press and the Conservation Fund.

Folke, C., C. S. Holling, and C. Perrings. 1996. Biological diversity, ecosystems, and the human scale. *Ecological Applications* 6:1018–1024.

Foreman, D., B. Dugelby, J. Humphrey, B. Howard, and A. Holdsworth. 2000. The elements of a wildlands network conservation plan: an example from the Sky Islands. *Wild Earth* 10:17–30.

Foreman, D., R. List, B. Dugelby, J. Humphrey, B. Howard, and A. Holdsworth. 2000. Healing the wounds: an example from the Sky Islands. *Wild Earth* 10:31–42.

Forman, R. T. T. 1997. *Land Mosaics: the Ecology of Landscapes and Regions.* Cambridge, UK: Cambridge University Press.

Forman, R. T. T., and S. K. Collinge. 1996. The "spatial solution" to conserving biodiversity in landscapes and regions. In R. M. DeGraaf and R. I. Miller, eds., *Conservation of Faunal Diversity in Forested Landscapes,* pp. 536–568. New York: Chapman and Hall.

Franklin, I. R. 1980. Evolutionary change in small populations. In M E. Soulé and B. A. Wilcox, eds., *Conservation Biology: An Evolutionary-Ecological Perspective,* pp. 135–150. Sunderland, MA: Sinauer Associates, Inc.

Franklin, J. F. 1993. Preserving biodiversity: species, ecosystems, or landscapes? *Ecological Applications* 3:202–205.

Franklin, J. F., D. Lindenmayer, J. A. MacMahon, A. McKee, J. Magnuson, D. A. Perry, R. Waide, and D. Foster. 2000. Threads of continuity. *Conservation Biology in Practice* 1:8–16.

Frissell, C. A. 1997. Ecological principles. In J. E. Williams, C. A. Wood, and M. P. Dombeck, eds., *Watershed Restoration: Principles and Practices,* pp. 96–115. Bethesda, MD: American Fisheries Society.

Frissell, C. A., W. J. Liss, C. E. Warren, and M. D. Hurley. 1986. A hierarchical framework for stream classification: viewing streams in a watershed context. *Environmental Management* 10:199–214.

Fujiwara, M. and H. Caswell. 2001. Demography of the endangered North Atlantic right whale. *Nature* 414:537–541.

Gallant, A. L., E. F. Binnian, J. M. Omernik, and M. B. Shasby. 1995. *Ecoregions of Alaska.* U.S. Geological Survey Professional Paper Number 1567, Washington, DC: U.S. Government Printing Office.

Garcia Fernandez, J. J. 1998. *Guide for the Preparation of Action Plans within the Framework of the Convention on Biodiversity.* New York, NY: United Nations Development Programme.

Gerber, L. R., and L. T. Hatch. 2002. Are we recovering? An evaluation of recovery criteria under the U.S. Endangered Species Act. *Ecological Applications* 12:668–673.

Gerber, L. R., and C. B. Schultz. 2001. Authorship and the use of biological information in endangered species recovery plans. *Conservation Biology* 15:1308–1314.

Gergeley, K. J., M. Scott, and D. Goble. 2000. A new direction for the U.S. National Wildlife Refuges: the National Wildlife Refuge System Improvement Act of 1997. *Natural Areas Journal* 20:107–118.

Gilpin, M. E., and M. E. Soulé. 1986. Minimum viable populations: processes of species extinction. In M. E. Soulé, ed., *Conservation Biology: The Science of Scarcity and Diversity,* pp. 19–34. Sunderland, MA: Sinaeur Associates, Inc.

Glowka, L., F. Burhenne-Guilman, H. Synge, J. A. McNeely, and L. Gundling. 1994. *A Guide to the Convention on Biological Diversity.* Environmental Policy and Law Paper No. 30, IUCN Environmental Law Centre. Gland, Switzerland: IUCN–The World Conservation Union.

Glynn, P. W., J. L. Mate, A. C. Baker, and M. O. Calderon. 2001. Coral bleaching and mortality in Panama and Ecuador during the 1997–1998 El Niño–Southern Oscillation event: spatial/temporal patterns and comparisons with the 1982–1983 event. *Bulletin of Marine Science* 69:79–109.

Goolsby, D. A., W. A. Battaglin, B. T. Aulenbach, and R. P. Hooper. 2000. Nitrogen flux and sources in the Mississippi River basin. *The Science of the Total Environment* 248:75–86.

Goolsby, D. A., W. A. Battaglin, G. B. Lawrence, R. S. Artz, B. T. Aulenbach, R. P. Hooper, D. R. Keeney, and G. J. Stensland. 1999. *Flux and Sources of Nutrients in the Mississippi–Atchafalaya River Basin—Topic 3.* Report for the integrated assessment on hypoxia in the Gulf of Mexico, NOAA Coastal Ocean Program Decision Analysis Series No. 17. Silver Spring, MD: NOAA Coastal Ocean Office.

Goreau, T., T. R. McClanahan, R. Hayes, and A. Strong. 2000. Conservation of coral reefs after the 1998 global bleaching event. *Conservation Biology* 14:5–15.

Groom, M., and P. Coppolillo. 2001. Ecologically functional populations. Unpublished manuscript. New York, NY: Wildlife Conservation Society.

Groom, M., and M. Pascual. 1998. The analysis of population persistence: an outlook on the practice of population viability analysis. In P. L. Fiedler and P. M. Kareiva, eds., *Conservation Biology for the Coming Decade,* pp. 4–27. New York: Chapman and Hall.

Groombridge, B., and M. Jenkins. 1998. *Freshwater Biodiversity: A Preliminary Global Assessment.* World Conservation Monitoring Centre, Cambridge, UK: World Conservation Press.

Grossman, D. H., P. Bourgeron, W.-D. N. Busch, D. Cleland, W. Platts, G. C. Ray, C. R. Roberts, and G. Roloff. 1999. Principles for ecological classification. In R. C. Szaro, N. C. Johnson, W. T. Sexton, and A. J. Malk, eds., *Ecological Stewardship: A Common Reference for Ecosystem Management Volume II,* pp. 353–393. Oxford, UK: Elsevier Science.

Grossman, D. H., D. Faber-Langendoen, A. S. Weakley, M. Anderson, P. Bourgeron, R. Crawford, K. Goodin, S. Landaal, K. Metzler, K. D. Patterson, M. Pyne, M. Reid, and L. Sneddon. 1998. *International Classification of Ecological Communities: Terrestrial Vegetation of the United States. Volume 1: The National Vegetation Classification System: Development, Status, and Applications.* Arlington, VA: The Nature Conservancy. Available online: www.conserveonline.org.

Groves, C. R., M. L. Klein, and T. F. Breden. 1995. Natural heritage programs: public private partnerships for biodiversity conservation. *Wildlife Society Bulletin* 23:784–790.

Groves, C. R., D. B. Jensen, L. L. Valutis, K. H. Redford, M. L. Shaffer, J. M. Scott, J. V. Baumgartner, J. V. Higgins, M. W. Beck, and M. G. Anderson. 2002a. Planning for biodiversity conservation: putting conservation science into practice. *BioScience* 52:499–512.

Groves, C. R., L. S. Kutner, D. M. Stoms, M. P. Murray, J. M. Scott, M. Schafale, A. S. Weakley, and R. L. Pressey. 2000b. Owning up to our responsibilities: who owns land important for biodiversity? In B. A. Stein, L. S. Kutner, and J. S. Adams, eds., *Precious Heritage: The Status of Biodiversity in the United States,* pp. 275–300. Oxford, UK: Oxford University Press.

Groves, C., L. Valutis, D. Vosick, B. Neely, K. Wheaton, J. Touval, and B. Runnels. 2000a. *Designing a Geography of Hope: A Practitioner's Handbook for Ecoregional Conservation Planning.* Arlington, VA: The Nature Conservancy. Available online: www.conserveonline.org.

Gucinski, H., M. J. Furniss, R. R. Ziemer, and M. H. Brookes, eds. 2001. *Forest Roads: A Synthesis of Scientific Information.* General Technical Report PNW-GTR-509, Pacific Northwest Research Station. Portland, OR: USDA Forest Service.

Guerry, A. D. and M. L. Hunter Jr. 2002. Amphibian distributions in a landscape of forests and agriculture: an examination of landscape composition and configuration. *Conservation Biology* 16:745–754.

Gurd, D. B., T. D. Nudds, and D. H. Rivard. 2001. Conservation of mammals in eastern North American wildlife reserves: how small is too small? *Conservation Biology* 15:1355–1363.

Hagen, R. T. 1999. *A Guide for Countries Preparing National Biodiversity Strategies and Action Plans.* Biodiversity Planning Support Programme, New York, NY: United Nations Development Programme.

Hale, P., and D. Lamb, eds. 1997. *Conservation Outside Nature Reserves.* Brisbane, AU: Centre for Conservation Biology, University of Queensland.

Halpern, B. 2003. The impact of marine reserves: do reserves work and does reserve size matter? *Ecological Applications* 13:S117–S137.

Halpern, B. S., and R. R. Warner. 2002. Marine reserves have rapid and lasting effects. *Ecology Letters* 5:361–366.

Halpin, P. N. 1997. Global climate change and natural-area protection: management responses and research direction. *Ecological Applications* 7:828–843.

Hann, W. J., J. L. Jones, M. G. Karl, et al. 1997. An assessment of landscape dynamics of the basin. In T. M. Quigley and S. J. Arbelbide, eds., *An Assessment of Ecosystem Components in the Interior Columbia Basin and Portions of the Klamath and Great Basins.* General Technical Report, PNW-GTR-405, Portland, OR: USDA, Forest Service.

Hannah, L., G. F. Midgley, T. Lovejoy, W. J. Bond, M. Bush, J. C. Lovett, D. Scott and F. I. Woodward. 2002. Conservation of biodiversity in a changing climate. *Conservation Biology* 16:264–268.

Hansen, A. J., T. A. Spies, F. J. Swanson, and J. L. Ohman. 1991. Conserving biodiversity in managed forests. *BioScience* 41:382–392.

Hanski, I. 1991. Single species metapopulation dynamics: concepts, models, and observations. In M. E. Gilpin and I. Hanski, eds., *Metapopulation Dynamics: Empirical and Theoretical Investigations*, pp.17–38 New York, NY: Academic Press.

———. 1999. *Metapopulation Ecology.* Oxford, UK: Oxford University Press.

Harding, J. S., E. F. Benfield, P. V. Bolstad, G. S. Helfman, and E. B. D. Jones III. 1998. Stream biodiversity: the ghost of land use past. *Proceedings of the National Academy of Science* 95:14843–14847.

Hargrove, B., T. Tear, and L. Valutis. 2002. *Geography of Hope Update #9: When and Where to Consider Restoration in Ecoregional Planning.* Arlington, VA: The Nature Conservancy.

Hargrove, W. W., and F. M. Hoffman. 1999. Using multivariate clustering to characterize ecoregion borders. *Computing in Science and Engineering* 1:18–25. Available online: http://research.esd.ornl.gov/~forrest/cise-1999/c4018.pdf.

———. 2000. An analytical assessment tool for predicting changes in a species distribution map following changes in environmental conditions. Fourth International Conference on Integrating GIS and Environmental Modeling (GIS/EM4): Problems, Prospects and Research Needs. September 2–8, 2000. Banff, Alberta, Canada. Available online: http://www.colorado.edu/research/cires/banff/upload/5/index.html.

Harris, L. D. 1984. *The Fragmented Forest: Island Biogeographic Theory and the Preservation of Biotic Diversity.* Chicago, IL: University of Chicago Press.

Harte, J. 1997. The central scientific challenge for conservation biology. In S. T. A. Pickett, R. S. Osterfield, M. Shacak, and G. E. Likens, eds. *The Ecological Basis of Conservation: Heterogeneity, Ecosystems, and Biodiversity*, pp. 379–383. New York, NY: Chapman and Hall.

Haskell, B. D., V. R. Leeworthy, P. C. Wiley, T. M. Beuttler, M. R. Haflich, J. Delaney, B. L. Richards, and E. Franklin. 2000. *Tortugas Ecological Reserve.* Final Supplemental Environmental Impact Statement/Final Supplemental Management Plan. Silver Spring, MD: National Oceanographic and Atmospheric Administration.

Haskell, B. D., B. G. Norton, and R. Costanza. 1992. What is ecosystem health and why should we worry about it? In R. Costanza, B. G. Noton, and B. D. Haskell, eds., *Ecosystem Health: New Goals for Environmental Management*, pp. 3–20. Washington, DC: Island Press.

Hastings, A., and L. W. Botsford. 1999. Equivalence in yield from marine reserves and traditional fisheries management. *Science* 284:1537–1538.

Haufler, J. B., C. A. Mehl, and G. J. Roloff. 1996. Using a coarse filter approach with species assessment for ecosystem management. *Wildlife Society Bulletin* 24:200–208.

Havlick, D. G. 2002. *No Place Distant.* Washington, DC: Island Press.

Hayden, B. P., G. C. Ray, and R. Dolan. 1984. Classification of coastal and marine environments. *Environmental Conservation* 11:199–207.

Heijnis, C. E., A. T. Lombard, R. M. Cowling, and P. G. Desmet. 1999. Picking up the pieces: a biosphere reserve framework for a fragmented landscape—The Coastal Lowlands of the Western Cape, South Africa. *Biodiversity and Conservation* 8:471–496.

Heyer, W. R., M. Donnelly, R. W. Heyer, D. Wake, and R. W. McDiarmid. 1994. *Measuring and Monitoring Biological Diversity: Standards Methods for Amphibians.* Washington, DC: Smithsonian Institution Press.

Higgins, J., M. Lammert, M. Bryer, M. DePhilip, and D. Grossman. 1998. *Freshwater Conservation in the Great Lakes Basin: Development and Application of an Aquatic Community Classification Framework.* Chicago, IL: The Nature Conservancy.

Hilderbrand, R. H., and J. L. Kershner. 2000. Conserving inland cutthroat trout in small streams: how much stream is enough? *North America Journal of Fisheries Management* 20:513–520.

Hillborn, R. 1987. Living with uncertainty in resource management. *North American Journal of Fisheries Management* 7:1–5.

Hilton-Taylor, C. 2000. *The 2002 IUCN Red List of Threatened Species.* Gland, Switzerland: IUCN–World Conservation Union.

Hobbs, R., and D. A. Norton. 1996. Towards a conceptual framework for restoration ecology. *Restoration Ecology* 8:260–267.

Hockings, M., S. Stolton, and N. Dudley. 2000. *Evaluating Effectiveness: A Framework for Assessing the Management of Protected Areas.* World Commission on Protected Areas, Best Practice Protected Area Guidelines Series No. 6. Cambridge, UK: IUCN–World Conservation Union.

Hoctor, T. S., M. H. Carr, and P. D. Zwick. 2000. Identifying a linked reserve system using a regional landscape approach: the Florida Ecological Network. *Conservation Biology* 14:984–1000.

Hocutt, C. H., and E. O. Wiley, eds. 1986. *The Zoogeography of North American Freshwater Fishes.* New York, NY: John Wiley and Sons.

Hoegh-Guldberg, O. 1999. Climate change, coral bleaching, and the future of the world's coral reefs. *Marine and Freshwater Research* 50:839–866.

Holdridge, L.R. 1967. Determination of world plant formations from simple climatic data. *Science.* 105:367–368.

Holling, C. S. 1986. The resilience of terrestrial ecosystems: local surprise and global change. In W. C. Clark and R. E. Munn, eds., *Sustainable Development of the Biosphere.* Cambridge, UK: Cambridge University Press.

Holthus, P. F., and J. E. Maragos. 1995. Marine ecosystem classification for the Tropical Island Pacific. In J. E. Maragos, M. N. Peterson, L. G. Eldredge, J. E. Bardach, and H. F. Takeuchi, eds., *Marine and Coastal Biodiversity in the Tropical Island Pacific Region, Volume 1: Species Systematics and Information Management Priorities,* pp. 239–280. Honolulu, HI: East–West Center.

Howard, P.C., T. R. B. Davenport, F. W. Kigenyi, P. Viskanic, M.C. Baltzer, C. J. Dickinson, J. Lwanga, R. A. Matthews, and E. Mupada. 2000. Protected area planning in the tropics: Uganda's national system of forest nature reserves. *Conservation Biology* 14:858–875.

Hudson, P. L, R. W. Griffiths, and T. J. Wheaton. 1992. Review of habitat classification schemes appropriate to streams, rivers and connecting channels in the Great Lakes drainage basin. In W.-D. N. Busch and P. G. Sly, eds., *The Development on an Aquatic Habitat Classification System for Lakes,* pp. 73–107. Boca Raton, FL: CRC Press.

Hughes, L. 2000. Biological consequences of global warming: is the signal already apparent? *Trends in Ecology and Evolution* 15:56–61.

Hunter, M. L., Jr. 1991. Coping with ignorance: the coarse filter strategy for maintaining biodiversity. In K. A. Kohm, ed., *Balancing on the Brink of Extinction: The Endangered Species Act and Lessons for the Future,* pp. 266–281. Washington, DC: Island Press.

———. 1996. Benchmarks for managing ecosystems: are human activities natural? *Conservation Biology* 10:695–697.

———. 2001. *Fundamentals of Conservation Biology, 2nd ed.* Malden, MA: Blackwell Science, Inc.

Hunter, M. L., G. L. Jacobson Jr., and T. Webb III. 1988. Paleoecology and the coarse filter approach to maintaining biological diversity. *Conservation Biology* 2:375–385.

Hunter, M. L., Jr., and P. Yonzon. 1993. Altitudinal distributions of birds, mammals, people, forests, and parks in Nepal. *Conservation Biology* 7:420–423.

Hutchings, J. A. 2000. Collapse and recovery of marine fishes. *Nature* 406:882–885.

Hutchinson, M. F., L. Belbin, A. O. Nicholls, H. A. Nix, J. P. McMahon, and K. D. Ord. 1996. *Volume Two: Spatial Modeling Tools, BioRap: Rapid Assessment of Biodiversity.* Canberra: The Australian BioRap Consortium.

Hwang, C. L., and K. Yoon. 1981. *Multiple Attribute Decision Making: Methods and Applications.* New York, NY: Springer-Verlag.

Iacobelli, T., K. Kavanaugh, and S. Rowe. 1993. *A Protected Areas Gap Analysis Methodology: Planning for the Conservation of Biodiversity.* Toronto, Ontario: World Wildlife Fund Canada.

Ilies, J. 1961. Versuch einer allgemein biozonotishcen Gleiderung der Fliessgewasser. Verhandlungen der Internationalen Veringung fur theroetische und angewandte. *Limnologie* 13:834–844.

State of Illinois Environmental Protection Agency, Bureau of Water. 1996. *Illinois Water Quality Report, Volume I, 1994–1995.* IEPA/BOW/96-060a.

Intergovernmental Panel on Climate Change (IPCC) Working Group II, 2001. Technical summary–climate change 2001: impacts, adaptation, and vulnerability. In J. J. McCarthy, O. F. Canziani, N. A. Leary, D. J. Dokken, and K. S. White, eds., *Climate Change 2001: Impacts, Adaptation, and Vulnerability,* pp. 19–73. Cambridge, UK: Cambridge University Press.

Interior Low Plateau Planning Team Plan. 2001. Interior Low Plateau Ecoregional Plan. Arlington, VA: The Nature Conservancy.

Jackson, J. A. 1994. Red-cockaded woodpecker (*Picoides borealis*). In A. Poole and F. Gill, eds., The Birds of North America, No. 85. Philadelphia: The Academy of Natural Sciences.

Jackson, J. B. C. 2001. What was natural in the coastal oceans? *Proceedings of the National Academy of Sciences* 98:5411–5418.

Jackson, J. B. C., M. X. Kirby, W. H. Berger, K. A. Bjorndal, L. W. Botsford, B. J. Bourque, R. H. Bradbury, R. Cooke, J. Erlandson, J. A. Estes, T. P. Hughes, S. Kidwell, C. B. Lange, H. S. Lenihan, J. M. Pandolfi, C. H. Peterson, R. S. Steneck, M. J. Tegner, and R. R. Warner. 2001. Historical overfishing and the recent collapse of coastal ecosystems. *Science* 293:629–638.

James, A., K. J. Gaston, and A. Balmford. 2001. Can we afford to conserve biodiversity? *BioScience* 51:43–52.

James, F. C. 1999. Lessons learned from a study of habitat conservation planning. *BioScience* 49:871–874.

Jamieson, G. S., and C. O. Levings. 2001. Marine protected areas in Canada: implications for both conservation and fisheries management. *Canadian Journal of Fisheries and Aquatic Sciences* 58:138–156.

Janis, B. A., ed. 1993. *Papua New Guinea Conservation Needs Assessment Volume 1.* Lander, MD: Corporate Press, Biodiversity Support Program, and Government of Papua New Guinea, Department of Environment and Conservation.

Jenkins, R. E. 1996. Natural Heritage Data Center Network: managing information for managing biodiversity. In R. C. Szaro and D. W. Johnston, eds., *Biodiversity in Managed Landscapes*, pp. 176–192. New York, NY: Oxford University Press.

Jennings, M. D. 2000. Gap analysis: concepts, methods, and recent results. *Landscape Ecology* 15:5–20.

Jeppson, P., and R. J. Whittaker. 2002. Ecoregions in context: a critique with special reference to Indonesia. *Conservation Biology* 16:42–57.

Johnson, K. N., F. Swanson, M. Herring, and S. Greene. 1999. *Bioregional Assessments: Science at the Crossroads of Management and Policy.* Washington, DC: Island Press.

Johnson, N. 1995. *Biodiversity in the Balance: Approaches to Setting Geographic Conservation Priorities.* Washington, DC: Biodiversity Support Program.

Jones, B. K., A. C. Neale, M. S. Nash, R. D. Van Remortel, J. D. Wickham, K. H. Ritters, and R. V. O'Neill. 2001. Predicting nutrient and sediment loadings to streams from landscape metric: a multiple watershed study from the United States Mid–Atlantic region. *Landscape Ecology* 16:301–312.

Jones, J. A., F. J. Swanson, B. C. Wemple, and K. Snyder. 2000. Effects of roads on hydrology, geomorphology, and disturbance patches in stream networks. *Conservation Biology* 14:76–85.

Kaiser, J. 2000. Rift over biodiversity divides ecologists. *Science* 289:1282–1283.

———. 2001. Bold corridor project confronts political reality. *Science* 293:2196–2199.

Kareiva, P., M. Marvier, and M. McClure. 2000. Recovery and management options for spring/summer chinook salmon in the Columbia river basin. *Science* 290:977–979.

Karl, J. W., P. J. Heglund, E. O. Garton, J. M. Scott, N. M. Wright, and R. L. Hutto. 2000. Sensitivity of species habitat-relationship model performance to factors of scale. *Ecological Applications* 10:1690–1705.

Karr, J. R., and E. W. Chu. 1995. Ecological integrity: reclaiming lost connections. In L. Westra and J. Lemon, eds., *Perspectives on Ecological Integrity*, pp. 34–48. Doredrecht, Netherlands: Kluwer Academic Publishers.

———. 1999. *Restoring Life in Running Waters: Better Biological Monitoring.* Washington, DC: Island Press.

Karr, J. R., and D. R. Dudley. 1981. Ecological perspective on water quality goals. *Environmental Management* 5:55–68.

Katzenbach, J. R., and D. K. Smith. 1993. *The Wisdom of Teams.* Boston, MA: Harper Business.

Kautz, R. S., and J. A. Cox. 2001. Strategic habitats for biodiversity conservation in Florida. *Conservation Biology* 15:55–77.

Keitt, T. H., D. L. Urban, and B. T. Milne. 1997. Detecting critical scales in fragmented landscapes. *Conservation Ecology* 11:1–20. Available online: www.consecol.org/vol11/iss1/art4.

Kerr, J. T. 1997. Species richness, endemism, and the choice of areas for conservation. *Conservation Biology* 11:1094–1100.

Kershaw, M., G. M. Mace, and P. H. Williams. 1995. Threatened status, rarity, and diversity as alternative selection measures for protected areas: a test using Afrotropical antelopes. *Conservation Biology* 9:324–334.

Ketchum, B. K. 1972. *The Water's Edge: Critical Problems of the Coastal Zone.* Cambridge, MA: Massachusetts Institute of Technology Press.

Khan, M. L., S. Menon, and K. S. Bawa. 1997. Effectiveness of the protected area network in biodiversity conservation: a case study of Meghalaya state. *Biodiversity and Conservation* 6:853–868.

Killeen, T. J., and T. S. Schulenberg, eds. 1998. *A Biological Assessment of Parque Nacional Noel Kempff Mercado, Bolivia.* RAP Working Paper 10, Washington, DC: Conservation International.

Kinlan, B. P., and S. D. Gaines. 2003. Propagule dispersal in marine and terrestrial environments: a community perspective. *Ecology.*

Kintsch, J. A., and D. L. Urban. 2002. Focal species, community representation, and physical proxies as conservation strategies: a case study in the Amphibolite Mountains, North Carolina, U.S.A. *Conservation Biology* 16:936–947.

Kirkpatrick, J. B. 1983. An iterative method for establishing priorities for the selection of nature reserves: an example from Tasmania. *Biological Conservation* 2:316–328.

Kirkpatrick, J. B., and M. J. Brown. 1994. A comparison of direct and environmental domain approaches to planning reservation of forest higher plant communities and species in Tasmania. *Conservation Biology* 8:217–224.

Kirkpatrick, J. B., and L. Gilfedder. 1995. Maintaining integrity compared with maintaining rare and threatened taxa in remnant bushland in subhumid Tasmania. *Biological Conservation* 74:1–8.

Klahr, T., R. K. Moseley, B. Butterfield, M. Bryer, D. Vanderschaaf, J. Kagan, S. Cooper, B. Hall, C. Harris, and B. Hargrove. 2000. *Middle Rockies–Blue Mountains Ecoregional Conservation Plan.* Arlington, VA: The Nature Conservancy. Available online: www.conserveonline.org.

Kleiman, D. G., R. P. Reading, B. J. Miller, T. W. Clark, J. M. Scott, J. Robinson, R. L. Wallace, R. J. Cabin, and F. Felleman. 2000. Improving the evaluation of conservation programs. *Conservation Biology* 14:356–365.

Knight, R. L., and K. J. Gutzwiller. eds., 1995. *Wildlife and Recreationists: Coexistence through Management and Research.* Washington, DC: Island Press.

Knowlton, N. 2001. Coral reef biodiversity—habitat size matters. *Science.* 292:1493–1494.

Kotar, J., J. A. Kovach, and C. T. Locey. 1988. *Field Guide to Forest Habitat Types of Northern Wisconsin.* Madison, WI: University of Wisconsin, Department of Forestry and Wisconsin Department of Natural Resources.

Kotliar, N. B. 2000. Application of the new keystone species concept to prairie dogs: how well does it work? *Conservation Biology* 14:1715–1721.

Krajina, V. J. 1965. Biogeoclimatic zones and classification of British Columbia. In V. J. Krajina, ed., *Ecology of Western North America*, pp. 1–17. Vancouver, British Columbia, Canada: University of British Columbia.

Küchler, A. W. 1964. *Potential Natural Vegetation of the Conterminous United States.* American Geographical Society Special Publication 36. New York, NY: American Geographical Society.

Kuhlmann, W. 1996. Wildlife's burden. In W. J. Snape III and O. A. Houck, eds., *Biodiversity and the Law*, pp. 189–201. Washington, DC: Island Press.

Lambeck, R. J. 1997. Focal species: a multi-species umbrella for nature conservation. *Conservation Biology* 11:849–856.

Lammert, M., J. Higgins, D. Grossman, and M. Bryer. 1997 *A Classification Framework for Freshwater Communities: Proceedings of The Nature Conservancy's Aquatic Community Classification Workshop.* Arlington, VA: The Nature Conservancy.

Lande, R. 1995. Mutation and conservation. *Conservation Biology* 9:782–791.

Landres, P. B., P. Morgan, and F. J. Swanson. 1999. Overview of the use of natural variability concepts in managing ecological systems. *Ecological Applications* 9:1179–1188.

Landres, P. B., J. Verner, and J. W. Thomas. 1988. Ecological uses of vertebrate indicator species: a critique. *Conservation Biology* 2:316–328.

Langdon, R., J. Andrews, K. Cox, S. Fiske, N. Kamman, and S. Warren. 1998. *A Classification of the Aquatic Communities of Vermont.* The Aquatic Classification Workgroup for The Nature Conservancy and the Vermont Biodiversity Project. Montpelier, VT: The Vermont Agency of Natural Resources.

Lawler, J. L., S. P. Campbell, A. D. Guerry, M. B. Kolozsvary, R. J. O'Connor, and L. C. N. Seward. 2002. The scope and treatments of threats in endangered species recovery plans. *Ecological Applications* 12:663–667.

Lawton, J. H. 1994. What do species do in ecosystems? *Oikos* 71:367–374.

Lawton, J. H., and R. M. May. 1995. *Extinction Rates.* Oxford, UK: Oxford University Press.

Leach, J. H., and R. C. Heron. 1992. A review of lake habitat classification. In W.-D. N. Busch and P. G. Sly, eds., *The Development of an Aquatic Habitat Classification System for Lakes*, pp. 27–57. Boca Raton, FL: CRC Press.

Lenihan, H. S. 1999. Physical–biological coupling on oyster reefs: how habitat structure influences individual performance. *Ecological Monographs* 69:251–275.

Lesica, P., and F. W. Allendorf. 1995. When are peripheral populations valuable for conservation? *Conservation Biology* 9:753–760.

Leslie, H., M. Ruckelshaus, I. Ball, S. Andelman, and H. Possingham. 2003. Using siting algorithms in the design of marine reserve networks. *Ecological Applications* 13:S185–S198.

Leslie, M., G. K. Meffe, J. L. Hardesty, and D. Adams. 1996. *Conserving Biodiversity on Military Lands: A Handbook for Natural Resource Managers.* Arlington, VA: The Nature Conservancy.

Levin, S. A. 1992. The problem of pattern and scale in ecology. *Ecology* 73:1943–1967.

Lewis, J. P. 2000. *The Project Manager's Desk Reference, 2nd. ed.* New York, NY: McGraw-Hill.

Lindenmayer, D. B., A. D. Manning, P. L. Smith, H. P. Possingham, J. Fischer, I. Oliver, and M. A. McCarthy. 2002. The focal-species approach and landscape restoration: a critique. *Conservation Biology* 16:338–345.

Lodge, D. M. 2001. Are non-native species harming the environment? *Congressional Quarterly Research* 11:807.

Loeb, S. L., and A. Spacie, eds. 1994. *Biological Monitoring of Aquatic Systems.* Boca Raton, FL: Lewis Publishers.

Lombard, A. T., R. M. Cowling, R. L. Pressey, and P. J. Mustart. 1997. Reserve selection in a species-rich and fragmented landscape on the Agulhas Plain, South Africa. *Conservation Biology* 11:1101–1116.

Lombard, A.T., R. M. Cowling, R. L. Pressey, A. G. Rebelo, and N. S. Cole. In press. Efficiency of land class data versus species locality data in conservation planning for the Cape Floristic Region. *Biological Conservation*.

Lorimer, C. G. 2001. Historical and ecological roles of disturbance in eastern North American forests: 9000 years of change. *Wildlife Society Bulletin* 29:425–439.

Lotka, A. J. 1956. *Elements of Mathematical Ecology.* New York, NY: Dover Publications.

Lubchenco, J., S. Gaines, R. Warner, S. Airame, and B. Simler. 2002. *The Science of Marine Reserves.* Corvallis, OR: Partnership for Interdisciplinary Study of Coastal Oceans.

Lugo, A. E., J. S. Baron, T. P. Frost, T. W. Cundy, and P. Dittberner. 1999. Ecosystem processes and functioning. In R. C. Szaro, N.C. Johnson, W.T. Sexton, and A. J. Malk, eds., *Ecological Stewardship: a Common Reference for Ecosystem Management*, pp. 219–254. Oxford, UK: Elsevier Science Ltd.

MacArthur, R. H., and E. O. Wilson. 1967. *The Theory of Island Biogeography.* Princeton, NJ: Princeton University Press.

MacNally, R., A. F. Bennett, G. W. Brown, L. F. Lumsden, A. Yen, S. Hinkley, P. Lillywhite, and D. Ward. 2002. How well do ecosystem-based planning units represent different components of biodiversity? *Ecological Applications* 12:900–912.

Maddock, A., and G. A. Benn. 2000. Identification of conservation-worthy areas in Northern Zululand, South Africa. *Conservation Biology* 14:155–166.

Maddock, A., and M. A. Du Plessis. 1999. Can species data only be appropriately used to conserve biodiversity? *Biodiversity and Conservation* 8:603–615.

Maddock, A. H., and M. J. Samways. 2000. Planning for biodiversity conservation based on the knowledge of biologists. *Biodiversity and Conservation* 9:1153–1169.

Maginnis, S., W. Jackson, and N. Dudley. 2001. *A Landscape Approach to Forest Conservation.* Gland, Switzerland: IUCN and WWF.

Maguire, L.A. 1994. Science, values, and uncertainty: a critique of the Wildlands Project. In R. E. Grumbine, ed., *Environmental Policy and Biodiversity,* pp. 267–272. Washington, DC: Island Press.

———. 1996. Making the role of values in conservation explicit: values and conservation biology. *Conservation Biology* 10:914–916.

Mangel, M., L. M. Talbot, G. K. Meffe, M. T. Agardy, D. L. Alverson, J. Barlow, D. B. Botkin, G. Budowski, T. Clark, J. Cooke, R. H. Crozier, P. K. Dayton, D. L. Elder, C. W. Fowler, S. Funtowicz, J. Giske, R. J. Hofman, S. J. Holt, S. R. Kellert, L. A. Kimball, D. Ludwig, K. Magnusson, B. S. Malayang, C. Mann, E. A. Norse, S. P. Northridge, W. F. Perrin, C. Perrings, R. M. Peterman, G. B. Rabb, H. A. Regier, J. E. Reynolds III, K. Sherman, M. P. Sissenwine, T. D. Smith, A. Starfield, R. J. Taylor, M. F. Tilman, C. Toft, J. R. Twiss Jr., J. Wilen, and T. P. Young. 1996. Principles for the conservation of wild living resources. *Ecological Applications* 6:338–362.

Margoluis, R., C. Margoluis, K. Brandon, and N. Salafsky. 2000. *In Good Company: Effective Alliances for Conservation.* Washington, DC: Biodiversity Support Program.

Margoluis, R., and N. Salafsky. 2001. *Is Our Project Succeeding? A Guide to Threat Reduction Assessment for Conservation.* Washington, DC: Biodiversity Support Program.

Margules, C. R. 1989. Introduction to some Australian developments in conservation evaluation. *Biological Conservation* 50:1–11.

Margules, C. R., and M. P. Austin. 1994. Biological models for monitoring species decline: the construction and use of data bases. *Philosophical Transactions of the Royal Society of London* 344:69–75.

Margules, C. R., A. O. Nicholls, and R. L. Pressey. 1988. Selecting networks of reserves to maximize biological diversity. *Biological Conservation* 43:63–76.

Margules, C. R., and R. L. Pressey. 2000. Systematic conservation planning. *Nature* 405:243–253.

Margules, C., and M. B. Usher. 1981. Criteria used in assessing wildlife conservation potential: a review. *Biological Conservation* 21:79–109.

Markham, A., N. Dudley, and S. Stolton, 1993. *Some Like It Hot: Climate Change, Biodiversity, and Survival of Species.* Gland, Switzerland: World Wildlife Fund.

Markham, A., and J. R. Malcolm Jr. 1996. Biodiversity and wildlife conservation: adaptation to climate change. In J. B. Smith, N. Bhatti, G. Menzhulin, R. Benioff, M. I. Budyko, M. Campos, B. Jallow, and F. Rijsberman, eds., *Adapting to Climate Change: Assessments and Issues*, pp. 384–401. New York: Springer-Verlag.

Marshall, R. M., S. Anderson, M. Batcher, P. Comer, S. Cornelius, R. Cox, A. Gondor, D. Gori, J. Humke, R. Paredes Aguilar, I. E. Parra, and S. Schwartz. 2000. *An Ecological Analysis of Conservation Priorities in the Sonoran Desert Ecoregion.* Arlington, VA: The Nature Conservancy, Sonoran Institute, and Instituto del Medio Ambiente y el Desarollo Sustenable del Estado de Sonora.

Master, L. L. 1991. Assessing threats and setting priorities for conservation. *Conservation Biology* 5:559–563.

Master, L. L., S. R. Flack, and B. A. Stein, eds. 1998. *Rivers of Life: Critical Watersheds for Protecting Freshwater Biodiversity.* Arlington, VA: The Nature Conservancy.

Master, L. L., B. A. Stein, L. S. Kutner, and G. A. Hammerson. 2000. Vanishing assets: conservation status of U.S. species. In B. A. Stein, L. S. Kutner, and J. S. Adams, eds., *Precious Heritage: The Status of Biodiversity in the United States*, pp. 93–118. Oxford, UK: Oxford University Press.

Maxwell, J. R., C. J. Edwards, M. E. Jensen, S. J. Paustain, H. Parrot, and D. M. Hill. 1995. *A Hierarchical Framework of Aquatic Ecological Units in North America (Nearctic Zone).* General Technical Report NC-176, North Central Forest Experimental Station. St. Paul, MN: USDA Forest Service.

Maybury, K. P., ed. 1999. *See the Forest and the Trees: Ecological Classification for Conservation.* Arlington, VA: The Nature Conservancy. Available online: www.conserve online.org.

McAllister, D. E., A. L. Hamilton, and B. Harvey. 1997. Global freshwater biodiversity: striving for the integrity of freshwater ecosystems. *Sea Wind* 11(3).

McCullough, D. R. 1996. *Metapopulations and Wildlife Conservation.* Washington, DC: Island Press.

McCune, B., and M. J. Mefford. 1995. PC-ORD. Multivatiate Analysis of Ecological Data, Version 2.0. Gleneden Beach, OR: MjM Software Design.

McDonald, K. A., and J. H. Brown. 1992. Using montane mammals to model extinctions due to global change. *Conservation Biology* 6:409–415.

McHarg, I. H. 1969. *Design with Nature.* Garden City, NY: Doubleday/The Natural History Press.

———. 1997. Natural factors in planning. *Journal of Soil and Water Conservation* 52:13–17.

McMahon, E. 2002. Speech to Greater Yellowstone Coalition Annual Meeting, June 2002. West Yellowstone, MT.

McMahon, G., S. M. Gregonis, S. W. Waltman, J. A. Omernik, T. D. Thorson, J. A. Freequf, A. H. Rorick, and J. E. Keys. 2001. Developing a spatial framework of common ecological regions for the conterminous United States. *Environmental Management* 28:293–316.

McNeely, J. A., M. Gadgil, C. Leveque, C. Padoch, and K. Redford. 1995. Human influences on biodiversity. In V. H. Heywood, ed., *Global Biodiversity Assessment*, pp. 711–822. United Nations Environment Program, Cambridge, UK: Cambridge University Press

Meffe, G. K., and C. R. Carroll. 1997. *Principles of Conservation Biology, 2nd ed.* Sunderland, MA: Sinauer Associates, Inc.

Mendel, L. C., and J. B. Kirkpatrick. 2002. Historical progress of biodiversity conservation in the protected area system of Tasmania, Australia. *Conservation Biology* 16:1520–1529.

Menge, B. A., E. L. Berlow, C. A. Blanchette, S. A. Navarrete, and S. B. Yamada. 1994. The keystone species concept: variation in interaction strength in a rocky intertidal habitat. *Ecological Monographs* 64:249–286.

Met Office. 2000. *Climate Change: An Update from the Hadley Centre*. Bracknell, UK: Hadley Centre for Climate Prediction and Research.

Meyer, M., and J. Booker. 1991. *Eliciting and Analyzing Expert Judgment, a Practical Guide*. London, UK: Academic Press.

Miller, K. R., and S. M. Lanou. 1995. *National Biodiversity Planning: Guidelines Based on Early Experiences Around the World*. Gland, Switzerland: World Resources Institute, United Nations Environment Programme, and the World Conservation Union.

Miller, R. R., J. D. Williams, and J. E. Williams. 1989. Extinctions of North American fishes during the past century. *Fisheries* 14:22–38.

Miller, S., J. V. Higgins, and J. Perot. 1998. *The Classification of Aquatic Communities in the Illinois River Watershed and Their Uses in Conservation Planning*. Peoria, IL: The Nature Conservancy.

Mills, E. L., J. H. Leach, J. L. Carlton, and C. L. Secor. 1993. Exotic species in the Great Lakes: a history of biotic crisis and anthropogenic introductions. *Journal of Great Lakes Research* 19:1–54.

Mills, L. S., M. E. Soulé, and D. F. Doak. 1993. The keystone species concept in ecology and conservation. *BioScience* 43:219–224.

Minello, T., and J. W. Webb Jr. 1997. Use of natural and created spartina alterniflora salt marshes by fishery species and other aquatic fauna in Galveston Bay, Texas, USA. *Marine Ecology Progress Series* 151:165–179.

Minckley, W. L., and J. E. Deacon, eds. 1991. *Battle Against Extinction. Native Fish Management in the American West*. Tucson, AZ: University of Arizona Press.

Mitsch, W. J., J. W. Day Jr., J. W. Gilliam, P. M. Groffman, D. L. Hey, G. W. Randall, and N. Wang. 2001. Reducing nitrogen loading to the Gulf of Mexico from the Mississippi River basin: strategies to counter a persistent ecological problem. *BioScience* 51:373–388.

Mitsch, W. J., and J. G. Gosselink. 1986. *Wetlands*. New York, NY: Van Nostrand Rheinhold.

Mittermeier, R. A., N. Myers, P. R. Gil, and C. G. Mittermeier. 1999. *Hotspots: Earth's Biologically Richest and Most Endangered Terrestrial Ecoregions*, Mexico City, Mexico: Cemex, S.A., and Conservation International.

Mittermeier, R. A., N. Myers, J. B. Thomsen, G. A. B. Da Fonseca, and S. Olivieri. 1998. Biodiversity hotspots and major tropical wilderness areas: approaches to setting conservation priorities. *Conservation Biology* 12:516–520.

Morgan, M. G., and M. Henrion. 1990. *Uncertainty: A Guide to Dealing with Uncertainty in Qualitative Risk and Policy Analysis*. New York, NY: Cambridge University Press.

Morris, W. F., P. L. Bloch, B. R. Hudgens, L. C. Moyle, and J. R. Stinchcombe. 2002. Population viability analyses in endangered species recovery plans: past use and future improvements. *Ecological Applications* 12:708–712.

Morris, W. F., and D. F. Doak. 2002. *Quantitative Conservation Biology: Theory and Practice of Population Viability Analysis.* Sunderland, MA: Sinauer Associates, Inc.

Morris, W., D. Doak, M. Groom, P. Kareiva, J. Fieberg, L. Gerber, P. Murphy, and D. Thomson. 1999. *A Practical Handbook for Population Viability Analysis.* Arlington, VA: The Nature Conservancy. Available online: www.conserveonline.org.

Moyle, P. B., and J. P. Ellison. 1991. A conservation-oriented classification system for the inland waters of California. *California Fish and Game* 77:161–180.

Moyle, P. B., and R. A. Leidy. 1992. Loss of biodiversity in aquatic ecosystems: evidence from fish fauna. In P. L. Fielder and S. K. Jain, eds., *Conservation Biology: The Theory and Practice of Nature, Conservation, Preservation, and Management*, pp. 128–169. New York, NY: Chapman and Hall.

Moyle, P. B., H. W. Li, and B. A. Barton. 1986. The Frankenstein effect: impact of introduced fishes on native fishes in North America. In R. H. Stroud, ed., *Fish Culture in Fisheries Management*, pp. 415–426. Bethesda, MD: American Fisheries Society.

Moyle, P. B., and P. J. Randall. 1998. Evaluating the Biotic Integrity of Watersheds in the Sierra Nevada, California. *Conservation Biology* 12:1318–1326.

Moyle, P. B., and R. M. Yoshiyama. 1994. Protection of aquatic biodiversity in California: a five-tiered approach. *Fisheries* 19:6–18.

Mumby, P. J., and A. R. Harborne. 1999. Development of a systematic classification scheme of marine habitats to facilitate regional management and mapping of Caribbean coral reefs. *Biological Conservation* 88:155–163.

Myers, N. 1979. *The Sinking Ark.* Elmsford, NY: Pergamon Press.

Myers, N., R. Mittermeier, C. G. Mittermeier, G. A. B. Da Fonseca, and J. Kent. 2000. Biodiversity hotspots for conservation priorities. *Nature* 403:853–858.

Nachlinger, J., K. Sochi, P. Comer, G. Kittel, and D. Dorfman. 2001. *Great Basin: An Ecoregion-Based Conservation Blueprint.* Arlington, VA: The Nature Conservancy.

Naeem, S., F. S. Chapin III, R. Costanza, P. R. Ehrlich, F. B. Golley, D. U. Hooper, J. H. Lawton, R. V. O'Neill, H. A. Mooney, O. E. Sala, A. J. Symstad, and D. Tilman. 1999. Biodiversity and ecosystem functioning: maintaining natural life support processes. *Issues in Ecology 4.*

Naiman, R. J., D. G. Lonzarich, T. J. Beechie, and S. C. Ralph. 1992. General principles of classification and the assessment of conservation potential in rivers. In P. J. Boon, P. Calow, and G. E. Petts, eds., *River Conservation and Management*, pp. 93–123. West Sussex, UK: John Wiley and Sons, Ltd.

Nalepa, T. F., and G. L. Fahnenstiel. 1995. *Dreissena polymorpha* in Saginaw Bay, Lake Huron ecosystem: overview and perspective. *Journal of Great Lakes Research* 21:411–416.

Namibian National Biodiversity Task Force. 1998. *Biological Diversity in Namibia—A Country Study.* Cape Town, South Africa: ABC Press.

National Research Council. 1999. *Sustaining Marine Fisheries.* Washington, DC: National Academy Press.

———. 2001. *Marine Protected Areas: Tools for Sustaining Ocean Ecosystems.* Washington, DC: National Academy Press.

Natural Resources Defense Council. 2001. *Priority Ocean Areas for Protection in the Mid-Atlantic.* New York, NY: Natural Resources Defense Council.

Neave, H. M., T. W. Norton, and H. A. Nix. 1996. Biological inventory for conservation evaluation I. Design of a field survey for diurnal, terrestrial birds in southern Australia. *Forest Ecology and Management* 85:107–122.

Neilson, R. P., and L. H. Wullstein. 1983. Biogeography of two southwest American oaks in relation to atmospheric dynamics. *Journal of Biogeography* 10:275–297.

Nicholls, A. O. 1989. How to make biological surveys go further with generalised linear models. *Biological Conservation* 50:51–75.

Nichols, W. F., K. T. Killingbeck, and P. V. August. 1998. The influence of geomorphological heterogeneity on biodiversity II. A landscape perspective. *Conservation Biology* 12:371–379.

Nix, H. A., D. P. Faith, M. F. Hutchinson, C. R. Margules, J. West, A. Allisoin, J. L. Kesteven, G. Natera, W. Slater, J. L. Stein, and P. Walker. 2000. *The BioRap Toolbox: A National Study of Biodiversity Assessment and Planning for Papua New Guinea.* Consultancy Report to the World Bank. Canberra, AU: Center for Resource and Environmental Studies, Australian National University.

Noble, S. J. 1996. *Volume Four: Tools for Storing and Mapping Spatial Data, BioRap: Rapid Assessment of Biodiversity.* Canberra, AU: The Australian BioRap Consortium.

Noon, B. R., R. H. Lamberson, M. S. Boyce, and L. I. Irwin. 1999. Population viability in ecosystem management. In R. C. Szaro, N. C. Johnson, W. T. Sexton, and A. J. Malk, eds., *Ecological Stewardship: A Common Reference for Ecosystem Management,* pp. 87– 134. Oxford, UK: Elsevier Science.

Northern Andes Ecoregional Team. 2001. *Biodiversity Vision for the Northern Andes Regional Complex.* World Wildlife Fund–U.S., Fundacion Natura. Washington, DC: FUDENA.

Noss, R., G. Wuerthner, K. Vance-Borland, and C. Carroll. 2001. *A Biological Conservation Assessment for the Greater Yellowstone Ecosystem.* Corvallis, OR: Conservation Science, Inc.

Noss, R. F. 1987. From plant communities to landscapes in conservation inventories: a look at The Nature Conservancy (USA). *Biological Conservation* 41:11–37.

———. 1990. Indicators for monitoring biodiversity: a hierarchical approach. *Conservation Biology* 4:355–364.

———. 1991. From endangered species to biodiversity. In K. A. Kohm, ed., *Balancing on the Brink of Extinction: The Endangered Species Act and Lessons for the Future,* pp. 227–246. Washington, DC: Island Press.

———. 1995. *Maintaining Ecological Integrity in Representative Reserve Networks.* Toronto: World Wildlife Fund Canada.

———. 1996a. Ecosystems as conservation targets. *Trends in Ecology and Evolution* 11:351.

———. 1996b. Protected areas: how much is enough? In R. G. Wright, ed., *National Parks and Protected Areas: Their Role in Environmental Protection,* pp. 91–120. Cambridge, MA: Blackwell Science.

———. 1996c. On attacking a caricature of reserves: response to Everett and Lehmkuhl. *Wildlife Society Bulletin* 24:777–779.

———. ed. 2000. *The Redwood Forest: History, Ecology, and Conservation of the Coast Redwoods.* Washington, DC: Island Press.

———. 2001. Beyond Kyoto: forest management in a time of rapid climate change. *Conservation Biology* 15:578–590.

———. 2002. Context matters: considerations for large-scale conservation. *Conservation Biology in Practice* 3:10–19.

Noss, R. F., C. Carroll, K. Vance-Borland, and G. Wuerthner. 2002. A multicriteria assessment of irreplaceability and vulnerability of sites in the Greater Yellowstone Ecosystem. *Conservation Biology* 16:895–908.

Noss, R. F., and A. Y. Cooperrider. 1994. *Saving Nature's Legacy: Protecting and Restoring Biodiversity.* Washington, DC: Island Press.

Noss, R. F., and B. Csuti. 1997. Habitat fragmentation. In G. Meffe, C. R. Carroll, and contributors, *Principles of Conservation Biology, 2nd ed.*, pp. 269–304. Sunderland, MA: Sinauer Associates, Inc.

Noss, R. F., E. Dinerstein, B. Gilbert, M. Gilpin, B. J. Miller, J. Terborgh, and S. Trombulak. 1999a. Core areas: where nature begins. In J. Terborgh and M. Soulé, eds., *Continental Conservation: Scientific Foundations of Regional Reserve Networks,* pp. 99–128. Washington, DC: Island Press.

Noss, R. F., and L. D. Harris. 1986. Nodes, networks, and MUMs: preserving diversity at all scales. *Environmental Management* 10:299–309.

Noss, R. F., E. T. Laroe III, and J. M. Scott. 1995. *Endangered Ecosystems of the United States: A Preliminary Assessment of Loss and Degradation.* USDI National Biological Service, Biological Report 28. Washington, DC: USDI National Biological Service.

Noss, R. F., M. A. O'Connell, and D. D. Murphy. 1997. *The Science of Conservation Planning: Habitat Conservation under the Endangered Species Act.* Washington, D. C.: Island Press.

Noss, R. F., H. B. Quigley, M. G. Hornocker, T. Merrill, and P. C. Pacquet. 1996. Conservation biology and carnivore conservation in the Rocky Mountains. *Conservation Biology* 10:949–963.

Noss, R. F., J. R. Strittholt, K. Vance-Borland, C. Carroll, and P. Frost. 1999b. A conservation plan for the Klamath-Siskiyou Ecoregion. *Natural Areas Journal* 19:392–410.

Nott, M. P., and S. L. Pimm. 1997. The evaluation of biodiversity as a target for conservation. In S. T. A. Pickett, R. S. Ostfeld, M. Shachak, and G. E. Likens, eds., *The Ecological Basis of Conservation: Heterogeneity, Ecosystems, and Biodiversity,* pp. 125–135. New York, NY: Chapman and Hall.

Novaro, A. J., M. C. Funes, and R. S. Walker. 2000. Ecological extinction of native prey of a carnivore asemblage in Argentine Patagonia. *Biological Conservation* 92:25–33.

Olson, D., E. Dinerstein, P. Canevari, I. Davidson, G. Castro, V. Morisset, R. Abell, and E. Toledo, eds. 1998. *Freshwater Biodiversity of Latin America and the Caribbean: A Conservation Assessment.* Washington, DC: World Wildlife Fund.

Olson, D. M., B. Chernoff, G. Burgess, I. Davidson, P. Canevari, E. Dinerstein, G. Castro, V. Morisset, R. Abell, and E. Toledo, eds. 1997. *Freshwater Biodiversity of Latin America and the Caribbean: A Conservation Assessment. Proceedings of a Workshop.* Washington, DC: World Wildlife Fund.

Olson, D. M., and E. Dinerstein. 1998. The Global 200: a representation approach to conserving the earth's most biologically valuable ecoregions. *Conservation Biology* 12:502–515.

Olson, D. M., E. Dinerstein, G. V. N. Powell, and E. D. Wikramanayake. 2002. Conservation biology for the biodiversity crisis. *Conservation Biology* 16:1–3.

Olson, D. M., E. Dinerstein, E. D. Wikramanayake, N. D. Burgess, G. V. N. Powell, E. C. Underwood, J. A. D'Amico, I. Itoua, H. E. Strand, J. C. Morrison, C. J. Loucks, T. F. Allnutt, T. H. Ricketts, Y. Kura, J. F. Lamoreux, W. W. Wettengel, P. Hedao, and K. R. Kassem. 2001. Terrestrial ecoregions of the world: a new map of life on earth. *BioScience* 51:933–938.

Omernik, J. M. 1987. Ecoregions of the Conterminous United States. *Annals of the Association of American Geographers* 77:118–125.

————. 1995. *Level III Ecoregions of the Continent.* Washington, DC: National Health and Environmental Effects Laboratory, Environmental Protection Agency.

Omernik, J. M., and R. G. Bailey. 1997. Distinguishing between watersheds and ecoregions. *Journal of the American Water Resources Association* 33:935–949.

O'Neill, R. V., A. R. Johnson, and R. V. King. 1986. *A Hierarchical Concept of Ecosystems.* Princeton, NJ: Princeton University Press.

Oregon Biodiversity Project. 1998. *Oregon's Living Landscape: Strategies and Opportunities to Conserve Bioodiversity.* Washington, DC: Defenders of Wildlife.

Orians, G. H. 1993a. Endangered at what level? *Ecological Applications* 3:206–208.

Orians, G. H. 1993b. Policy implications of global climate change. In P. M. Kareiva, J. G. Kingsolver, and R. B. Huey, eds., *Biotic Interactions and Global Change,* pp. 467–479. Sunderland, MA: Sinauer Associates, Inc.

Page, L. M., and B. M. Burr. 1991. *A Field Guide to Freshwater Fishes, North America North of Mexico.* Boston, MA: Houghton Mifflin Company.

Paine, R. T. 1966. Food web complexity and species diversity. *American Naturalist* 100:65–75.

————. 1969. A note on trophic complexity and community stability. *American Naturalist* 103:91–93.

Palumbi, S. R. 1994. Genetic divergence, reproductive isolation, and marine speciation. *Annual Review of Ecology and Systematics* 25:547–572.

Parks, S. A., and A. H. Harcourt. 2002. Reserve size, local human density, and mammalian extinctions in U. S. Protected Areas. *Conservation Biology* 16:800–808.

Parmesan, C., N. Ryrholm, C. Stefanescu, J. K. Hill, C. D. Thomas, H. Descimon, B. Huntley, L. Kaila, J. Kullberg, T. Tammaru, J. Tennent, J. A. Thomas, and M. Warren. 1999. Poleward shift of butterfly species' ranges associated with regional warming. *Nature* 399:579–583.

Pauly, D., V. Christensen, J. Dalsgaard, R. Froese, and F. J. Torres. 1998. Fishing down marine food webs. *Science* 279:860–863.

Paustian, S. J., ed. 1992. *Channel Type Users Guide for the Tongass National Forest, Southeast Alaska.* Report R10-TP-26. Southeast Alaska Region, Anchorage, AK: USDA Forest Service.

Peck, S. 1998. *Planning for Biodiversity: Issues and Examples.* Washington, DC: Island Press,

Pekelman, D., and S. K. Sen. 1974. Mathematical programming models for the determination of attribute weights. *Management Science* 20:1217–1229.

Pence, G. Q. K., M. A. Botha, and J. K. Turpie. In press. Evaluating combinations of on- and off-reserve conservation strategies for the Agulhas Plain, South Africa: a financial perspective. *Biological Conservation.*

Peres, C. A., and J. W. Terborgh. 1995. Amazonian nature reserves: an analysis of the defensibility status of existing conservation units and design criteria for the future. *Conservation Biology* 9:24–46.

Pernetta, J., R. Leemans, D. Elder, and S. Humphrey, eds. 1994. *Impacts of Climate Change on Ecosystems and Species.* Gland, Switzerland: World Conservation Union.

Peters, R. L. 1992. Conservation of biotic diversity in the face of climate change. In R. L. Peters and T. E. Lovejoy, eds., *Global Warming and Biological Diversity,* pp. 15–30. New Haven, CT: Yale University Press.

Peters, R. L., and J. D. S. Darling. 1985. The greenhouse effect and nature reserves. *BioScience* 35:707–717.

Peters, R. S., D. M. Waller, B. Noon, S. T. A. Pickett, D. Murphy, J. Cracraft, R. Kiester, W. Kuhlmann, O. Houck, and W. J. Snape III. 1997. Standard scientific procedures for implementing ecosystem management on public lands. In S. T. A. Pickett, R. S. Ost-feld, M. Shachak, and G. E. Likens, eds., *The Ecological Basis of Conservation: Heterogeneity, Ecosystems, and Biodiversity*, pp. 320–336. New York: Chapman and Hall.

Peterson, A. T. , M. A. Ortega-Herta, J. Bartley, V. Sanchez-Cordero, J. Soberon, R. H. Buddemeier, and D. R. B. Stockwell. 2002. Future projections for Mexican faunas under global climate change scenarios. *Nature* 416:626–629.

Pfister, R. D., and S. F. Arno. 1980. Classifying forest habitat based on potential climax vegetation. *Forest Science* 26:52–70.

Pflieger, W. L. 1989. *Aquatic Community Classification System for Missouri. Aquatic Series No. 19.* Jefferson City, MO: Missouri Department of Conservation.

Pharo, E. J., A. J. Beatties, and D. Binns. 1999. Vascular plant diversity as a surrogate for bryophyte and lichen diversity. *Conservation Biology* 13:282–292.

Pickett, S. T. A., and J. N. Thompson. 1978. Patch dynamics and the design of nature reserves. *Biological Conservation* 13:27–37.

Pickett, S. T. A., V. T. Parker, and P. L. Fiedler. 1992. The new paradigm in ecology: implications for conservation biology above the species level. In P. L. Fielder and S. K. Jain, eds., *Conservation Biology: The Theory and Practice of Nature Conservation, Preservation, and Management*, pp. 66–88. New York, NY: Chapman and Hall.

Pielou, E. C., 1991. *After the Ice Age: The Return of Life to Glaciated North America*. Chicago, IL: University of Chicago Press.

Pimentel, D., L. Westra, and R. F. Noss. 2000. *Ecological Integrity: Intergrating Environment, Conservation, and Health*. Washington, DC: Island Press.

Pimm, S. L., M. Ayres, A. Balmford, G. Branch, K. Brandon, T. Brooks, R. Bustamante, R. Costanza, R. Cowling, L. M. Curran, A. Dobson, S. Farber, G. A. B. da Fonseca, C. Gascon, R. Kitching, J. McNeely, T. Lovejoy, R. A. Mittermeier, N. Myers, J. A. Patz, B. Raffle, D. Rapport, P. Raven, C. Roberts, J. P. Rodriguez, A. B. Rylands, C. Tucker, C. Safina, C. Samper, M. L. J. Stiassny, J. Supriatna, D. H. Wall, and D. Wilcove. 2001. Can we defy nature's end? *Science* 293:2207–2208.

Pimm, S. L., and J. H. Lawton. 1998. Planning for biodiversity. *Science* 279:2068–2069.

Pineda, J. 1999. Circulation and larval distribution in internal tidal bore warm fronts. *Limnology and Oceanography* 44:1400–1414.

Plathong, S., G. J. Inglis, and M. E. Huber. 2000. Effects of self-guided snorkeling trails on corals in a tropical marine park. *Conservation Biology* 14:1821–1830.

Poff, N. L. 1997. Landscape filters and species traits: towards mechanistic understanding and prediction in stream ecology. *Journal of the North American Benthological Society* 16:391–409.

Poff, N. L., J. D. Allan, M. B. Bain, J. R. Karr, K. L. Prestegaard, B. D. Richter, R. E. Sparks, and J. C. Stromberg. 1997. The natural flow regime: a paradigm for river conservation and restoration. *BioScience* 47:769–784.

Poiani, K. A., J. V. Baumgartner, S. C. Buttrick, S. L. Green, E. Hopkins, G. D. Ivey, K. P. Sutton, and R. D. Sutter. 1998. A scale-independent site conservation planning framework in The Nature Conservancy. *Landscape and Urban Planning* 43:143–156.

Poiani, K A., B. D. Richter, M. G. Anderson, and H. E. Richter. 2000. Biodiversity conservation at multiple scales: functional sites, landscapes, and networks. *BioScience* 50:133–146.

Possingham, H., I. Ball, and S. Andelman. 2000. Mathematical mothods for identifying representative reserve networks. In S. Ferson and M. A. Burgman, eds., *Quantitative Methods in Conservation Biology*, pp. 291–306. New York, NY: Springer-Verlag.

Povilitis, T. 2002. What is a natural area? *Natural Areas Journal* 21:70–74.

Powell, G. V. N., J. Barborak, and M. Rodriquez S. 2000. Assessing representativeness of protected natural areas in Costa Rica for conserving biodiversity: a preliminary gap analysis. *Biological Conservation* 93:35–41.

Power, M. E., D. Tilman, J. A. Estes, B. A. Menge, W. J. Bond, L. S. Mills, G. Daily, J. C. Castilla, J. Lubchenco, and R. T. Paine. 1996. Challenges in the quest for keystones. *BioScience* 46:609–620.

Pregitzer, K.S., and B.V. Barnes. 1982. The use of ground flora to indicate edaphic factors in upland ecosystems of the McCormick Experimental Forest, Upper Michigan. *Canadian Journal of Forest Research* 12:661–672.

Prendergast, J. R., R. M. Quinn, and J. H. Lawton. 1999. The gaps between theory and practice in selecting nature reserves. *Conservation Biology* 13:484–492.

Prendergast, J. R., R. M. Quinn, J. H. Lawton, B. C. Eversham, and D. W. Gibbons. 1993. Rare species, the coincidence of diversity hotspots, and conservation strategies. *Nature* 365:335–337.

Press, D., D. F. Doak, and P. Steinberg. 1996. The role of local government in the conservation of rare species. *Conservation Biology* 10:1538–1548.

Pressey, R. L. 1994. *Ad Hoc* reservations: forward to backward steps in developing representative reserve systems? *Conservation Biology* 8:662–668.

———. 1997. Priority conservation areas: towards an operational definition for regional assessments. In J. J. Pigram and R. C. Sundell, eds., *National Parks and Protected Areas: Selection, Delimitation, and Management*, pp. 337–357. Armidale, AU: Centre for Water Policy Research, University of New England.

———. 1998. Algorithms, politics, and timber: an example of the role of science in a public, political negotiation process over new conservation areas in production forests. In R. Wills and R. Hobbs, eds., *Ecology for Everyone: Communicating Ecology to Scientists, the Public, and Politicians*, pp. 73–87. Sydney, AU: Surrey, Beatty, and Sons.

———. 1999. Systematic conservation planning for the real world. *Parks* 9:1–6.

Pressey, R. L., and M. Bedward. 1991. Mapping the environment at different scales: benefits and costs for nature conservation. In C. R. Margules and M. P. Austin, eds., *Nature Conservation: Cost Effective Biological Surveys and Data Analysis*, pp. 7–13. Melbourne, AU: Commonwealth Scientific and Industrial Research Organisation.

Pressey, R. L., S. Ferrier, T. C. Hager, C. A. Woods, S. L. Tully, and K. M Weinman. 1996a. How well protected are the forests of north–eastern New South Wales—analyses of forest environments in relation to formal protection measures, land tenure, and vulnerability to clearing. *Forest Ecology and Management* 85:311–333.

Pressey, R. L., S. Ferrier, C. D. Hutchinson, D. P. Sivertsen, and G. Manion. 1995. Planning for negotiation: using an interactive geographic information system to explore alternative protected area networks. In D. E. Saunders, J. L. Craig, and E. M. Mattiske, eds., *Nature Conservation 4: The Role of Networks*, pp. 23–33. Sydney, AU: Surrey, Beatty, and Sons.

Pressey, R. L., T. C. Hager, K. M. Ryan, J. Schwarz, S. Wall, S. Ferrier, and P. M. Creaser. 2000. Using abiotic data for conservation assessments over extensive regions: quantitative methods applied across New South Wales, Australia. *Biological Conservation* 96:55–82.

Pressey, R. L, C. J. Humphries, C. R. Margules, R. I. Vane-Wright, and P. H. Williams. 1993. Beyond opportunism: key principles for systematic reserve selection. *Trends in Ecology and Evolution* 8:124–128.

Pressey, R. L., I. R. Johnson, and P. D. Wilson. 1994. Shades of irreplaceability: towards a measure of the contribution of sites to a reservation goal. *Biodiversity and Conservation* 3:242–262.

Pressey, R. L., and V. S. Logan. 1994. Level of geographic subdivision and its effects on assessments of reserve coverage: a review of regional studies. *Conservation Biology* 8:1037–1046.

———. 1998. Size of selection units for future reserves and its influence on actual vs. targeted representation of features: a case study in western New South Wales. *Biological Conservation* 85:305–319.

Pressey, R. L., and A. O. Nicholls. 1989. Application of a numerical algorithm to the selection of reserves in semi-arid New South Wales. *Biological Conservation* 50:263–278.

Pressey, R. L., H. P. Possingham, and J. R. Day. 1997. Effectiveness of alternative heuristic algorithms for identifying indicative minimum requirements for conservation reserves. *Biological Conservation* 80:207–219.

Pressey, R. L., H. P. Possingham, V. S. Logan, J. R. Day, and P. H. Williams. 1999. Effects of data characteristics on the results of reserve selection algorithms. *Journal of Biogeography* 26:179–191.

Pressey, R. L., H. P. Possingham, and C. R. Margules. 1996b. Optimality in reserve selection algorithms: when does it matter and how much? *Biological Conservation* 76:259–267.

Pressey, R. L., and K. H. Taffs. 2001a. Sampling of land types by protected areas: three measures of effectiveness applied to western New South Wales. *Biological Conservation* 101:105–117.

———. 2001b. Scheduling conservation action in production landscapes: priority areas in western New South Wales defined by irreplaceability and vulnerability to vegetation loss. *Biological Conservation* 100:355–376.

Pressey, R. L., G. L. Whish, T. W. Barrett, and M. E. Watts. 2002. Effectiveness of protected areas in north-eastern New South Wales: recent trends in six measures. *Biological Conservation* 106:57–69.

Pringle, C. M. 2000. Managing riverine connectivity in complex landscapes to protect "remnant natural areas." Plenary lecture, *Verhandlungen Internationals Verein. Limnol.* 27:1149–1164.

Pritchard, D. W. 1967. What is an estuary? In G. H. Lauff, ed., *Estuaries*, pp. 3–5. Washington, DC: AAAS Press.

Pulliam, H. R. 1988. Sources, sinks, and population regulation. *American Naturalist* 132:652–661.

Quigley, T. M., R. W. Haynes, and R. T. Graham, eds. 1996. *Integrated Scientific Assessment for Ecosystem Management in the Interior Columbia Basin.* General Technical Report PNW-GTR-382, Pacific Northwest Research Station. Portland, OR: USDA Forest Service.

Ramsar Convention Bureau. 2001. *Ramsar Convention on Wetlands.* Gland, Switzerland: Ramsar Convention Bureau. Available online: www.ramsar.org.

Rao, M., and P. J. K. McGowan. 2002. Wild meat, food security, livelihoods, and conservation. *Conservation Biology* 16:580–583.

Rao, M., A. Rabinowitz, and S. T. Khaing. 2002. Status review of the protected-area system in Myanmar, with recommendations for conservation planning. *Conservation Biology* 16:360–368.

Rapport, D. J., R. Costanza, and A. J. McMichael. 1998. Assessing ecosystem health. *Trends in Ecology and Evolution* 13:397–402.

Ray, G. C., and B. P. Hayden. 1992. Coastal zone ecotones. In F. di Castri and A. J. Hansen, eds., *Landscape Boundaries: Consequences for Biodiversity and Ecological Flows.* New York, NY: Springer-Verlag.

Redford, K. H. 1992. The empty forest. *BioScience* 42:412–422.

————. 2000. Natural areas, hunting, and nature conservation in the Neotropics. *Wild Earth* 10:41–48.

Redford, K. H., P. Coppolillo, E. W. Sanderson, G. A. B. Da Fonseca, E. Dinerstein, C. Groves, G. Mace, S. Maginnis, R. A. Mittermeier, R. Noss, D. Olson, J. G. Robinson, A. Vedder, and M. Wright. In press. Mapping the conservation landscape. *Conservation Biology.*

Redford, K. H., and P. Feinsinger. 2001. The half-empty forest: sustainable use and the ecology of interactions. In *Conservation of Exploited Populations,* J. D. Reynolds, G. M. Mace, K. H. Redford, and J. G. Robinson, eds., pp. 371–399. Cambridge, UK: Cambridge University Press.

Redford, K. H., and B. D. Richter. 1999. Conservation of biodiversity in a world of use. *Conservation Biology* 13:1246–1256.

Redford, K. H., and S. E. Sanderson. 2000. Extracting humans from nature. *Conservation Biology* 14:1362–1364.

Reed, J. M., L. S. Mills, J. B. Dunning Jr., E. S. Menges, K. S. McKelvey, R. Frye, S. R. Beissinger, M.-C. Anstett, and P. Miller. 2002. Emerging issues in population viability analysis. *Conservation Biology* 16:7–19.

Regan, H. M., M. Colyvan, and M. A. Burgman. 2002. A taxonomy and treatment of uncertainty for ecology and conservation biology. *Ecological Applications* 12:618–628.

Reid, T. S., and D. D. Murphy. 1995. Providing a regional context for local conservation action. *BioScience Supplement* S:84–90.

Reid, W. V., and K. R. Miller. 1989. *Keeping Options Alive: The Scientific Basis for Conserving Biodiversity.* Washington, DC: World Resources Institute.

Revelle, R. R. D., and H. E. Suess. 1957. Carbon dioxide exchange between atmosphere and ocean and the question of an increase of atmospheric CO_2 during the past decades. *Tellus* 9:18–27.

Revenga, C., J. Brunner, N. Henninger, K. Kassem, and R. Payne 2000. *Pilot Analysis of Global Ecosystems: Freshwater Ecosystems.* Washington, DC: World Resources Institute.

Revenga, C., S. Murray, J. Abramovitz, and A. Hammond. 1998. *Watersheds of the World: Ecological Value and Vulnerability.* Washington, DC: World Resources Institute and Worldwatch Institute.

Reyers, B., D. H. K. Fairbanks, K. J. Wessels, and A. S. Van Jaarsveld. 2002. A multicriteria approach to reserve selection: addressing long-term biodiversity maintenance. *Biodiversity and Conservation* 11:769–793.

Ricciardi, A., and J. B. Rasmussen. 1999. Extinction rates of North American freshwater fauna. *Conservation Biology* 13:1220–1222.

Richards, R. A., and P. J. Rago. 1999. A case history of effective fishery management: Chesapeake Bay striped bass. *North American Journal of Fisheries Management* 19:356–375.

Richter, B. D., J. V. Baumgartner, R. Wigington, and D. P Braun. 1997b. How much water does a river need? *Freshwater Biology* 37:231–249.

Richter, B. D., J. V. Baumgartner, J. Powell, and D. P. Braun. 1996. A method for assessing hydrologic alteration within ecosystems. *Conservation Biology* 10:1163–1174.

Richter, B. D., D. P. Braun, M. A. Mendelson, and L. L. Master. 1997a. Threats to imperiled freshwater fauna. *Conservation Biology* 11:1081–1093.

Richter, B. D., R. Mathews, D. L. Harrison, and R. Wigington. In press. Ecologically sustainable water management: managing river flows for ecological integrity. *Ecological Applications.*

Ricketts, T., E. Dinerstein, D. M. Olson, C. J. Loucks, W. Eichbaum, D. DellaSala, K. Kavanagh, P. Hedao, P. Hurley, K. M. Carney, R. Abell, and S. Walters. 1999. *Terrestrial Ecoregions of North America: A Conservation Assessment.* Washington, DC: Island Press.

Roberts, C. M. 1997. Connectivity and management of Caribbean coral reefs. *Science* 278:1454–1457.

————. 1998. Sources, sinks, and the design of marine reserve networks. *Fisheries* 23:16–19.

Roberts, C. M., and J. P. Hawkins. 1999. Extinction risk in the sea. *Trends in Ecology and Evolution* 15:241–246.

————. 2000. *Fully Protected Marine Reserves: A Guide.* Washington, DC: World Wildlife Fund.

Roberts, C. M., C. J. McClean, J. E. N. Veron, J. P. Hawkins, G. R. Allen, D. E. McAllister, C. G. Mittermeier, F. W. Schueler, M. Spalding, F. Wells, C. Vynne, and T. B. Werner. 2002. Marine biodiversity hotspots and conservation priorities for tropical reefs. *Science* 295:1280–1284.

Rodrigues, A. S., and K. J. Gaston. 2002. Rarity and conservation planning across geopolitical units. *Conservation Biology* 16:674–682.

Rodrigues, A. S. L., R. Tratt, B. D. Wheeler, and K. J. Gaston. 1999. The performance of existing networks of conservation areas in representing biodiversity. *Proceedings of the Royal Society of London* 266:1453–1460.

Roff, J. C., and M. E. Taylor. 2000. National frameworks for marine conservation: a hierarchical geophysical approach. *Aquatic Conservation* 10:209–223.

Rosenberg, D. K., B. R. Noon, and E. C. Meslow. 1997. Biological corridors: form, function, and efficacy. *BioScience* 47:677–687.

Rosser, A. M. and S. A. Mainka. 2002. Overexploitation and species extinction. *Conservation Biology* 16:584–586.

Roth, N. E., J. D. Allan, and D. L. Erickson. 1996. Landscape influences on stream biotic integrity values assessed at multiple spatial scales. *Landscape Ecology* 113:141–156.

Rothley, K. D. 1999. Designing bioreserve networks to satisfy multiple, conflicting demands. *Ecological Applications* 9:741–750.

Rowe, J. S. 1984. Forestland classification: limitations of the use of vegetation. In J. G. Bockheim, ed., *Forest Land Classification: Experiences, Problems, Perspectives,* pp. 132–147. Proceedings of the Symposium, March 18–20, 1984. Madison, WI: University of Wisconsin.

Ruckelshaus, M., P. McElhany, and M. J. Ford. 2002. Recovering species of conservation concern—are populations expendable? In P. Kareiva and S. Levin, eds., *The Importance of Species,* pp. 305–329. Princeton, NJ: Princeton University Press.

Ruckelshaus, M., P. McElhany, M. McClure, and S. Heppell. In press. How many populations are needed for persistence of listed salmon species? Exploring the effects of spatially correlated catastrophes. In R. Ackakaya and M. Burgman, eds., *Species Conservation and Management: Case Studies.* New York, NY: Oxford University Press.

Saaty, T. L. 1994. *The Fundamentals of Decision Making and Priority Theory with the Analytic Hierarchy Process.* Volume VI, AHP Series, RWS Publications.

Sala, O. E., F. S. Chapin III, J. J. Armesto, E. Berlow, J. Bloomfield, R. Dirzo, E. Huber-Sannwald, L. Huenneke, R. B. Jackson, A. Kinzig, R. Leemans, D. M. Lodge, H. A. Mooney, M. Oesterheld, N. L. Poff, M. T. Sykes, B. H. Walker, M. Walker, and D. H. Wall. 2000. Biodiversity scenarios for the year 2100. *Science* 287:1770–1774.

Sala, O. E., F. S. Chapin III, R. H. Gardner, W. K. Lauenroth, H. A. Mooney, and P. S. Ramakrishnan. 1999. Global change, biodiversity and ecological complexity. In B. Walker, W. Steffen, J. Caadell, and J. Ingram, eds., *The Terrestrial Biosphere and Global Change: Implications for Natural and Managed Ecosystems*, pp. 304–328. Cambridge, UK: Cambridge University Press.

Salafsky, N., and R. Margoluis. 1999. Threat reduction assessment: a practical and cost-effective approach to evaluating conservation and development projects. *Conservation Biology* 13:830–841.

Salm, R. V., S. E. Smith, and G. Llewellyn. 2001. Mitigating the impact of coral bleaching through marine protected area design. In H. Z. Schuttenberg, ed., *Coral Bleaching: Causes, Consequences and Response*. Selected papers presented at the 9th International Coral Reef Symposium on "Coral Bleaching: Assessing and Linking Ecological and Socioeconomic Impacts, Future Trends and Mitigation Planning," pp. 81–88. Coastal Management Report #2230. Kingston, RI: Coastal Resources Center, University of Rhode Island.

Samson, F. B. 2002. Population viability analysis, management, and conservation planning at large scales. In S. R. Beissinger and D. R. McCullough, eds., *Population Viability Analysis*, pp. 425–441. Chicago, IL: University of Chicago Press.

Samson, F. B., F. Noon, F. Knopf, W. Ostlie, G. Plumb, S. Larson, and C. Sieg. 2000. *Terrestrial Assessment: A Broad-Scale Look at Species Viability on the Northern Great Plains*. Unpublished report, Chadron, NE: Nebraska National Forest, USDA Forest Service.

Sanderson, E. W., K. H. Redford, C. L. B. Chetkiewicz, R. A. Medellin, A. Rabinowitz, J. G. Robinson, and A. B. Tabor. 2002b. Planning to save a species: the jaguar as a model. *Conservation Biology* 16:58–72.

Sanderson, E. W., K. H. Redford, A. Vedder, P. B. Coppolillo, and S. E. Ward. 2002a. A conceptual model for conservation planning based on landscape species requirements. *Landscape and Urban Planning* 58:41–56.

Saunders, D. L., J. J. Meeuwig, and A. C. Vincent. 2002. Freshwater protected areas: strategies for conservation. *Conservation Biology* 16:30–41.

Saxon, E. 1983. Parks and reserves. In J. Messer and G. Moseley, eds., *What Future for Australia's Arid Lands*, pp. 165–168. Hawthorn, Victoria, AU: Australian Conservation Foundation.

Sayre, R., E. Roca, G. Sedaghatkish, B. Young, S. Keel, R. Roca, and S. Sheppard. 2000. *Nature in Focus: Rapid Ecological Assessment*. Washington, DC: The Nature Conservancy and Island Press.

Scheffer, M., S. Carpenter, J. A. Foley, C. Folke, and B. Walker. 2001. Catastrophic shifts in ecosystems. *Nature* 413:591–596.

Schlesinger, W. H., J. S. Clark, J. E. Mohan, and C. D. Reid 2001. Environmental change: effects on biodiversity. In M. E. Soulé and G. H. Orians, eds., *Conservation Biology: Research Priorities in the Next Decade*, pp. 175–223. Washington, DC: Island Press.

Schlosser, I. J. 1991. Stream fish ecology: a landscape perspective. *Bioscience* 41:704–712.

Schneider, S. H., and R. Londer. 1984. *The Coevolution of Climate and Life*. San Francisco, CA: Sierra Club Books.

Schumaker, N. H. 1998. *A User's Guide to the PATCH Model*. Report No. EPA/600/R-98/135, Corvallis, OR: Environmental Research Laboratory, Environmental Protection Agency.

Schupp, D. H. 1992. *An Ecological Classification f Minnesota Lakes with Associated Fish Communities*. Investigational Report 417. St. Paul, MN: Minnesota Department of Natural Resources.

Schwartz, M. W. 1999. Choosing the appropriate scale of reserves for conservation. *Annual Review of Ecology and Systematics* 30:83–108.

Schwartzman, S., A. Moreira, and D. Nepstad. 2000. Rethinking tropical forest conservation: perils in parks. *Conservation Biology* 14:1351–1357.

Scott, J. M. 1999. A representative biological reserve system for the United States. *Society for Conservation Biology Newsletter* 6(2).

Scott, J. M., R. J. F. Abbitt, and C. R. Groves. 2001b. What are we protecting? The United States conservation portfolio. *Conservation Biology in Practice* 2:18–19.

Scott, J. M., and B. Csuti. 1997. Noah worked two jobs. *Conservation Biology* 11:1255–1257.

Scott, J. M., B. Csuti, J. D. Jacobi, and J. E. Estes. 1987. Species richness: a geographic approach to protecting future biological diversity. *BioScience* 37:782–788.

Scott, J. M., F. Davis, B. Csuti, R. Noss, B. Butterfield, C. Groves, H. Anderson, S. Caicco, F. D'erchia, T. Edwards, J. Ulliman, and R. G. Wright. 1993. Gap analysis: a geographic approach to protection of biological diversity. *Wildlife Monographs* 123.

Scott, J. M., F. W. Davis, G. McGhie, and C. Groves. 2001a. Nature reserves: do they capture the full range of America's biological diversity? *Ecological Applications* 11:999–1007.

Scott, J. M., P. J. Heglund, M. L. Morrison, J. B. Haufler, M. G. Raphael, W. A. Wall, and F. B. Samson, eds. 2002. *Predicting Species Occurrences: Issues of Accuracy and Scale.* Washington, DC: Island Press.

Scott, T. A., and J. E. Sullivan. 2000. The selection and design of multiple-species habitat preserves. *Environmental Management* 26: S37–S53.

Scott, W. B., and E. J. Crossman. 1973. *Freshwater Fishes of Canada.* Bulletin No. 184. Ottawa, Canada: Fisheries Research Board of Canada.

Seaber, P. R., F. P. Kapinos, and G. L. Knapp. 1987. Hydrologic Unit Maps: U.S. Geological Survey Water-Supply Paper 2294. Reston, VA: U.S. Geological Survey.

Seabloom, E. W., A. P. Dobson, and D. M. Stoms. In press. Extinction rates under nonrandom patterns of habitat loss. *Proceedings of the National Academy of Science.*

Secretariat of the Convention on Biological Diversity.. 2000. *Sustaining Life on Earth: How the Convention on Biological Diversity Promotes Nature and Human Well-Being.* United Nations Environment Programme. Available online: www.cbd.org.

———. 2001. *The Handbook on the Convention of Biological Diversity.* London, UK: Earthscan Publications.

Seelbach, P. S., M. J. Wiley, J. C. Kotanchik, and M. E. Baker. 1997. *A Landscape-Based Ecological Classification System for River Valley Segments in Lower Michigan* (MI-VSEC version 1.0). Research Report Number 2036. Lansing, MI: State of Michigan Department of Natural Resources, Fisheries Division.

Seymour, R., and M. Hunter. 1999. Principles of ecological foresty. In M. L. Hunter Jr. ed., *Maintaining Biodiversity in Forest Ecosystems,* pp. 22–64. Cambridge, UK: Cambridge University Press.

Shafer, C. L. 1995. Values and shortcomings of small reserves. *BioScience* 45:80–88.

———. 1999a. History of selection and system planning for U.S. natural area national parks and monuments: beauty and biology. *Biodiversity and Conservation* 8:189–204.

———. 1999b. U.S. National Park buffer zones: historical, scientific, social, and legal aspects. *Environmental Management* 23:49–73.

———. 1999c. National park and reserve planning to protect biological diversity: some basic elements. *Landscape and Urban Planning* 44:123–153.

———. 2001. Inter-reserve distance. *Biological Conservation* 100:215–227

Shaffer, M. L. 1981. Minimum population sizes for species conservation. *BioScience* 31:131–134.

Shaffer, M. L., and B. L. Stein. 2000. Safeguarding our precious heritage. In B. A. Stein, L. S. Kutner, and J. S. Adams, eds., *Precious Heritage: The Status of Biodiversity in the United States,* pp. 301–322. Oxford, UK: Oxford University Press.

Shaffer, M. L., J. M. Scott, and F. Casey. 2002. Noah's options: initial cost estimates of a national system of habitat conservation areas in the United States. *BioScience* 52:439–443.

Shaklee, J., and P. Bentzen. 1998. Genetic identification of stocks of marine fish and shellfish. *Bulletin of Marine Science* 62:589–621.

Shemske, D. W., B. C. Husband, M. H. Ruckelhaus, C. Goodwillie, I. M. Parker, and J. G. Bishop. 1994. Evaluating approaches to the conservation of rare and endangered plants. *Ecology* 75:584–606.

Shrader-Frechette, K. S., and E. D. McCoy. 1993. *Method in Ecology: Strategies for Conservation.* Cambridge, UK: Cambridge University Press.

Shugart, H. H., and D. C. West. 1981. Long term dynamics of forest ecosystems. *American Scientist* 69:647–652.

Simberloff, D. 1997. Flagships, umbrellas, and keystones: is single species management passé in the landscape era? *Biological Conservation* 83:247–257.

Simberloff, D., J. A. Farr, J. Cox, and D. W. Mehlman. 1992. Movement corridors: conservation bargains or poor investments? *Conservation Biology* 6:493–505.

Simberloff, D. J., D. Doak, M. Groom, S. Trombulak, A. Dobson, S. Gatewood, M. E. Soulé, M. Gilpin, C. Martínez del Rio, and L. Mills. 1999. Regional and continental restoration. In M. E. Soulé and J. Terborgh, eds., *Continental Conservation: Scientific Foundations of Regional Reserve Networks.* Washington, DC: Island Press and The Wildlands Project.

Simon, T. P., ed. 1999. *Assessing the Sustainability and Biological Integrity of Water Resources Using Fish Communities.* Boca Raton, FL: CRC Press.

Slowinski, R., ed. 1998. *Fuzzy Sets in Decision Analysis, Operations Research, and Statistics.* Amsterdam, Netherlands: Kluwer Academic Publishers.

Smart, J. M., A. T. Knight, and M. Robinson. 2000. *A Conservation Assessment for the Cobar Peneplain Biogeographic Region—Methods and Opportunities.* Hurtsville, AU: New South Wales National Parks and Wildlife Service.

Smit, B., and O. Pilifosova. 2001. Adaptation to climate change in the context of sustainable development and equity. In J. J. McCarthy, O. F. Canziani, N. A. Leary, D. J. Dokken, and K. S. White, eds., *Climate Change 2001: Impacts, Adaptation, and Vulnerability,* pp. 879–912. Cambridge, UK: Cambridge University Press.

Smith, P. G. R., and J. B. Theberge. 1986. A review of criteria for evaluating natural areas. *Environmental Management* 10:715–734.

Smith, R. K., P. L. Freeman, J. V. Higgins, K. S. Wheaton, T. W. FitzHugh, A. A. Das, and K. J. Ernstrom. 2002. *Priority Areas for Freshwater Conservation Action: A Biodiversity Assessment of the Southeastern United States.* Arlington, VA: The Nature Conservancy.

Smith, T. B., S. Kark, C. J. Schneider, R. K. Wayne, and C. Moritz. 2001. Biodiversity hotspots and beyond: the need for preserving environmental transitions. *Trends in Ecology and Evolution* 16:431.

Smith, T. B., M. W. Bruford, and R. K. Wayne. 1993. The preservation of process: the missing element of conservation programs. *Biodiversity Letters* 1:164–167.

Society for Ecological Restoration. 2002. *The SER Primer on Ecological Restoration.* Available online: http://www.ser.org/reading.php?pg=primer2.

Soulé, M. E. 1980.Thresholds for survival: maintaining fitness and evolutionary potential. In *Conservation Biology: An Evolutionary–Ecological Perspective*, eds. M. E. Soulé and B. A. Wilcox, pp. 151–170, Sunderland, MA: Sinaeuer Associates, Inc.

———. 1987.Where do we go from here? In M. E. Soulé, ed., *Viable Populations for Conservation*, pp. 175–183. Cambridge, UK: Cambridge University Press.

———. 1996. Are ecosystem processes enough? *Wild Earth* 6:59–60.

Soulé, M. E., and M. A. Sanjayan. 1998. Conservation targets: do they help? *Science* 279:2060–2061.

Soulé, M. E., and D. Simberloff. 1986. What do genetics and ecology tell us about the design of nature reserves? *Biological Conservation* 35:19–40.

Soulé, M. E., and J. Terborgh. 1999a. Conserving nature at regional and continental scales—a scientific program for North America. *BioScience* 49:808–818.

———. eds. 1999b. *Continental Conservation: Scientific Foundations of Regional Reserve Networks*. Washington, DC: Island Press.

Southern Appalachians Man and the Biosphere Cooperative. 1996. *Southern Appalachians Assessment Summary Report*. Available online: www.sunsite.utk.edu/samab/saa.

Southern Rocky Mountains Ecoregional Assessment Team. 2001. *Southern Rocky Mountains: An Ecoregional Assessment and Conservation Blueprint*. Arlington, VA: The Nature Conservancy.

Spellerburg, I. F. 1992. *Evaluation and Assessment for Conservation: Ecological Guidelines for Determining Priorities for Nature Conservation*. London, UK: Chapman and Hall.

Spencer, C. N., B. R. McClelland, and J. A. Stanford. 1991. Shrimp stocking, salmon collapse, and eagle displacement. *BioScience* 41:14–21

Sprugel, D. G. 1991. Disturbance, equilibrium, and environmental variability: what is natural vegetation in a changing environment? *Biological Conservation* 58:1–18.

Stanfield, L.W., and R. Kuyvenhoven. 2002. Protocol for applications used in the Aquatic Landscape Inventory System: an application for delineating, characterizing, and classifying valley segments. Glenora, Canada: Ontario Ministry of Natural Resources.

Stanford, J. A. 1994. *Instream Flows to Assist the Recovery of Endangered Fishes of the Upper Colorado River Basin*. Biological Report Number 24. Washington, DC: USDI National Biological Survey.

Stattersfield, A. J., M. J. Crosby, A. J. Long, and D. C. Wege. 1998. *Endemic Bird Areas of the World: Priorities for Biodiversity Conservation*. Cambridge, UK: Birdlife International.

Stein, B., and F. Davis. 2000. Discovering life in America: tools and techniques of biodiversity inventory. In B. A. Stein, L. S. Kutner, and J. S. Adams, eds., *Precious Heritage: The Status of Biodiversity in the United States,* pp. 19–53. Oxford, UK: Oxford University Press.

Stein, B. A., J. S. Adams, L. L. Master, L. E. Morse, and G. A. Hammerson. 2000. A remarkable array: species diversity in the United States. In B. A. Stein, L. S. Kutner, and J. S. Adams, eds., *Precious Heritage: The Status of Biodiversity in the United States,* pp. 55–92. Oxford, UK: Oxford University Press.

Steiner, F. R. 2000. *The Living Landscape: An Ecological Approach to Landscape Planning*. 2nd ed. NewYork: McGraw-Hill.

Steinitz, C., C. Adams, L. Alexander, J. DeNormandie, R. Durant, L. Eberhart, J. Felkner, K. Hickey, A. Mellinger, R. Narita, T. Slattery, C. Viellard, Y. Wang, and E. M. Wright. 1997. *An Alternative Future for the Region of Camp Pendleton, California*. Cambridge, MA: Harvard University, Graduate School of Design.

Stephenson, J. R., and G. M. Calcarone. 1999. *Southern California Mountains and Foothills Assessment: Habitat and Species Conservation Issues.* General Technical Report GTR-PSW-172, Pacific Southwest Research Station. Albany, CA: USDA Forest Service.

Stine, P. A., F. W. Davis, B. Csuti, and J. M. Scott. 1996. Comparative utility of vegetation maps of different resolutions for conservation planning. In R. C. Szaro and D. W. Johnston, eds., *Biodiversity in Managed Landscapes,* pp. 210–220. New York, NY: Oxford University Press.

Stolton, S., M. Hockings, and N. Dudley. 2002. *Reporting Progress on Management Effectiveness in Protected Areas: A Simple Site-Level Tracking Tool Developed for the World Bank and WWF.* Gland, Switzerland: World Wildlife Fund International.

Stoms, D. M. 1992. Effects of habitat map generalization in biodiversity assessment. *Photogrammetric Engineering and Remote Sensing* 58:1587–1591.

———. 1994. Scale dependence of species richness maps. *Professional Geographer* 46:346–358.

———. 2000. GAP management status and regional indicators of threats to biodiversity. *Landscape Ecology* 15:21–33.

Stoms, D. M., F. W. Davis, K. L. Direse, K. M. Cassidy, and M. P. Murray. 1998. Gap analysis of the vegetation communities of the Intermountain Semi-Desert Ecoregion. *Great Basin Naturalist* 58:199–216.

Striplen, C., and S. DeWeerdt. 2002. Old science, new science: incorporating traditional ecological knowledge in contemporary management. *Conservation Biology in Practice* 3:20–27.

Strittholt, J. R., and D. A. Dellasala. 2001. Importance of roadless areas in biodiversity conservation in forested ecosystems: case study of the Klamath–Siskiyou Ecoregion of the United States. *Conservation Biology* 15:1742–1754.

Strittholt, J. R., R. F. Noss, P. A. Frost, K. Vance-Borland, C. Carroll, and G. Heilman. 1999. *A Science-Based Conservation Assessment of the Klamath–Siskiyou Ecoregion.* Corvallis, OR: Earth Design Consultants and the Conservation Biology Institute.

Sugden, A., and E. Pennisi. 2000. Diversity digitized. *Science* 29:2305.

Sullivan Sealey, K. M., and G. Bustamante. 1999. *Setting Geographic Priorities for Marine Conservation in Latin America and the Caribbean.* Arlington, VA: The Nature Conservancy.

Sutherland, W. J. 2000. *The Conservation Handbook: Research, Management, and Policy.* Oxford, UK: Blackwell Science, Ltd.

Swearer, S. E., J. E. Caselle, D. W. Lea, and R. R. Warner. 1999. Larval retention and recruitment in an island population of a coral-reef fish. *Nature* 402:799–804.

Swetnam, T. W., C. D. Allen, and J. L. Betancourt. 1999. Applied historical ecology: using the past to manage for the future. *Ecological Applications* 9:1189–1206.

Szaro, R. C., and D. W. Johnston, eds. 1996. *Biodiversity in Managed Landscapes: Theory and Practice.* New York, NY: Oxford University Press.

Tarlock, A. D. 1993. Local government protection of biodiversity: what is its niche? *University of Chicago Law Review* 60:555.

Tear, T. H., J. M. Scott, P. H. Hayward, and B. Griffith. 1993. Status and prospects for success of the Endangered Species Act: a look at recovery plans. *Science* 262:976–977.

———. 1995. Recovery plans and the Endangered Species Act: are criticisms supported by data? *Conservation Biology* 9:182–195.

Terborgh, J. 1976. Island biogeography and conservation: strategy and limitations. *Science* 193:1029–1030.

———. 2000. The fate of tropical forests. *Conservation Biology* 14:1358–1361.

Thackway, R., and I. D. Cresswell. 1995. *An Interim Biogeographic Regionalization for Australia: A Framework for Setting Priorities in the National Reserves System Cooperative Program.* Canberra, AU: Australian Nature Conservation Agency.

———. 1997. A bioregional framework for planning the national system of protected areas in Australia. *Natural Areas Journal* 17:241–247.

Theberge, J. B. 1989. Guidelines to drawing ecologically sound boundaries for national parks and nature reserves. *Environmental Management* 13:695–702.

The Nature Conservancy. 1997. *Designing a Geography of Hope: Guidelines for Ecoregion-Based Conservation in The Nature Conservancy.* Arlington, VA: The Nature Conservancy.

———. 1998. *Ecoregional Planning in the Northern Tallgrass Prairie.* Arlington, VA: The Nature Conservancy. Available online: www.conserveonline.org.

———. 2000a. *Great Lakes Ecoregional Plan: A First Iteration.* Arlington, VA: The Nature Conservancy.

———. 2000b. *The Five-S Framework for Site Conservation: A Practitioner's Handbook for Site Conservation Planning and Measuring Conservation Success.* Arlington, VA: The Nature Conservancy. Available online: www.conserveonline.org.

———. 2000c. *Tools for GIS Analysis.* Arlington, VA: The Nature Conservancy. Available online: www.freshwaters.org.

———. 2001. *Conservation by Design: A Framework for Mission Success.* Arlington, VA: The Nature Conservancy.

The Nature Conservancy of China. 2002. *Yunnan Great Rivers Project: Northwestern Yunnan Ecoregional Conservation Plan.* Arlington VA: The Nature Conservancy.

Theobald, D. 2000. Fragmentation by inholdings and exurban development. In R. L. Knight, F. W. Smith, S. W. Buskirk, W. H. Romme, and W. L. Baker, eds., *Forest Fragmentation in the Southern Rocky Mountains,* pp. 155–174. Boulder, CO: University Press of Colorado.

Theobald, D. M., N. T. Hobbs, T. Bearly, J. A. Zack, T. Shenk, and W. E. Riebsame. 2000. Incorporating biological information in local land-use decision making: designing a system for conservation planning. *Landscape Ecology* 15:35–45.

Thomas, J. W., E. D. Forsman, J. B. Lint, E. C. Meslow, B. R. Noon, and J. Verner. 1990. *A Conservation Strategy for the Northern Spotted Owl: Report of the Interagency Scientific Committee to Address the Conservation of the Northern Spotted Owl.* Portland, OR: USDA Forest Service, USDI Bureau of Land Management, Fish and Wildlife Service, National Park Service.

Thomsett, M. C. 1990. *The Little Black Book of Project Management.* New York, NY: American Management Association.

Tilman, D., R. M. May, C. H. Lehman, and M. A. Nowak. 1994. Habitat destruction and the extinction debt. *Nature* 371:65–66.

Titus, J. G. 1998. Rising seas, coastal erosion, and the takings clause: how to save wetlands and beaches without hurting property owners. *Maryland Law Review* 57:1279–1399.

Tonn, W. M. 1990. Climate change and fish communities: a conceptual framework. *Transactions of the American Fisheries Society* 119:337–352.

Toth, L. A. 1995. Principles and guidelines for restoration of river/floodplain ecosystems—Kissimmee River, Florida. In J. Cairns, ed., *Rehabilitating Damaged Ecosystems,* pp. 49–73. Boca Raton, FL: Lewis Publishers/CRC Press.

Touval, J. 2000. *The Nature Conservancy Colombia Country Program Five-Year Strategy: 2000–2005.* Arlington, VA: The Nature Conservancy.

Tracy, C. R., and P. F. Brussard. 1994. Preserving biodiversity: species in landscapes. *Ecological Applications* 4:205–207.

Trautman, M. B. 1981. *The Fishes of Ohio*. Columbus, OH: Ohio State University Press.

Trimble, S. W. 2001. Geomorphology, hydrology, and soils. In D. Egan and E. A. Howell, eds., *The Historical Ecology Handbook: A Restorationist's Guide to Reference Ecosystems*, pp. 317–334. Washington, DC: Island Press.

Trombulak, S. C., and C. A. Frissell. 2000. Review of ecological effects of roads on terrestrial and aquatic communities. *Conservation Biology* 14:18–30.

Turner, I. M., and R. T. Corlett. 1996. The conservation value of small, isolated fragments of lowland tropical rain forest. *Trends in Ecology and Evolution* 11:330–333.

Turpie, J. K., L. E. Beckley, and S. M. Katua. 2000. Biogeography and the selection of priority areas for conservation of South African coastal fishes. *Biological Conservation* 92:59–72.

Twilley, R. R. 1998. Mangrove Wetlands. In M. G. Messina and W. H. Conner, eds., *Southern Forested Wetlands: Ecology and Management*. Boca Raton, FL: Lewis Publishers.

Udvardy, M. D. F. 1975. *A Classification of the Biogeographical Provinces of the World*. IUCN Occasional Paper No. 18, Morges, Switzerland: International Union for the Conservation of Nature and Natural Resources.

United Nations Environment Programme. 1995. *Global Biodiversity Assessment*. Cambridge, UK: Cambridge University Press.

———. 1999. *GEO-2000: UNEP's Millennium Report on the Environment*. London, UK: Earthscan.

United Nations Framework Convention on Climate Change (UNFCCC). 1992. Available online: http://unfccc.int.

UNESCO (United Nations Educational, Scientific and Cultural Organization). 1973. *International Classification and Mapping of Vegetation, Series 6, Ecology and Conservation*. Paris: UNESCO.

Urban, D. L., R. V. O'Neill, and H. H. Shugart. 1987. Landscape ecology. *BioScience* 37:119–127.

U.S. Army Corps of Engineers. 2000. *President signs Everglades restoration bill*. U.S. Army Corps of Engineers. Available online: www.saj.usace.army.mil.

U.S. Fish and Wildlife Service. 2001. Oregon Silverspot Butterfly (*Speyeriua zerene hippolyta*) Revised Recovery Plan. Portland, OR: U.S. Fish and Wildlife Service.

U.S.G.S. (United States Geological Survey). 1982. Hydrologic unit map of the United States. Washington, DC: U.S. Government Printing Office.

———. 2001. Land processes distributed active archive center. Available online: http://edcdaac.usgs.gov/gtopo30/gtopo30.html.

Usher, M. B. 1986. Wildlife conservation evaluation: attributes, criteria, and values. In M. B. Usher, ed., *Wildlife Conservation Evaluation*, pp. 3–44. London, UK: Chapman and Hall.

Valles, H. 2001. Larval supply to a marine reserve and adjacent fished area in the Soufriere Marine Management Area, St. Lucia, West Indies. *Journal of Fish Biology* 59:152–177.

Valutis, L., and R. Mullen. 2000. The Nature Conservancy's approach to prioritizing conservation action. *Environmental Science and Policy* 3:341–346.

Van Jaarsveld, A. S., S. Frietag, S. L. Chown, C. Muller, S. Koch, H. Hull, C. Bellamy, M. Kruger, S. Endrody-Younga, M. W. Mansell, and C. H. Scholtz. 1998. Biodiversity assessment and conservation strategies. *Science* 279:2106–2108.

Vardon, M. J., C. Missi, M. Cleary, and G. J. W. Webb. 1997. Aboriginal use and conservation of wildlife in northern Australia: a cultural necessity. In P. Hale and D. Lamb, eds.,
Conservation Outside Nature Reserves, pp. 241–245. Brisbane, Queensland, AU: Centre
for Conservation Biology, University of Queensland.

Vitousek, P. M., H. A. Mooney, J. Lubchenco, and J. M. Melillo. 1997. Human domination
of Earth's ecosystems. *Science* 277:494–499.

Walker, B. H., W. L. Steffen and J. Langridge. 1999. Interactive and integrated effects of
global change on terrestrial ecosystems. In B. H. Walker, W. Steffen, J. Canadell, and J.
Ingram, eds., *The Terrestrial Biosphere and Global Change,* pp. 329–375. Cambridge, UK:
Cambridge University Press.

Walters, C. J., and C. S. Holling. 1990. Large-scale management experiments and learning
by doing. *Ecology* 71:2060–2068.

Waples, R. S. 2002. Definition and estimation of effective population size in the conservation of endangered species. In S. R. Beissinger and D. R. McCullough, eds., *Population Viability Analysis,* pp.147–168. Chicago, IL: University of Chicago Press.

Ward, T. J., M. A. Vanderklift, A. O. Nicholls, and R. A. Kenchington. 1999. Selecting
marine reserves using habitats and species assemblages as surrogates for biological
diversity. *Ecological Applications* 9:691–698.

Ward, J. V., and J. A. Stanford. 1989a. Riverine ecosystems: the influence of man on catch-
ment dynamics and fish ecology. In D. P. Dodge, ed., *Proceedings of the International Large
River Symposium,* pp. 56–64. Canadian Special Publications in Fisheries and Aquatic
Sciences 106.

Ward, J. V., and Stanford, J. A. 1989b. The four-dimensional nature of lotic ecosystems.
Journal of North American Benthological Society 8:2–8.

Water Research Commission. 2001. State of Rivers Report. Letaba and Luvuvhu River
Systems. Report no. TT 165/01. Pretoria, South Africa: Water Research Commission.

Watling, L., and E. A. Norse. 1998. Disturbance of the seabed by mobile fishing gear: a
comparison to forest clearcutting. *Conservation Biology* 12:1180–1197.

Weber, T., and J. Wolf. 2000. Maryland's green infrastructure—using landscape assessment
tools to identify a regional conservation strategy. *Environmental Monitoring and Assessment* 63:265–277.

Wessels, K. J., S. Frietag, and A. S. van Jaarsveld. 1999. The use of land facets as biodiversity surrogates during reserve selection at a local scale. *Biological Conservation* 89:21–38.

White, D., P. G. Minotti, M. J. Barczak, J. C. Sifneos, K. E. Freemark, M. V. Santelmann,
C. F. Steinitz, A. R. Kiester, and E. M. Preston. 1997. Assessing risks to biodiversity
from future landscape change. *Conservation Biology* 11:349–360.

White, P. S., J. Harrod, W. H. Romme, and J. Betancourt. 1999. Disturbance and temporal
dynamics. In R. C. Szaro, N. C. Johnson, W. T. Sexton, and A. J. Malk, eds., *Ecological
Stewardship: A Common Reference for Ecosystem Management,* pp. 281–312. Oxford, UK:
Elsevier Science, Ltd.

Whitney, G. G., and J. DeCant. 2001. Government land office surveys and other early
surveys. In D. Egan and E. A. Howell, eds., *The Historical Ecology Handbook: A Restorationist's Guide to Reference Ecosystems,* pp. 147–176. Washington, DC: Island Press.

Whittaker, R. H. 1975. *Communities and Ecosystems.* 2nd ed. New York, NY: Macmillan
Publishing Co.

Wieland, R. G. 1993. *Marine and Estuarine Habitat Types and Associated Ecological Communities of Mississippi Coast.* Jackson, MS: Mississippi Department of Wildlife, Fisheries,
and Parks.

Wiens, J. A. 1989. Spatial scaling in ecology. *Functional Ecology* 3:385–397.

———. 1996. Wildlife in patchy environments: metapopulations, mosaics, and management. In D. R. McCullough, ed., *Metapopulations and Wildlife Conservation,* pp. 53–84. Washington, DC: Island Press.

———. 1997. The emerging role of patchiness in conservation biology. In S. T. A. Pickett, R. S. Ostfeld, M. Shacak, and G. E. Likens, eds., *The Ecological Basis of Conservation: Heterogeneity, Ecosystems, and Biodiversity,* pp. 93–107. New York, NY: Chapman and Hall.

———. 2002. Predicting species occurrences: progress, problems, and prospects. In J. M. Scott, P. J. Heglund, M. L. Morrison Jr., J. B. Haufler, M. G. Raphael, W. A. Wall, and F. B. Samson, eds., *Predicting Species Occurrences: Issues of Accuracy and Scale,* pp. 739–749. Washington, DC: Island Press.

Wiken, E.B., compiler. 1986. *Terrestrial Ecozones of Canada.* Ecological Land Classification Series No. 19. Ottawa, Canada: Environment Canada, Lands Directorate.

Wikramanayake, E. D., C. Carpenter, H. Strand, and M. McKnight. 2001. *Ecoregion-Based Conservation in the Eastern Himalaya: Identifying Important Areas for Biodiversity Conservation.* Washington, DC: World Wildlife Fund.

Wikramanayake, E. D., E. Dinerstein, J. G. Robinson, U. Karanth, A. Rabinowitz, D. Olson, T. Mathew, P. Hedao. M. Conner, G. Hemley, and D. Bolze. 1998. An ecology-based method for defining priorities for large mammal conservation: the tiger as case study. *Conservation Biology* 12:865–878.

Wilcove, D. 1993. Getting ahead of the extinction curve. *Ecological Applications* 3:218–220.

Wilcove, D. S., D. Rothstein, J. Dubow, A. Phillips, and E. Losos. 1998. Quantifying threats to imperiled species in the United States. *BioScience* 48:607–615.

Wilkie, D. S., and J. T. Finn. 1996. *Remote Sensing Imagery for Natural Resources Monitoring: A Guide for First-Time Users.* New York, NY: Columbia University Press.

Williams, P., D. Gibbons, C. Margules, A. Rebelo, C. Humphries, and R. Pressey. 1996. A comparison of richness hotspots, rarity hotspots, and complementary areas for conserving diversity of British birds. *Conservation Biology* 10:155–174.

Williams, P. H. 1996. *WORLDMAP4WINDOWS: Software and Help Document.* London, UK: Privately distributed.

Williams, P. H. 1998. Key sites for conservation: area-selection methods for biodiversity. In G. M Mace, A. Balmford, and J. R. Ginsberg, eds., *Conservation in a Changing World,* pp. 211–250. Cambridge, UK: Cambridge University Press.

Willson, M. F., S. M. Gende, and B. H. Marston. 1998. Fishes and the forest: expanding perspectives on fish-wildlife interactions. *Bioscience* 48:455–462.

Willson, M. F., and K. C. Halupka. 1995. Anadramous fish as keystone species in vertebrate communities. *Conservation Biology* 9:489–497.

Wilson, D. E., F. R. Cole, J. D. Nichols, R. Rudran, and M. S. Foster, eds. 1996. *Measuring and Monitoring Biological Diversity: Standard Methods for Mammals.* Washington, DC: Smithsonian Institution Press.

Wilson, E. O., ed. 1988. *Biodiversity.* Washington, DC: National Academy Press.

———. 1992. *The Diversity of Life.* Cambridge, UK: Belknap Press of Harvard University Press.

Wing, S. R., and E. S. Wing. 2001. Prehistoric fisheries in the Caribbean. *Coral Reefs* 20:1–8.

Wolff, W. J. 2000. Causes of extirpations in the Wadden Sea, an estuarine area in the Netherlands. *Conservation Biology* 14:876–885.

Woodroffe, R., J. Ginsberg, D. Macdonald, and IUCN/SSC Canid Specialist Group. 1997. *The African Wild Dog: Status Survey and Conservation Action Plan.* Available online: www.iucn.org/themes/ssc/sgs/sgs.html.

World Commission on Environment and Development. 1987. *Our Common Future.* New York, NY: Oxford University Press.

World Commission on Protected Areas. 2001. *WCPA Draft Strategic Plan.* Gland, Switzerland: IUCN Program on Protected Areas. Available online: http://wcpa.iucn.org/.

World Conservation Monitoring Centre. 1992. *Global Biodiversity: Status of the Earth's Living Resources.* London, UK: Chapman and Hall.

———. 2001. *Freshwater Biodiversity: A Preliminary Global Assessment.* WCMC Biodiversity Series No. 8. Available online: www.unep-wcmc.org/information_services/publications/freshwater/toc.htm.

World Conservation Union. 1994. *Guidelines for Protected Area Management Categories.* Gland Switzerland: IUCN–World Conservation Union.

World Resources Institute. 2001. *Global Forest Watch: A World Resources Initiative.* Washington, DC: World Resources Institute. Available online: www.igc.org/wri/gfw.

World Resources Institute (WRI), The World Conservation Union (IUCN), and the United Nations Environment Programme (UNEP). 1992. *Global Biodiversity Strategy: Guidelines for Action to Save, Study, and Use Earth's Biotic Wealth Sustainably and Equitably.* Washington, DC: WRI, IUCN, and UNEP.

World Wildlife Fund. 2000. *Stakeholder Collaboration: Building Bridges for Conservation.* Washington, DC: Ecoregional Conservation Strategies Unit, World Wildlife Fund.

World Wildlife Fund–United States. 1997. *A Report on the Chihuahuan Desert Conservation Priority-Setting Workshop.* Washington, DC: World Wildlife Fund.

Wright, R. G., M. M. Murray, and T. Merrill. 1998. Ecoregions as a level of ecological analysis. *Biological Conservation* 86:207–213.

Young, T. P. 2000. Restoration ecology and conservation biology. *Biological Conservation* 92:72–83.

Zacharias, M. A., D. E. Howes, J. R. Harper, and P. Wainwright. 1998. The British Columbia marine ecosystem classification: rationale, development, and verification. *Coastal Management* 26:105–124.

Zacharias, M. A., and J.C. Roff. 2000. A hierarchical ecological approach to conserving marine biodiversity. *Conservation Biology* 14:1327–1334.

Zedler, J. B. 2000. Progress in wetland restoration ecology. *Trends in Ecology and Evolution* 15:402–407.

Zonneveld, I. S. 1989. The land unit—a fundamental concept in landscape ecology, and its application. *Landscape Ecology* 3:67–86.

Index

Abiotic units, 109–11, 167, 347–51, *figs. 12.1–2*
Aboriginal groups, 173
Accountability, 44–45, *fig. 3.1*
Accuracy assessments, 142–43
Action Plans (of SSC), 153–54
Adaptive management, 175
Adequacy, 226–27, 228
Ad hoc conservation efforts, 81, 217
Aggregation sites, 330–31
AHP (Analytic Hierarchy Process), 284–86, *fig. 9.3*
Algorithms for site or area selection, 229–37, 239–40; freshwater conservation and, 315–16; marine conservation and, 336–37
Alliances, 59–60
Alpha diversity, 9–10, *fig. 1.2*
Alternative goals, 173–74
American eels (*Anguilla rostrata*), 304
American swallow-tailed kite (*Elanoides forficatus*), 394
AntiEnvironment Movement, 58, 279–80
Aquatic ecosystems. *See* Freshwater biodiversity; Marine environments
Aquatic Gap Program, 301
Arctic Coast Ecoregion, 352–53, 357
Arctic National Wildlife Refuge, 219
ArcView, 307
Area-limited species, 94, 95
Area selection. *See* Selection of conservation areas
Area-sensitive species, 198
Assessments, 69–70, 133–45. *See also* threats assessment

Association of Biodiversity Information, 37, 63, 104
Atlantic eelgrass limpet (*Lottia alveus alveus*), 322
At-risk species, 91–92
Audience, 50
Australian National University, 231

Babbitt, Bruce, 5, 399
Bailey's ecoregional classification, 26, *fig. 2.1*
Baker, William, 179–80
Balance, 226
"balance of nature," 178
Ball, Ian, 336
Barndoor skate (*Raja laevis*), 322
"battle zones," 336
BCIS (Biodiversity Conservation Information System), 73
Benchmarks, 185–87
Benedict, Mark, 3
Bennett, Andrew, 249
Beta diversity, 9–10, *fig. 1.2*, 240, 242
Biodiversity: conservation plans at multiple scales for, 369–97; definition of, 7–13, 81, *fig. 1.1*; hierarchy of, 8–9, *fig. 1.1*, 82–85
Biodiversity health factor, 263–64, 376
Biodiversity indicators, 96–97
Biodiversity Ranking (B Rank) system, 222, *table 8.1*
Biodiversity Support Program. *See* BSP
Biodiversity value, 263–67, *table 9.1*
Bioinformatics, 70
"Bioinformatics for Biodiversity," 70

Biological distinctiveness, 262, 264, *table 9.1*
Biological integrity, 184–85
Biological inventories, 139–45, *fig. 5.6*
Biological legacies, 200
BioPlan, 68
BioRap, 231
Bioregional assessments, 69–70
Bioregions, 26
Biosphere reserves, 14, 30, 218–19
Biotic attributes of freshwater biodiversity,
    296–98
BirdLife International, 17, 27
Black bear (*Ursus americanus*), 94, 250, 252,
    394
"Black box" effect, 240
Black-footed ferret (*Mustela nigripes*), 157
Black-tailed prairie dog (*Cynomys ludovi-
    cianus*), 195
Blue crab (*Callinectes sapidus*), 329
Bottom-up classification of ecological systems,
    306–8
B Rank. *See* Biodiversity Ranking
Braulio-Carillo National Park (Costa Rica),
    251
Brook trout (*Salvelinus fontinalis*), 304
Brown, Jim, 254
Brundtland Commission, 151, 152
BSP (Biodiversity Support Program): Bul-
    garia, 51; Colombia, 59–60; Papua New
    Guinea, 142
Budgets for conservation projects, 53, *table
    3.2*, 395–97
Building blocks for planning, 41–78
Burden of proof, 174–76
Bushmeat trade, 386–87

CABS (Center for Applied Biodiversity Sci-
    ence), 67–68
California Condor (*Gymnogyps californianus*),
    64
California Legacy Project, 232, 284
Callicott, J. Baird, 180
Cape Canaveral, 403
CAPE (Cape Action Plan for the Environ-
    ment), 42–43, 50, 174, 397
CARA (Conservation and Recovery Act),
    17–18
Caribou (*Rangifer tarandus*), 352, 357
CBD (Convention on Biological Diversity), 6,
    10, 15–16, 398, 402–3; in Colombia, 60;
    conservation areas and, 218; ecosystem
    approach of, 379, 381; precautionary prin-
    ciple and, 174; priority-setting exercises
    and, 27; WCMC and, 66. *See also* NBSAPs
Central Appalachian Forest Ecoregion, 111,
    234, *fig. 8.3*

Central Shortgrass Prairie Ecoregion, 136–38,
    fig. 5.5, 396
Central Tallgrass Prairie Ecoregion, 170, 199,
    table 7.2
CH2M Hill, 50
Chacoan fairy armadillo (*Burmeisteria retusus*),
    84
Chaco region, 84
Chaotic systems (moisture regimes), 361–62,
    tables 12.2-4
Chesapeake striped bass (*Morone saxatilis*), 323
CI (Conservation International), 18, 27,
    67–68; in Brazil, 60; expert workshops and,
    126–30; hotspots and, 17; priority-setting
    and, 261, 398; RAP and, 135, 138–39
Classifications of ecosystems, 100–109, *table
    4.2*; for freshwater conservation, 294–96,
    302–3; for marine conservation, 329
Clean Water Act (U.S.), 184
Clearinghouse Mechanism, 68–69
Climate change, 345–65; assessment of exist-
    ing conservation areas and, 356; conserva-
    tion goals and, 354–55; conservation
    targets and, 351–53; ecoregional planning
    and, 347–51, *figs. 12.1-2*; information gaps
    and, 353–54; persistence and, 356–57, *table
    12.1*; portfolio of conservation areas and,
    353, 357–62, *fig. 12.3*; threats assessment
    and priority-setting, 363–64, *table 12.5*
Clinton administration, 388
*Closing the Gaps in Florida's Wildlife Habitat
    Conservation System*, 394
Closure, 55, 57
Coarse-and-fine-filter approach, 85–90, *fig. 4.2*
Coarse-scale targets, 227, 232–33
CODA, 231
Cognitive biases, 131
Coherence, 227
Collaborative relationships, 57–60
Collinge, Sharon, 253
Columbia Plateau Ecoregion: Gap Analysis
    Programs and, 118–21, *figs. 5.1-2, table 5.2*;
    historic or NRV analyses and, 170–71;
    species targets and, 150–51, 163; suitability
    indices and, 201–3, *table 7.3, fig. 7.3. See
    also* Southern Rocky Mountains Ecore-
    gion
Commitment (of team members), 44–45, *fig.
    3.1*
Commitments in area selection algorithms,
    240
Committee of Scientists, 150
Common wombat (*Vombatus irsinus*), 163
Community targets, 100–109, *table 4.3*; con-
    servation goals and, 159–65, 171. *See also*
    Conservation targets

Complementarity, 223, 227, 237, 258

Completeness, 227

Compositionalism, 180, *fig. 7.1*

Composition of biodiversity, 8–9, *fig. 1.1*

Comprehensiveness, 226, 228

Condition (for EO ranks), 189–90

Conference of the Environment and Development. *See* Earth Summit

Congo Basin, 234, *fig. 8.4*

Connectivity, 76–77, 240, 247–54

Conservation areas, 217–20, *fig. 8.1*, 356; evaluation of, 115–25; Five-S Framework and, 374–77; Four-R Framework and, 30–33; in freshwater environments, 313–16; in marine environments, 333–39; selection of, 220–37, *table 8.2, fig. 8.2*; setting priorities for action among, 260–87; seven-step planning process and, 34–40, *figs. 2.4-5*; six-step process for freshwater conservation and, 302–17. *See also* Networks of conservation areas; Portfolio of conservation areas

Conservation biology, 16–20, 181, *fig. 7.1*, 237

Conservation Biology, 219, 386

Conservation Data Centers (CDCs), 62–63, 297

Conservation goals, 36, 146–77, *fig. 6.1*; climate change and, 354–55; guiding principles in setting, 153–77; for marine environments, 331–33, *fig. 11.2*; networks of conservation areas and, 255–57, *table 8.3*; for freshwater environments, 309–11, *tables 10.1-2*

Conservation International. *See* CI

Conservation Needs Assessment, 142

Conservation planning: building blocks for, 41–78; at crossroads, 398–403; emergence as a discipline, 13–15; Five-S Framework and, 374–77; for freshwater ecosystems, 291–318, *fig. 10.1*; key components of, 371–72; landscape species approach to, 377–79, *fig. 13.4*; for marine environments, 319–44; principles of, 28–29. *See also* Ecoregional plans

Conservation Priorities Setting Program, 67–68

Conservation Process (TNC), 21–22, *fig. 1.4*, 372–74, *fig. 13.1*

Conservation Science Program (WWF-US), 65–66

Conservation targets: ability to persist of, 178–215, 356–57, *table 12.1*; in Four-R Framework, 31–33; in seven-step planning process, 34–40; climate change and, 351–53; in freshwater environments, 303–9; in marine environments, 327–31,

table 11.2; selection of, 81–113; setting goals for, 146–77, *fig. 6.1*; in six-step planning process, 303–9

Conservation value, 262, 263–67, *table 9.1*

Consistency, 227

Consortiums, 59

Contractual agreements, 59

Control as element of project management, 52

Convention on Biological Diversity. *See* CBD

Cooperative Fish and Wildlife Research Units (U.S. states), 64

Core areas, 218, 228

"correlated fates," 155–57, *figs. 6.2-3*

Corridors, 247–54, *fig. 8.9*

Cost effectiveness, 227

Costs in area selection algorithms, 240

Cottonwood (*Populus* sp.), 179

Covey, Stephen, 34

Cowling, Richard, 180

C-Plan, 231, 391

Creek chub (*Semotilus atromaculatus*), 304

Crested caracara (*Caracara plancus*), 394

Critical threats, 262

CSIRO (Commonwealth Scientific and Industrial Research Organization), 231

Csuti, Blair, 220–21

Cumberland Southern Ridge and Valley Ecoregion, 310, *table 10.2, fig. 10.6*

Data. *See* Information

Data issues in conservation area selection, 232–37

"dead zone," 321, *fig. 11.1*

Decision support tools, 284–86, *fig. 9.3*

Defenders of Wildlife, 50

Definition as element of project management, 52

Delphi workshops, 336

Demographic uncertainty, 183

DEMs (digital elevation models), 294, 303

Design of conservation area management, 220–21

Design with Nature (McHarg), 179

Desired future conditions, 150

Devils Hole pupfish (*Cyprinodon diabolis*), 163, 225

Diamond, Jared, 243

Dinerstein, Eric, 26, 65

Direct utilitarian values, 11

Disjunct targets, 163

Dispersal-limited species, 94

Distribution patterns, 164–65, *table 6.1*

Disturbance-based approach to assessing ecological integrity, 196–98, *fig. 7.2*

Douglas fir (*Pseudotsuga menziesii*), 25

East Gulf Coastal Plain Ecoregion, 383
Ecological Applications, 72, 84–85, 153
Ecological Drainage Units (EDUs), 111, 167, 303, 309–11, *fig. 10.2, fig. 10.6*
"ecological extinction," 195
Ecological integrity, 150, 182, 184–85, 196–210, *tables 7.2-3*; freshwater conservation and, 311–13; marine conservation and, 334–35
Ecological Land Units (ELUs), 110–11, 167–68
Ecological processes as conservation targets, 111–12, 179–80
Ecological restoration, 210–13, *figs. 7.8-9*
Ecological Society of America, 7, 13, 30
Ecological systems, 233, 305–9
Ecology, 16–20
ECOMAP, 26
Ecoregional plans, 25–28, *figs. 2.1-2*, 387–95; climate change and, 345–65, *figs. 12.1-2*; cost of implementing, 395–97. *See also* Conservation planning; Regional planning
Ecosystem as term, 100, 327–29
Ecosystem classifications, 100–109, *table 4.2*
Ecosystem geography, 24
Ecosystem targets, 100–109, *table 4.4*; conservation goals and, 159–65, 171. *See also* Conservation targets
EDUs. *See* Ecological Drainage Units
Effective population size (N^e), 188–89
Efficiency, 227, 229, 237, 239
EFPs (ecologically functional populations), 194–96
Eglin Air Force Base, 383–84
Empty forest metaphor, 86, 193
Endangered species, 4–6, 90–91, *table 4.1*, 153, 170, 383, 399
Endangered Species Act (U.S.), 4–5, 13; conservation goals and, 147; identifying conservation targets and, 34–35; recovery plans and, 153; spatial scales and, 84–85; species targets and, 91, *table 4.1*; threat assessment and, 268
Endemic species, 92, *table 4.1*, 163
Environmental gradients, 240
Environmental Law Institute, 17
Environmental Resources Information Network (Australia), 231
Environmental uncertainty, 183
Environmental units, 109–11, 167
EO (element occurrence) ranks, 189–92, *table 7.1*, 200
Equilibrium paradigm, 178
Equity, 227
Esthetic values, 11–12
Ethical values, 11–12

Eucalyptus radiata, 141–42, *fig. 5.6*
European Union, 17, 91
Evaluation, 77–78; of existing conservation areas, 115–25; in seven-step planning process, 36–37
Expectations, 49
Expert opinion (for PVAs), 189–92
Expert workshops, 126–33, *fig. 5.3*
Extent, 28, *fig. 2.3*
Extinction, probability of, 155–57, *figs. 6.2-3*
Extinction crisis, 4–7; marine diversity and, 322
"Extinction debt," 5

Feasibility, 273–74, 279–80
FEMAT (Forest Ecosystem Management Assessment Team), 69
Fencerow corridors, 248
50/500 rule, 188
Fine-and-coarse-filter approach, 85–90, *fig. 4.2*
Fisher (*Martes pennanti*), 250
Five-S Framework for Site Conservation (TNC), 374–77, *figs. 13.2-3*
Five-S Handbook, 377
Flagship species, 92–93, *table 4.1*
Flexibility, 223–24, *table 8.2*, 237, 239, 258
Florida Fish and Wildlife Conservation Commission, 393–95
Florida manatee (*Trichechus manatus latirostris*), 329
Florida panther (*Felis concolor*), 252, 394
Floristic-based classifications, 104, *figs. 4.3-4*
Focal species, 94–95, *table 4.1*, 105, *fig. 4.3*; PVAs and, 194; in Wildlands Project, 194, 228
Folke, Carl, 13
Forest Conservation Program (IUCN), 249, 379–82, *fig. 13.5*
Forman, Richard, 253
Four-R Framework, 30–33, 257–58, 403
Fragmentation, 245–46
Fragmented Forest, The (Harris), 245
Freshwater biodiversity, 291–318, *fig. 10.1*; Biotic attributes of, 296–98; global and continental analyses of, 300–302; physical attributes of, 294–96; regional planning process for, 302–17; threats assessment and, 298–300
Functionalism, 180, *fig. 7.1*
Functionality, 227, 228, 257
Functional sites, landscapes, and networks, 241–42
Function of biodiversity, 8–9, *fig. 1.1*
Fundación ProSierra Nevada de Santa Marta, 60
Future threats assessment, 282–84
Gamma diversity, 9–10, *fig. 1.2*
Gantt, Henry, 53
Gantt chart, 53, *fig. 3.3*

Gamma diversity, 9–10, *fig. 1.2*
Gantt, Henry, 53
Gantt chart, 53, *fig. 3.3*
Gap Analysis Handbook, 109, 119, 122
Gap Analysis Programs (U.S.), 16, 17, 64–65, *fig. 3.5*; Aquatic Gap Program, 301; conservation area assessment and, 36; conservation areas and, 218; conservation goals and, 152; ecosystem targets and, 106–7, 110; evaluation of existing conservation areas and, 115–22, *figs. 5.1-2, tables 5.1-2*; information management and, 76; predictive models and, 142, 193; priority-setting exercises and, 27; REAs and, 134, 136; umbrella species and, 93
GEF (Global Environmental Facility), 16, 60, 68
General Land Office (GLO), 172
Genetic uncertainty, 183
GIS (Geographic Information System), 37, 62; conservation goals and, 174, 400; ELUs and, 110, 167; expert workshops and, 127–29; freshwater conservation and, 294, 304, 307, 311; in Gap Analysis Programs, 64–65, 115, *fig. 3.5*; information management and, 74; landscape species approach and, 379, *fig. 13.4*; PATCH program and, 194; predictive models and, 143; REAs and, 134; spatial surrogates and, 201, 209; in TIGER files, 71
GLMs (Generalized Linear Models), 140–42, *fig. 5.6*
Global 200, 237
Global Biodiversity Assessment, 6
Global Biodiversity Strategy, 20
Global Environmental Facility. *See* GEF
Goals. *See* Conservation goals
"good, fast, and cheap," 52, *fig. 3.2*
Gradient systems (moisture regimes), 361–62, *tables 12.2-4*
Grain, 28, *fig. 2.3*
Gray brocket deer (*Mazama gouazoubira*), 84
Great Barrier Reef Marine Park Authority, 231, 323, 334
Greater Yellowstone Ecosystem, 272–74, *fig. 9.1*
Great Lakes Ecoregion, 268–71, *table 9.2*
Green infrastructure hubs, 252
Green River (Ky.), 385
Green turtle (*Chelonia mydas mydas*), 331
Grizzly (*Ursus arctos*), 83, 85, 93, 119
Guanacaste National Park (Costa Rica), 212
Guanaco (*Lama guanicoe*), 195
Gulf Coastal Plain Ecosystem Partnership, 384

Habitat as term, 327–29
Habitat-based approaches to prediction, 193–94

Habitat Conservation Plans (under ESA), 148
Habitat Evaluation Procedures (HEP), 96
Habitat patches, 244–47
Harlequin duck (*Histrionicus histrionicus*), 191–92, *table 7.1*
Harris, Larry, 245
Health indicators, 96
Heinemayer, Kim, 72
Heuristic algorithms, 230
Hierarchical classification structure, 104–5, *fig. 4.4*
Hierarchy of biodiversity, 8–9, *fig. 1.1*, 82–85, *fig. 4.1*
HI (Habitat Index), 312
Historical context on conservation goals, 151–53, 168–73, *fig. 6.7*
Home ranges, 198
Horneyhead chub (*Nocomis biguttatus*), 304
Hotspots, 17–18; aspects of scale and, 76; CABS and, 67; Cape Floral Kingdom (South Africa), 42–43; Cerrado region (Brazil), 18; freshwater conservation and, 313; spatial scale of, 27
Hypothetico-deductive science, 175
Hypoxia, zone of, 321, *fig. 11.1*

IBI (Index of Biotic Integrity), 311–12
ICBMP (Interior Columbia Basin Ecosystem Management Project): ecological integrity and, 205–7, *fig. 7.7*, 210; historic or NRV analyses and, 170, *fig. 6.7*
IMADES (Sonora State Institute for Environment and Sustainable Development), 60, 387–89, *fig. 13.6*
Imperiled species, 90–91, *table 4.1*
Index of Hydrologic Alteration, 169
Indicator species, 96–97, 100
Indigenous groups, 58, 72, 173, 186
Indirect utilitarian values, 11
Information, assembling of, 61–72, *table 3.3*
Information gaps, 35, 63; assessments of, 133–39; climate change and, 353–54; expert workshops and, 126–33, *fig. 5.3*; inventories of, 139–45
Information management, 72–77
Institutional expectations, 49
Instituto Alexander von Humbolt, 60
Integrated Coastal Zone Management Programs, 343
Integrated Natural Resources Management Plan, 383
Integration, 227
Integrity. *See* Ecological integrity
Intergovernmental Panel on Climate Change, 359

Interior Columbia River Basin. *See* ICBMP
International Association of Fish and Wildlife
  Agencies, 18
International Classification of Ecological
  Communities, 104
International Data Base, 71
Intrinsic values, 11–12
Inventories, 139–45, *fig. 5.6*
Inventory and Monitoring Program, 142
Investment, 49
Irreplaceability, 224–26, *fig. 8.2*, 266–67,
  272–73, *fig. 9.1*, 278–79
Island biogeography, 243, *fig. 8.8*
Iterative algorithms, 230–31
IUCN Red List of Threatened Species, The, 4, 66,
  90–91, 386
IUCN (World Conservation Union), 66; con-
  servation areas and, 218, 221; conservation
  goals and, 151, 152, 154, 395–96; evalua-
  tion of existing conservation areas and,
  122–25, *table 5.3*; Forest Conservation Program, 249, 379–82, *fig. 13.5*; wide-ranging
  species and, 168

Jaguar (*Panthera onca*), 95, 220
Johnson, Lyndon, 403
Johnson, Nels, 29
Juniper (*Juniperus* sp.), 136

Kankakee Sands conservation area (Indiana,
  U.S.), 212
Karoo Succulent Desert (South Africa), 220
Kemp's Ridley turtle (*Lepidochelys kempii*), 329
Keystone species, 95–96, *table 4.1*, 195, 228,
  329–30
Klamath–Siskiyou Ecoregion, 246, 250, *fig.
  8.10*

Landscape approach to forest conservation,
  249, 379–82, *fig. 13.5*
Landscape context (for EO ranks), 190
Landscape ecology, 17
Landscape integrity, 264, *table 9.1*; freshwater
  conservation and, 313–14, *fig. 10.5*
Landscape mosaic corridors, 248–49
Landscapes, 25
Landscape-scale conservation areas, 37,
  241–43
Landscape species, 93
Landscape species approach to planning,
  377–79
Land survey system, 172
Land system classifications, 109–10
Land trusts, 18–19, *fig. 1.3*
La Selva Biological Station (Costa Rica), 251

Leadership, 48
Legacy program (DOD), 384
Lesser rhea (*Rhea pennata*), 195
Leverage, 274, 280–81
Limited targets, 163–64
Linkages, 247–54, *fig. 8.10*
Linkages in the Landscape (Bennett), 249
Little Black Book of Project Management, The
  (Thomsett), 52
Living Landscape, The (Steiner), 33, 201
Local levels of conservation, 3; collaborative
  relationships with, 60; land trusts, 18–19,
  fig. 1.3
Local opportunity, 274
Loggerhead turtle (*Caretta caretta*), 335
Longleaf pine (*Pinus palustris*), 179, 383
"low-hanging fruit," 274, 336
Lynx (*Felis lynx*), 163

MAB (Man and the Biosphere Program), 14,
  219
Management Effectiveness Task Force, 124–25
Management Indicator Species (MIS), 96
Man and the Biosphere program, 14, 30
Margules, Chris, 39
Marine environments, 319–44; assessment of
  existing conservation areas in, 333–34;
  assessment of population viability and ecological integrity in, 334–35; conservation
  goals for, 331–33, *fig. 11.2*; conservation
  targets in, 327–31, *table 11.2*; priority-set-
  ting and, 342–43; regional planning for,
  323–27, *table 11.1*; selection of conserva-
  tion areas in, 335–39, *figs. 11.5-6*; threats
  assessment and, 320–23, *fig. 11.1*
Marine Life Protection Act (Calif.), 336
Marine Protected Areas (MPAs), 326, 333–34
MARXAN, 336
Masks in area selection algorithms, 240
Matrix-forming communities, 164–65, *table 6.1*
MCDA (multicriteria decision analysis),
  284–86
McHarg, Ian, 179
McMahon, Edward, 3
Memorandum of Understanding (MOU), 385
Mesoamerican Biological Corridor, 252
Metalite, 74
Metapopulation theory, 157–58, *fig. 6.4*
Middle Rocky Mountain–Blue Mountain
  Ecoregion, 315
Millennium Ecosystem Assessment, 70
Ministry of Sustainable Resource Manage-
  ment (B.C.), 326
Minumum dynamic area, 196–98, 200, *fig. 7.2*
Mississippi Basin, 321, *fig. 11.1*, 340

Moisture regimes, 361–62, *tables 12.2-4*
Mojave black-collared lizard (*Crotaphytus bicinctores*), 163
Motivational biases, 130–31
MOU (Memorandum of Understanding), 385
Movement among populations, 155–57, *figs. 6.2-3*
MPAs (Marine Protected Areas), 326, 333–34
Multiarea conservation strategies, 383–87
Multiobjective models, 232
Myers, Norman, 260

National Audubon Society, 218
National Focal Points, 68
National Forest Management Act (U.S.), 20
National Forest Policy Statement (Australia), 390–93
National Inventory of Dams, 298
National Marine Fisheries Service, 153
National Monuments, 388
National Nature Reserves (U.K.), 14
National Reserves Program (Australia), 14–15
Natura 2000, 17, 91
Natural Areas Inventory (Fla.), 383
Natural catastrophes, 183–84
Natural Heritage Programs (NHPs), 13–14, 37, 62–63; ecological integrity assessments and, 198–201; ecosystem targets and, 106–7; EO ranks and, 189–92; fine-and-coarse-filter approach in, 86–87; freshwater conservation and, 297; imperiled species and, 91, *table 4.1*
"natural insurance capital," 13
Naturalness, 185–87, 220
Natural Resources Defense Council, 336
The Nature Conservancy. *See* TNC
Nature Conservancy (Canada), 60
Nature Conservation Review, A, 14
NatureServe, 37, 63; ecological integrity assessments and, 198–201; ecosystem targets and, 104, 107; EO ranks and, 189–92; freshwater conservation and, 297; imperiled species and, 91, *table 4.1*
NBII (National Biological Information Infrastructure), 71
NBSAPs (National Biodiversity Strategies and Action Plans), 16, 18, 19, 51, 68, 123, 147
Networks of conservation areas, 216, 243–47; corridors and linkages in, 247–54; freshwater conservation and, 316–17, *fig. 10.6*; marine conservation and, 339–42; measuring success of, 255–57, *table 8.3*; semi-natural matrix in, 254–55; in seven-step planning process, 37–38, *fig. 2.5*

New World bias, 172–73, 186
NOAA (National Oceanographic and Atmospheric Administration), 326, 333–34, 335
Noah's species approach, 86
"Noah worked two jobs," 220–21, 401
Nonequilibrium paradigm, 16, 179
Non-native species, 299–300
Northern Appalachians Ecoregion, 197, *fig. 7.2*
Northern Gulf of Mexico Ecoregion, 328, *table 11.2*, 336, 340
Northern hog sucker (*Hypentelium nigricans*), 304
Northern spotted owl (*Strix occidentalis caurina*), 243, 398
Northern Tallgrass Prairie Ecoregion, 106, *table 4.3*, 170, 210, 229, 385
NRV (natural range of variability), 168–71, 185

Okefenokee National Wildlife Refuge, 250, *fig. 8.9*
Olson, David, 26, 65
"opportunistic" approach, 81
Opportunity, 281
Opposition stakeholders, 58
Optimization algorithms, 230–31
Oregon Biodiversity Project, 117–18, *table 5.1*, 171–72, *fig. 6.8*
Oregon silverspot butterfly (*Speyeria zerene hippolyta*), 153
Organizational expectations, 49
Osceola National Forest, 250, *fig. 8.9*
Overfishing, 322

Paine, Robert, 95
Panda (*Ailuropoda melanoleuca*), 92
Pantanal (Brazil) Ecosystem, 308–9, *fig. 10.4*
Partnerships, 59
Partners in Flight, 91, 168
Patch communities, 164–65, *table 6.1*
PATCH program, 194
Pattern-process dichotomy, 180
Patterns of distribution, 164–65, *table 6.1*
Peer review, 77–78
Peer review workshops, 57
Peripheral populations, 97, 100, 163–64
Persistence: of conservation targets, 178–215, 356–57, *table 12.1*; as selection principle, 226–27, 228, 237, 239, 257
physical attributes of freshwater biodiversity, 294–96
Physiognomic-based classifications, 104, *figs. 4.3-4*
Pinhook Swamp, 249–50, *fig. 8.9*
Pinyon pine (*Pinus edulis*), 136

Plan implementation, 240–41
Plant associations, 233
Polar bear (*Ursus maritimus*), 352
Pollution, 299
Ponderosa pine (*Pinus ponderosa*), 25, 136, 179
Population indicators, 96
Population Reference Bureau, 71
Portfolio of conservation areas, 216–43, *figs. 8.6-7*; climate change and, 353, 357–62, *fig. 12.3*; freshwater conservation and, 313–16; marine conservation and, 336, *fig. 11.3*
Possingham, Hugh, 336
Potential restoration areas, 211
Prairie dog (*Cynomys* sp.), 96, 157, 195
Precautionary principle, 174–76
Predicting Species Occurrences (Scott), 142
Predictive models, 142–43, 313
Preferences in area selection algorithms, 240
Prendergast, John, 28–29, 231, 237
Preserves, 218
Pressey, Bob, 39, 81, 217, 263, 400
Priority-setting, 260–87; criteria for, 263–84, *table 9.1*; decision support tools for, 284–86, *fig. 9.3*; exercises in spatial scales and, *fig. 2.3*; marine conservation and, 342–43; in seven-step planning process, 38; spatial scales exercises and, 27–28; climate change and, 363–64, *table 12.5*
Problem definition, 42–43
Process-limited species, 94
Products, 49–50
Project management, 49–57, *figs. 3.2-4, table 3.2*
Protected areas, 16, 218
Public opinion, American, 10–11
Puget Sound/Georgia Straits Ecoregion, 337, 343
Puma (*Felis concolor*), 84
PVAs (population viability analyses), 148, 155–59, 182–83, 187–94, 198, 334–35

Quantitative PVAs, 187–88

"ranchette" development, 283
Rangewide distribution of targets, 162–64
Rapid Assessment Program (RAP), 135, 138–39
Rapid Ecological Assessments (REAs), 109, 134–38, *fig. 5.4*, 389
Rare species, 92
Rarity-based algorithms, 231
Recovery plans, 153–54
Recreational values, 11–12
Red-cockaded woodpecker (*Picoides borealis*), 250, 383
Red list. See IUCN Red List
Redundant (Four-R Framework), 31–32, 257

Redwoods Ecosystem, 203–5, *figs. 7.4-6*
Reef Condition Index, 139
Regional corridors, 248–49
Regional Forestry Agreements (Australia), 390–93, *fig. 13.7*
Regional planning: for freshwater conservation, 302–17; for marine conservation, 323–27, *table 11.1*
Representation, 149, 228, 237
Representative (Four-R Framework), 30–31, 257
Representativeness, 226, 227, 228
Reproductive success, 158
Reserve designs, 243–47; corridors and linkages in, 247–54
Resilient (Four-R Framework), 31, 257
Resource-limited species, 94
Restoration. See Ecological restoration
Restorative (Four-R Framework), 32–33, 257
Restricted targets, 163–64
RFAs (Regional Forestry Agreements-Australia), 390–93, *fig. 13.7*
Right whale (*Eubalaena glacialis*), 330, 335
Riverina Bioregion (Australia), 108, *table 4.4*
Roadless lands, 246–47
Rules of thumb (for PVAs), 188–89

Sage grouse (*Centrocercus urophasianus*), 150–51
Salmon (*Oncorhynchus* sp.), 86
Sanctuaries, 218
Saturated systems (moisture regimes), 361–62, *tables 12.2-4*
Saving Nature's Legacy (Noss and Cooperrider), 220–29
Science, 70–71
Scientific Advisory Group, 391
Scott, J. Michael, 64, 142, 220–21
Sea otter (*Enhydra lutris*), 331
Secondary stakeholders, 58
Selection of conservation areas, 20, 220–21; algorithms for, 229–37, 239–40, 315–16, 336–37; principles for, 221–29, *tables 8.1-2, fig. 8.2*, 237–41
Selection units, 235–37, *fig. 8.5*, 266
Semi-natural matrix, 254–55
Sensitive species, 91
Seven Habits of Highly Effective People, The (Covey), 34
Seven-step planning process, 34–40, *fig. 2.4*
Severity of threat, 263
Shafer, Craig, 20
Shaffer, Mark, 396
Shelford, Victor, 30
"shovel brigade," 280
Sink habitats, 158

Sinking Ark, The (Myers), 260–61
Sink populations, 158
Sites, 24–25, 218
Sites (algorithm), 231, 336–37
Site selection, 20; algorithms for, 229–37, 239–40, 315–16, 336–37
Situation (Five-S Framework), 375
Six-step planning process for freshwater conservation, 302–17
Size (for EO ranks), 189
Skills (of team members), 44–45, *fig. 3.1*
Sky Islands conservation plan, 211
Slimy sculpin (*Cottus cognatus*), 304
SLOSS (Single Large or Several Small Reserves), 241, 243, *fig. 8.8*
Snake River Birds of Prey National Conservation Area, 163
Society for Conservation Biology, 10, 72, 94
Society for Ecological Restoration, 210
Socioeconomic data, 71–72
Sockeye salmon (*Oncorhynchus nerka*), 5
Soil surveys, 172
Sonoran Desert Ecoregion, 60, 387–89, *fig. 13.6*
Sonoran Institute, 60, 387–89, *fig. 13.6*
Source habitats, 158
Source populations, 158
Sources: in Five-S Framework, 375–77, *figs. 13.2-3*; in threats assessment, 268, 272
Southern Rocky Mountains Ecoregion: conservation goals and, 255, *table 8.3*; ecological integrity assessments and, 201; information management and, 74; portfolio of conservation areas for, 238, *fig. 8.7*; species targets and, 165, *table 6.1*; subunits of ecoregions and, 166, *fig. 6.6*; threats assessment and, 274–79, *tables 9.3-4*, 283. *See also* Columbia Plateau Ecoregion
Southern three-banded armadillo (*Tolypeutes matacus*), 84
Spatial data units, 233–37
Spatial scales, 9–10, *fig. 1.2*; of biodiversity, 83–84, *fig. 4.1*; information management and, 75–77, *fig. 3.6*; in priority-setting exercises, 27–28, *fig. 2.3*
"spatial solution," 253
Spatial surrogates, 201, 208–10
Special elements, 228
Species-area relationships, 159–62, *fig. 6.5*; climate change and, 355
Species Survival Commission (SSC), 153–54, 168
Species targets, 90–100, *table 4.1*, 165, *table 6.1*; ecoregional planning and, 352–53; for freshwater conservation, 309–11, *table 10.1*; marine conservation and, 329

SPEXAN, 336
Spiritual values, 11–12
Spotted owl (*Strix occidentalis*), 69
Stakeholders, collaborative relationships with, 57–60
Starfish (*Pisaster* sp.), 95
State of the Hotspots Program, 67
Steiner, Frederick, 33, 201
Stellar sea cow (*Hydrodamalis gigas*), 322, 330
Stonecat (*Noturus flavus*), 304
Strategic Habitat Conservation Areas (SHCAs), 393–95
Strategies (Five-S Framework), 375–77, *figs. 13.2-3*
Stresses: in Five-S Framework, 374–77, *figs. 13.2-3*; in threats assessment, 268, 272
Structure of biodiversity, 8–9, *fig. 1.1*
Subunits of ecoregions, 165–68, *figs. 6.6-7*
Success (Five-S Framework), 375–77, *figs. 13.2-3*
Suitability indices, 201–10, *table 7.3, figs. 7.3-6*; freshwater conservation and, 313–14, *fig. 10.5*; marine conservation and, 335
Sustainable development, 15
Synchronization of conservation planning, 51
Systems (Five-S Framework), 374–77, *figs. 13.2-3*

Team building, 43–49, *fig. 3.1*; charters for, 45–47, *table 3.1*
Temporal scales, 75–77, *fig. 3.6*
Texas pipefish (*Sygnathus affinis*), 322
Thomas, Jack Ward, 243
Thomsett, Michael, 52
Threat abatement, 376
Threatened species, 4–6, 90–91, *table 4.1*, 153, 170
Threats assessment, 262, 268–79, *tables 9.2-4, fig. 9.1*; climate change and, 363–64, *table 12.5*; freshwater conservation and, 298–300; future, 282–84; marine conservation and, 320–23, *fig. 11.1*; in seven-step planning process, 38
Tiger (*Panthera tigris*), 95
TIGER (Topologically Integrated Geographic Encoding and Referencing), 71
TNC (The Nature Conservancy): ad hoc conservation efforts and, 217; algorithms and, 239; alliances of, 50–51, 59–60; audience for plans of, 50; biodiversity value and, 262, 263–64; bioregional assessments and, 69; closure to conservation planning and, 57; conservation areas and, 218, 221; conservation area selection and, 222, *table 8.1*, 227, 233; conservation goals and, 148, *fig. 6.1*, 152, 165, *table 6.1*, 396–97; Conservation

TNC (*continued*)
  Process, 21–22, *fig. 1.4*, 372–74, *fig. 13.1*;
    ecological integrity assessments and, 198;
    ecological restoration and, 212; ecoregional
    planning process of, 26–27, 398; ecosystem
    targets and, 104, 106, 110–11, 385; evalua-
    tion of existing conservation areas and,
    114–15; expert workshops and, 126–30;
    Five-S Framework and, 374–77, *figs. 13.2-3*;
    freshwater conservation and, 296, 298,
    301–3; functional sites and, 241; Gantt chart
    for projects of, 53–54, *fig. 3.3*; information
    management and, 73–74; investment in time
    and money, 49; leverage and, 281; marine
    conservation and, 323, 328, *table 11.2*,
    336–37, 343; Mianus Gorge (N.Y.) and,
    81–82; population viability assessments and,
    194; REAs and, 134, *fig. 5.4*; representation
    and, 237, *fig. 8.6*; seven-step planning process
    and, 33, 36–37; Sonoran Desert Ecoregion
    and, 60, 387–89, *fig. 13.6*; suitability indices
    and, 201; team building and, 43–47; threats
    assessment and, 262, 274–79, *tables 9.3-4*;
    U.S. Army Corps and, 385; U.S. Department
    of Defense and, 149, 383–84; Yunnan Great
    Rivers Project and, 389–90. *See also* Natural
    Heritage Programs
Top-down classification of ecological systems,
    308–9, *fig. 10.4*
Tortugas ecological reserves, 335–36
Traditional ecological knowledge (TEK), 72
"train wrecks," 5, 399
TRA (Threat Reduction Assessment), 278
"triage strategy," 260–61
Type I and II errors, 175

U.K. Wildlife Trusts, 14
Umbrella species, 93–94, *table 4.1*, 377
Uncertainty, categories of, 183–84
U.N. Conference on the Environment and
    Development. *See* Earth Summit
UNDP (United Nations Development Pro-
    gramme), 16, 68, 147
UNEP (United Nations Environment Pro-
    gramme), 6, 16, 66, 74, 300
UNESCO, 14, 30, 218–19
Urgency of threats, 262–63
U.S. Army Corps of Engineers, 51, 281, 385
U.S. Bureau of Land Management, 20, 51, 57,
    91, 151
U.S. Bureau of Reclamation, 51
U.S. Census Bureau, 71
U.S. Congress, 17
U.S. Department of Defense, 149, 383–84,
    387–89, *fig. 13.6*

U.S. Environmental Protection Agency, 235,
    fig. 8.5
U.S. Fish and Wildlife Service, 20, 51, 64, 96,
    125, 153, 168
U.S. Forest Service, 20, 51, 57, 91, 96, 150,
    166, 241, 246–47
U.S. Geological Survey, 64, 71, 74, 294
U.S. Marine Mammal Commission, 7
U.S. National Estuarine Research Reserve
    (NERR), 326
U.S. National Marine Sanctuaries, 323
U.S. National Oceanographic and Atmos-
    pheric Administration. *See* NOAA
U.S. National Park Service, 125, 142, 179
U.S. National Wildlife Refuge system, 20
USNVC (U.S. National Vegetation Classifica-
    tion), 104–7, *figs. 4.3-4*, *table 4.3*, 233
Utilitarian values, direct and indirect, 11

Value. *See* Conservation value
Values, 150
Vegetation classification. *See* USNVC
Vegetation maps, historic, 171–72
Viability, 150; freshwater conservation and,
    311–13. *See also* PVAs
Vole (*Muridae* sp.), 352
Vulnerability, 263, 272–73, *fig. 9.1*, 278–79

Watson, Bob, 359
WCPA (World Commission on Protected
    Areas), 124–25, *table 5.4*, 218, 226
Web Site urls: BCIS, 73; bushmeat trade, 387;
    CABS, 67, 68; CBD, 68; endangered
    species, 4, 91; Gap Analysis Programs, 65,
    109, 122; Great Barrier Reef Marine Park
    Authority, 323; ICBMP and, 70; Interna-
    tional Data Base, 71; IUCN, 154; landbirds
    (U.S.), 91; Land Trust Alliance, 18–19; Latin
    American and Caribbean species, 63;
    MAB, 14, 219; Metalite, 74; Millennium
    Ecosystem Assessment, 70; National Inven-
    tory of Dams, 298; NatureServe, 63; NBII,
    71; non-native species, 299–300; Popula-
    tion Reference Bureau, 71; RAP, 138; *Sci-
    ence*, 70–71; SSC, 154; TIGER, 71; UNDP,
    68; U.S. Fish and Wildlife Service, 154; U.S.
    National Park Service, 142; WCMC, 66,
    124; Wildlife Trusts, 14; World Heritage
    Sites, 219; Zebra mussel, 299–300
West Indian monk seal (*Monachus tropicalis*),
    322
White abalone (*Haliotis sorensi*), 322, 330
"white data," 63
Whittaker, Robert, 9
W-IBI (watershed index of biotic integrity),
    312

Wide-ranging species, 168, 193–94, 250
Widespread targets, 163–64
Wild dog (*Lycaon pictus*), 154
Wildlands Project (U.S.), 194, 210–11, 218, 228
Wildlife Conservation Society (WCS), 377–79, *fig. 13.4*
Wildlife habitat-relationship modeling, 193–94
Wildlife Society Bulletin, 219
Wildlife Trusts. *See* U.K. Wildlife Trusts
Wild meat use, 386–87
Wilson, E. O., 7, 11
Wise Use Movement, 279
Wolf (*Canis lupus*), 194, 357
Wolverine (*Gulo gulo*), 168, 352
Working groups, 44–45
Work plans, 53, *fig. 3.4*
World Bank, 16, 60, 125, 142
World Commission on Protected Areas. *See* WCPA
World Conservation Monitoring Center (WCMC), 6, 66–67, 124, *table 5.4*, 294, 300–301
World Conservation Union. *See* IUCN
World Heritage Sites, 219
WORLDMAP, 231
World Parks Congress, 151

World Resources Institute (WRI), 29, 59, 68, 301
World Wide Fund for Nature–South Africa, 42, 50
WWF (World Wildlife Fund): biodiversity value and, 264–66, *table 9.1*; biological distinctiveness and, 262; in Brazil, 60; bushmeat trade and, 386–87; in Canada, 110; ecoregional assessments and, 26–27, *fig. 2.2*, 65–66; evaluation of existing conservation areas and, 123–24, 125; expert workshops and, 126–30; flagship species and, 92–93; freshwater conservation and, 301, 303; hotspots and, 17; landscape approach to forest conservation and, 379–82, *fig. 13.5*; marine conservation and, 323, 336; priority-setting and, 261, 398; representation and, 237; spatial data units and, 234, *fig. 8.4*; stakeholders and, 58–59; threats assessment and, 262–63

Yunnan golden monkey (*Rhinopithecus bieti*), 163
Yunnan Great Rivers Project, 389–90

Zebra mussel (*Dressena polymorpha*), 299–300
Zoogeographic Subregions, 303